D1212674

The Development of
Mathematical Thinking

DEVELOPMENTAL PSYCHOLOGY SERIES

SERIES EDITOR
Harry Beilin

Developmental Psychology Program
City University of New York Graduate School
New York, New York

The list of titles in this series continues on the last page of this volume.

The Development of Mathematical Thinking

Edited by

HERBERT P. GINSBURG

Graduate School of Education and Human Development
University of Rochester
Rochester, New York

1983

ACADEMIC PRESS
A Subsidiary of Harcourt Brace Jovanovich, Publishers
New York   London
Paris   San Diego   San Francisco   São Paulo   Sydney   Tokyo   Toronto

ACADEMIC PRESS, INC.
111 Fifth Avenue, New York, New York 10003

United Kingdom Edition published by
ACADEMIC PRESS, INC. (LONDON) LTD.
24/28 Oval Road, London NW1 7DX

Library of Congress Cataloging in Publication Data
Main entry under title:

The Development of mathematical thinking.

(Developmental psychology)
Includes index.
1. Mathematics--Study and teaching--Psychological
aspects. 2. Cognition in children. I. Ginsburg,
Herbert. II. Series.
QA135.5.D48 1982   155.4'13   82-8693
ISBN 0-12-284780-6   AACR2

PRINTED IN THE UNITED STATES OF AMERICA

83 84 85 86   9 8 7 6 5 4 3 2 1

To the children who can love to learn mathematics

Contents

Herbert P. Ginsburg

Introduction

CHAPTER **1**

Protocol Methods in Research on Mathematical Thinking

Herbert P. Ginsburg
Nancy E. Kossan
Robert Schwartz
David Swanson

vii

CHAPTER **2**

The Acquisition of Early Number Word Meanings: A Conceptual Analysis and Review

Karen C. Fuson
James W. Hall

CHAPTER **3**

A Developmental Theory of Number Understanding

Lauren B. Resnick

CHAPTER **4**

Development of Children's Problem-Solving Ability in Arithmetic

Mary S. Riley
James G. Greeno
Joan I. Heller

CHAPTER **5**

On the Representation of Procedures in Repair Theory

Kurt Van Lehn

CHAPTER **6**

Complex Mathematical Cognition

Robert B. Davis

CHAPTER **7**

The Development of Numerical Cognition: Cross-Cultural Perspectives

Geoffrey B. Saxe
Jill K. Posner

CHAPTER **8**

Children's Psychological Difficulties in Mathematics

Barbara S. Allardice
Herbert P. Ginsburg

CHAPTER **9**

The Many Faces of Piaget

Guy Groen
Caroline Kieran

Contributors

Numbers in parentheses indicate the pages on which the authors' contributions begin.

BARBARA S. ALLARDICE (319), Learning Development Center, Rochester Institute of Technology, Rochester, New York 14623

ROBERT B. DAVIS (253), Curriculum Laboratory, University of Illinois, Urbana, Illinois 61801

KAREN C. FUSON (49), School of Education, Northwestern University, Evanston, Illinois, 60201

HERBERT P. GINSBURG (1, 319), Graduate School of Education and Human Development, University of Rochester, Rochester, New York 14627

JAMES G. GREENO (153), Learning Research and Development Center, University of Pittsburgh, Pittsburgh, Pennsylvania 15260

GUY GROEN (351), Department of Educational Psychology, McGill University, Montreal, Quebec, Canada

JAMES W. HALL (49), School of Education, Northwestern University, Evanston, Illinois, 60201

JOAN I. HELLER (153), Department of Physics, University of California, Berkeley, Berkeley, California 94720

CAROLINE KIERAN (351), Department of Educational Psychology, McGill University, Montreal, Quebec, Canada

NANCY E. KOSSAN (7), Office of Financial Systems, Holyoke Center, Harvard University, Cambridge, Massachusetts 02138

JILL K. POSNER (291), Casamance Project, USAID/Daker Air Pouch, Agency for International Development, Washington, D.C. 20523

LAUREN B. RESNICK (109), Learning Research and Development Center, University of Pittsburgh, Pittsburgh, Pennsylvania 15260

MARY S. RILEY[1] (153), Learning Research and Development Center, University of Pittsburgh, Pittsburgh, Pennsylvania 15260

GEOFFREY B. SAXE (291), Program in Education, Graduate Center, City University of New York, New York, New York 10036

ROBERT SCHWARTZ (7), 2125 North Stowell Avenue, Milwaukee, Wisconsin, 53211

DAVID SWANSON (7), American Board of Internal Medicine, Philadelphia, Pennsylvania 19104

KURT VAN LEHN (197), Cognitive and Instructional Systems, Xerox Corporation, Palo Alto, California 94304

---

[1]Present address: Program in Cognitive Science, Center for Human Information Processing, University of California, San Diego, La Jolla, California 92093.

Preface

This book is intended to introduce the student and researcher in developmental and educational psychology, in mathematics education, and in cognitive science to an exciting area of theory and research—the development of mathematical thinking. The book approaches the subject from a variety of perspectives, including those of the cognitive developmental psychologist, the cross-cultural researcher, the mathematics educator, the cognitive scientist, the philosopher, the psychometrician, and the clinician. The book shows that these perspectives, stressing as they do different kinds of ideas, methods, and values, can all aid in clarifying our understanding of that most complex phenomenon, mathematical cognition. Each chapter was specially prepared for this volume.

Many people contribute to an undertaking of this sort—primarily of course the authors, who offered not only their talents, but also a deep commitment to the project. Thanks are due also to Arthur Baroody who took the time to make useful comments on several of the papers; to Harry Beilin for his editorial support; to Edie Chamberlin for her patient and skillful secretarial help; and to Kathy Gannon for preparing the subject index.

The Development of
Mathematical Thinking

HERBERT P. GINSBURG

Introduction

In the 1970s, developmental psychologists rediscovered the topic of mathe-matical thinking, which once had been an important area for research, especially in educational psychology. In 1922, Thorndike wrote an insightful *Psychology of Arithmetic,* focusing strongly on educational problems and applications. Al-though his theory was stated in behaviorist terms, it continues to offer important insights for today's researchers. Indeed, Thorndike's work was only one exam-ple of several major research efforts carried out, with results still of interest today, by such educational psychologists as Buswell and Judd (1925). As is the case with too many important developments in psychology, work on the psychol-ogy of number lay dormant for many years, perhaps because the theory of the time had reached its limits. Then, in the United States, interest in Piaget emerged in the 1950s and 1960s, and developmental psychologists were captivated by the complexities of conservation, including the conservation of number. In the 1960s and 1970s, the developmentalists probed and prodded conservation, seriation, and classification every which way, perhaps with mixed results. But in the process of assimilating Piaget, some psychologists discovered a broader set of problems involving mathematical thinking and used Piagetian and other cogni-tive concepts and methods to investigate the new subject matter. In the 1970s, considerable research on mathematical thinking was undertaken. Investigations employed the clinical interview and talking aloud procedures, as well as more traditional paradigms, to investigate phenomena ranging from the 3-year-old's concept of *more* to the adolescent's understanding of the limit.

1

In the 1980s, research into mathematical thinking continues to enjoy considerable popularity and indeed attracts increasing interest in developmental and educational psychology and cognitive science. There are many reasons for this. Mathematical cognition is a splendid example of intellectual activity. It originates in the perceptions and sensorimotor activities of infancy and develops into the abstract concepts and mental procedures of adulthood. In the course of development, mathematical cognition assumes many faces, including perception (e.g., the discrimination of quantity), language (e.g., the grammar of the counting words), problem solving (e.g., word problems), procedural knowledge (e.g., mental calculation), and comprehension (e.g., the part–whole schema). Mathematical cognition involves everyday, out-of-school activities as well as instruction in the context of formal education. It involves informal, intuitive knowledge as well as written, codified abstractions. It involves rote repetition as well as the highest forms of human creativity. So the first reason for studying mathematical cognition is that it embraces a good deal of the complexity of the human mind.

Second, the researcher enjoys some tactical advantages in the study of mathematical cognition. The subject matter can be well-defined, even formalized. This makes it possible to develop theories that aim for similar precision and to employ formal mathematical models of the learning and thinking processes. Furthermore, mathematical cognition can be investigated with a variety of techniques, ranging from the traditional procedures of the laboratory (e.g., matching-from-sample and reaction-time methods) to recently popular methods like Piaget's clinical interview and Newell and Simon's talking aloud technique.

Third, research into mathematical cognition offers opportunities for important contributions to education. Our recently obtained knowledge already points the way to major innovations in diagnostic testing and the education of the "learning disabled." These are no small contributions; no doubt others will follow.

The general aim of this book is to introduce the reader—researchers in developmental and educational psychology, cognitive science, and mathematics education—to key topics in mathematical cognition. The chapters in this volume review, criticize, and analyze relevant literature and propose new syntheses, hypotheses, and theories. Some chapters focus on empirical studies, others on theoretical issues, and one on methods. The chapters are diverse, hopefully capturing key themes in the emerging field. No simple conclusion or grand unifying theory is possible; the field is in flux and our knowledge is limited.

The book begins with an examination of the novel *protocol methods*—the talking aloud technique and the clinical interview procedure—which have promoted substantial progress in research on mathematical cognition. Ginsburg, Kossan, Schwartz, and Swanson describe the methods and their aims and evaluate the rationales underlying the methods. In doing this, the authors deal with the central issues of introspection, reliability, and validity. The general theme is that protocol methods are especially valuable for cognitive research and in many respects are superior to standard procedures.

Fuson and Hall (Chapter 2) offer an analysis of the early use and understanding of number words. As is now well known, the child exhibits basic numerical skills and understandings at a remarkably young age. Fuson and Hall's chapter sheds new light on the complexities of this early development. The authors propose that early number develops within at least six different environmental contexts and involves varied activities, ranging from saying the number words in proper sequence to estimating quantities. The young child must learn that the meanings and uses of number differ in the various contexts. Once this differentiation has been successfully accomplished, the child must integrate different aspects of number so that he or she can shift meanings rapidly and with minimal confusion. The manner in which this integration is accomplished is an important topic for future research.

Resnick (Chapter 3) offers a new theory dealing primarily with the manner in which the young child's number concepts become elaborated during the early years of school. Her general theme is that key aspects of understanding—basic concepts or schemes—strongly underlie the development of procedural knowledge like mental addition or written calculation. She describes several stages. In the first stage the child in the late preschool years uses a kind of mental number line as a schema to direct elementary counting and quantity comparisons. In the second stage, the child begins to acquire a part–whole schema, which permits new accomplishments in mental calculation. And in the third stage, in the later elementary school years, the child elaborates on the part–whole schema to develop knowledge concerning the decimal structure of numbers and to achieve a deeper understanding of written calculation. Resnick's theory not only provides new insight into the child's knowledge and techniques of written arithmetic, but through its discussion of the part–whole schema also provides an interesting link between Piaget's theory and recent cognitive science notions.

Riley, Greeno, and Heller (Chapter 4) provide a detailed model of the child's solution (or attempts at solution) of arithmetic word problems. These authors concur with Resnick in proposing that conceptual knowledge, and not just procedural skill, plays a central role in such problem solving. The information-processing model proposes that problem schemas or interpretations organize plans, monitor actions, and decide among possible procedures. The model is described with precision and elegance and appears extremely successful as an explanation of key classes of word problem solving. The authors also offer stimulating speculations concerning the interplay between conceptual and procedural knowledge in the course of development. Needless to say, the work of Riley, Greeno, and Heller has important implications for educational practice in the areas of curriculum and testing.

Chapter 5 by VanLehn, which presents a formal computational theory of subtraction in the school context, may be read at several different levels. On one level, the chapter describes the nature and learning of *bugs*—systematic error procedures—underlying children's computations in elementary subtraction. VanLehn shows how instruction results in *core procedures,* how these may break down when confronted by certain environmental demands (e.g., new types of carrying procedures), and how the child creates systematic bugs when he or she attempts to *repair* the breakdowns. At this level, the chapter provides both a thorough, cognitive analysis of computational errors and, what is rarer still, a convincing theory of their acquisition. The theory of bugs and repairs has important implications for diagnostic procedures and the training of teachers. At another level, the chapter is metatheoretical, examining the nature of and criteria for formal models of mathematical thought. For example, VanLehn argues that the very language one uses to describe mental representation has a profound impact on a theory's success. He also discusses how cognitive theories can account for key constraints that characterize knowledge at all stages of development. Taken as a whole, the VanLehn chapter is an especially convincing demonstration of the power of formal theories in accounting for fundamental aspects of knowledge and learning.

In Chapter 6, Davis proposes an information-processing account of complex mathematical cognition, including aspects of the calculus. The chapter stresses the role of flexible methods of data collection: Many advances in the field have resulted from intensive case studies, interviews, and the like. Davis also demonstrates that our models of knowledge representation and processing must be extraordinarily complex to account for the intricacies of higher forms of mathematical thought. Davis makes creative use of information-processing notions to explore some of these intricacies. The chapter's emphasis on the conceptual and creative aspects of higher mathematics reinforces the notion that genuine mathematics is based on very humane forms of insight.

Saxe and Posner (Chapter 7) broaden our perspective by discussing the manner in which the development of number is embedded in social life. The authors begin by describing two major and conflicting theoretical influences on cross-cultural research into cognitive development, namely Piaget's notion of the self-directed equilibration of logical thought and Vygotsky's historical–cultural determinism. Saxe and Posner use the Vygotsky-Piaget contrast as a framework for organizing cross-cultural research into number development. The empirical evidence shows how economic and cultural factors shape mathematical thought and how certain aspects of number appear to develop spontaneously, regardless of cultural factors. Evidence reveals a complex situation: Although economic and cultural forces shape some aspects of mathematical cognition, the individual, regardless of culture, engages in self-directed activities that construct his or her own concept of number and produce cognitive universals. Both Piaget and Vygotsky seem to have perceived part of the truth. The authors conclude with a synthesis and extension of the two approaches—a new schema for examining the relations between culture and cognition.

Allardice and Ginsburg (Chapter 8) add a further dimension to this book by examining learning difficulties in children's mathematics. After analyzing the traditional concept of *learning disabilities,* the authors review both environmental and neurological theories of ''cognitive deficits'' in mathematical thought. There appears to be little evidence supporting either view. The chapter then examines the few existing studies that deal with cognitive factors underlying school failure in mathematics. The authors conclude that the vast majority of children experiencing difficulty with school mathematics suffer from no serious or deep-seated cognitive or motivational disorders and that there is a clear need to change the direction of research in this area.

Chapter 9 by Groen and Kieran examines the relevance of Piaget's theory for understanding children's knowledge of academic mathematics. The relevance is not immediately obvious since, the authors argue, Piaget's theory is primarily a genetic epistemology, designed to answer certain philsophical questions concerning the development of knowledge. Piaget's psychological investigations must be understood within the framework of genetic epistemology. Moreover, the psychological aspect of Piaget's work has evolved over many years, particularly with respect to the notion of *stages.* Piaget's empirical research on the child's concept of number, conducted several decades ago, needs to be reinterpreted in the light of changes in his theorizing. Only then can we clarify the relevance of Piagetian research for the understanding of the child's academic mathematics.

REFERENCES

Buswell, G. T., & Judd, C. H. Summary of educational investigations relating to arithmetic. *Supplementary Educational Monographs,* 1925, 27, 1–212.
Thorndike, E. L. *The psychology of arithmetic.* New York: Macmillan, 1922.

HERBERT P. GINSBURG
NANCY E. KOSSAN
ROBERT SCHWARTZ
DAVID SWANSON

CHAPTER **1**

Protocol Methods in Research on Mathematical Thinking[1]

Current research on the development of mathematical cognition employs a variety of data collection methods, ranging from the chronometric measurements of the laboratory (Groen & Parkman, 1972) to naturalistic observations gathered in African tailor shops (Lave, 1980). There is heavy emphasis on *protocol methods*—the talking aloud procedure and the clinical interview technique—which typically involve the gathering of a rich corpus of data, ultimately represented in written protocols. The protocols are used as the basis for inferences about underlying cognitive processes (or structures, strategies, operations, etc., depending on the terminology and concepts of the theory used) involved in intellectual activities. Although increasingly popular, protocol methods have not yet received thorough methodological analysis. Little is known about their fundamental natures, the rationales underlying their use, and their reliability and validity. This chapter offers an examination of these issues. First we present examples of the various protocol methods and describe their aims and rationales.

[1]Preparation of this chapter was partially supported by a grant to Herbert P. Ginsburg from the National Institute of Education, NIE-G-78-0163.

7

THE DEVELOPMENT OF MATHEMATICAL THINKING

EXAMPLES, AIMS, RATIONALES

Talking Aloud

EXAMPLE

Newell and Simon (1972) have made extensive use of the talking aloud method, in which the subject is instructed to say everything that comes into his or her head while solving a challenging problem. Aside from the initial instruction to talk aloud, the researcher seldom intervenes beyond presenting the relevant tasks. Consider a condensed version of a talking aloud protocol involving cryptarithmetic.

The subject is presented with the problem DONALD + GERALD = ROBERT and, given that D is 5, is asked to determine the numerical values of each letter.

INTERVIEWER (I): *I will give you that D is 5 in this problem. Please talk.*
SUBJECT (S): *Well D . . . giving D = 5 automatically makes T a zero. Could you make T a zero?*
I: *T is a zero.*
S: *Because 5 plus 5 is equal to 10. And that's simple from the problem. And looking at the leftmost column, you can see that R is either 1 or 2 greater than G, but that doesn't seem to help very much at this point. In the second column having the two L's equal, and also the two A's equal in the third column, doesn't seem to help too much at this point either. . .* [Newell & Simon, 1972, p. 329].

In talking aloud procedures, the data are mainly the subject's verbalizations, with rare comments by the researcher. Occasionally, behavioral observations may also be made. The protocol preserves the verbalizations and relevant behaviors.

AIMS AND RATIONALE

The aims of investigators who employ the talking aloud method are to elicit and describe the integrated activities constituting complex problem solving in intelligent adults and to identify the cognitive processes and internal symbolic mechanisms underlying adult problem solving. It is argued that both of these aims are best achieved by obtaining data from a situation in which the subject deals with a challenging problem and in which there is a minimum of investigator intervention.

The first aim of the talking aloud method is to elicit complex forms of problem solving, not isolated responses. Underlying this aim is the theoretical proposition that problem solving is a complex, sequential activity, involving such components as the formulation and testing of hypotheses and the execution of multistep algorithms. Problem solving involves far more than isolated responses or reaction times. In the talking aloud procedures utilizing challenging problems, subjects seem to exhibit the complex problem-solving activities that are of theoretical interest. Thus, in our example, the subject performs calculations, makes hypotheses, evaluates their utility, and in general engages in a complex chain of reasoning over time. The subject's behavior is obviously far richer and more complex than that obtained through eye movement or reaction time paradigms. Presumably the talking aloud procedure allows the investigator to elicit behavior *representative* of complex problem solving.

The second aim of the talking aloud procedure is to identify the internal symbolic mechanisms that underlie problem solving. Frequently during ordinary problem solving there are long periods in which subjects exhibit little overt behavior, but seem to be thinking. Given instructions to talk aloud, subjects report complex thought processes, such as rejected hypotheses and alternative solutions attempted. These verbal reports seem to offer insight into the working and structure of the mind. Using these verbal reports, the investigator extracts the meaning of the utterances, proposes statements about the knowledge and operations to be attributed to the subject, and constructs the web of inference that supports these attributions (Newell & Simon, 1972).

The two aims just described—eliciting complex problem solving and identifying underlying cognitive processes—are both said to require data obtained from a situation in which the subject deals with a challenging problem and in which there is a minimum amount of investigator intervention. It is obvious that the problem must be challenging: Complex problem-solving behavior cannot be elicited by trivial, uninteresting problems. Investigator intervention must be minimal for several reasons. First, since we have limited knowledge about problem-solving activities and underlying cognitive processes, a major aim is to learn more about them; therefore it is necessary to elicit behavior in an open-ended fashion and to avoid biasing or limiting what may occur. We need rich data to construct interesting theories of cognition. Second, in the Newell and Simon view, problem solving is a process strongly dependent on the individual's history and idiosyncratic knowledge base. The data collection method must be sufficiently sensitive to reveal what the individual brings to the problem-solving encounter. The method must be open-ended in order to permit the emergence of individual differences in problem solving and the identification of the individual's knowledge base.

In brief, the talking aloud procedure permits investigation of the complex activities that constitute problem solving and the exploration of the internal symbolic mechanisms that underlie it. Since the investigator plays a limited role in the protocol collection, the data can be viewed as relatively unbiased accounts of the individual subject's problem-solving performance and may shed light on individual differences in cognitive processes.

Clinical Interview

The clinical interview method takes two forms—verbal and revised. We first consider examples of each and then their aims and rationales.

EXAMPLES

The verbal clinical interview method, as originally used by Piaget, involves flexible questioning of individual children on a totally verbal level; concrete objects do not serve to illustrate or represent the problem posed by the interview. Consider the following example provided by Gelman (1980, p. 65):

I: *What's the biggest number you can think of?*
S: *A billion.*
I: *Is that the biggest number?*
S: *No.*
I: *What is?*
S: *I don't know.*
I: *Why don't you know?*
S: *Numbers never end.*
I: *What if somebody told you that a googol is the biggest number?*
S: *I'd say that numbers don't end.*
I: *Could you prove it?*
S: *No.*
I: *What if you add 1 to a googol?*
S: *A googol and 1.*

In this case, the data are entirely verbal, consisting of a series of questions by the interviewer and answers by the subject. The protocol simply preserves the content of the verbal interaction.

The revised clinical interview method was developed when Piaget concluded that the verbal method was sometimes inadequate, particularly in the case of younger children who often have difficulty in dealing with verbal problems. Piaget attempted to improve the method by using concrete objects to illustrate the problem to be solved. Presumably, these concrete supports permit better understanding of the problem and provide the child with an opportunity to manipulate the objects, thereby revealing his or her thought. Hence, in the revised clinical method, the data of interest are both verbalizations and aspects of nonverbal behavior. The classical example is the conservation of number, as described by Piaget and Szeminska (1952):

> Hab (5:3) began by putting 9 sweets opposite the 6 of the model, but made the row the same length. *That's it.*—Are they the same?—*I'm not sure.*—Where are there more?—*There* (row of 9, close together).—What must we do then?—(She put 6 opposite 6 of the model and removed the rest).—(The 6 of the model were then closed up).—Are they the same?—*No.*—Are there as many here (model) as there?—*No, there* (copy) *there are more.*—Is there more to eat on one side than on the other, or are they both the same?—*I shall have more to eat.*—Make them both the same, then—(She removed 2, then made the one–one correspondence, and finally put the two back when she found they were missing!) [p. 79].

In the revised clinical method, the interviewer presents problems and questions in a flexible manner in which the question may be contingent on the previous response. The data are the subject's verbalizations and his or her problem-solving behavior (as when Hab puts out nine sweets opposite the model's six).

AIMS AND RATIONALE

Piaget developed the clinical method in order to achieve three aims central to the study of cognitive development: (*a*) the elicitation of intellectual activities, (*b*) the specification of the nature and organization of cognitive processes (e.g., providing a logico-mathematical model of the child's thought, as in the case of the concrete operations), and (*c*) the evaluation of the child's level of cognitive competence.

One aim of cognitive developmental research is to elicit the intellectual activities used by children in a variety of contexts. Hence, the clinical method attempts to give children the opportunity to engage in various intellectual activities, like speculating on the origins of the sun, the fairness of a set of rules, or the conservation of an equivalence relation. Often, these activities involve a complex, sequential chain of reasoning; measurement of single, isolated responses is inadequate.

A second aim of cognitive developmental research involves the specification of the nature and organization of cognitive processes. Once relevant intellectual activities have been elicited, the cognitive processes underlying them need to be examined in detail and described as precisely as possible. Thus, after we observe that the child believes that a set of rules cannot be modified or that an equivalence relation is conserved over a visible transformation, we need to identify the cognitive processes presumably underlying these activities. Piaget usually describes these processes in terms of the "structures" of intelligence, general mental capacities presumably underlying performance on a wide range of tasks. These structures are portrayed in terms of logico-mathematical models.

Other theorists, of course, characterize the cognitive processes in terms of other representational models.

Note that elicitation of intellectual activities and specification of cognitive processes operate at several levels of specificity. At the beginning stages of research, we have a low level of specificity in which the investigator wishes to obtain preliminary indications of the relevant phenomena and of the cognitive processes underlying them. This stage of research may be said to involve *discovery*.[2] Piaget felt that in general our knowledge of cognition is so limited that it is necessary to begin most research investigations with the aim of discovering relevant phenomena and processes. Thus, the investigator may seek to establish whether 4-year-olds are capable of some form of mental addition or whether unschooled individuals know some number facts. Next, it is necessary to get a rough idea of the kind of cognitive processes children employ in these situations. Thus, does the 4-year-old add by some form of counting, or does the unschooled individual seem to "remember" the number facts? Such discovery requires extensive exploration of children's thought, and for this task Piaget designed the clinical interview procedure, an unstructured and open-ended method intended to give the child the opportunity to display his or her "natural inclination." Piaget (1929) describes the essence of the method as follows: "The clinical examination is dependent on direct observation . . . the practitioner lets himself be led . . . and takes account of the whole of the mental context [p. 8]." While beginning with a set problem, the clinical interview is designed to permit a kind of naturalistic observation of unanticipated results and a flexible exploration of their meaning. This is intended to satisfy the aim of discovery.

At a more precise level of research, which we may call *specification,* the investigator wishes to go beyond the discovery stage to obtain further empirical data of interest and to develop precise descriptions of cognitive process. Thus, if the 4-year-old is thought to add by counting, the investigator must develop a precise account (model) of the counting process (e.g., in the manner of Groen & Parkman, 1972). Or if the unschooled individual is thought to "remember," one must develop a relevant model of recall.

The accurate specification of structures is a complex, inferential process. Often, a given behavior may be produced by several different underlying structures or processes. To take a simple example, the answer "10" to the problem $5 + 5 = ?$ may be obtained by memory (which itself needs to be explained), by counting on the fingers, or by an "invented strategy" like "$4 + 4 = 8$ and then 9, 10 because there are 2 left over." Distinguishing among these different

---

[2]The term *discovery* has many different uses in discussions of methodology. For example, it is often used to distinguish the phase of generating or inventing (i.e., discovering) hypotheses from the subsequent testing of hypotheses. We are using *discovery* here to contrast the more exploratory phase of a study with later, more explicit and detailed theorizing.

strategies and describing the one actually used are best accomplished not by the standardized presentation of the problem $5 + 5 = ?$, but by the flexible questioning involved in the clinical method. This questioning must allow the examiner to test alternative hypotheses concerning the nature of the underlying processes and to evaluate the significance of the subject's verbalizations concerning his or her method of solution. (These may be misleading since often the subject does not have direct access to his or her cognitive processes.) The examiner's testing and evaluation both require considerable flexibility. In particular, they require the on-the-spot invention of critical tests that must be contingent on the subject's immediately preceding behavior. The clinical interview procedure affords the flexibility needed to accomplish the difficult aim of specifying cognitive processes.

Note that discovery and specification differ from each other mainly in degree. In both cases, the investigator must elicit relevant behaviors and develop an account of the underlying cognitive processes. Furthermore, discovery and specification appear to be cyclical processes. The researcher often engages in some discovery activity, followed by specification, which again leads to the need for further discovery, and so on. What is considered discovery and what specification is relative within the context of some body of theory.

The third aim of cognitive developmental research is the evaluation of cognitive competence. Piaget's theory aims at establishing underlying competence—the child's highest ability at his current stage of development—and does not often deal with typical performance—the child's ordinary behavior on a particular occasion.

The investigation of competence focuses on three areas: subjective equivalence, seriousness of response, and strength of belief. The examiner needs to determine whether (a) the child understands the question in the way intended, (b) the child takes the task seriously, and (c) the child's response is deeply held. All of these activities are best achieved through the flexibility of the clinical interview.

1. To measure competence in a given area, we must guarantee that all subjects perceive a problem's basic features and requirements in the way intended. Yet, presented with the *objectively* identical, standard problem, different children may understand it in very different terms. If standardized instructions do not produce the common, intended understanding, then a clarification or modification of the instructions may be required for many children. Thus, deliberately varying objective conditions—the unstandardized instructions of the clinical interview—may be needed to produce "subjective equivalence" of understanding. Similarly, in cross-cultural research, ". . . [subjective] equivalence is best approximated by the use of culture-specific materials that are equivalent along cultural dimensions but are not carbon copies of each other [Scribner & Cole, 1976]."

2. A second activity useful in the evaluation of competence is the assessment of the "seriousness" of an answer. In his early work, Piaget (1929) found that sometimes young children engage in "romancing." Faced with the necessity to respond to a problem they do not fully understand, children may *invent* an answer that does not reflect a well-thought-out belief. Asked to justify the answer, children may then invent a rationale that, in fact, originally was unconnected with the production of the answer. (Similarly, in cross-cultural research, respondents sometimes give answers that are thought to please or amuse foreigners, but that are not indicative of actual beliefs). To determine whether the child is romancing, Piaget found it necessary to engage in various forms of flexible questioning—for example, posing the same problem several times in different ways in order to investigate the consistency of response.

3. A third approach to competence is similar and appears in Piaget's later work where he attempts to determine the strength of the child's belief in a particular explanation. If the child seems to have achieved successful conservation, the examiner challenges the child's response through the use of countersuggestion. If the child then modifies his or her response, the examiner may conclude that the apparent competence is not genuine, or at least that it is fragile. Such information is vital for determining the child's stage.

In general, then, to establish different aspects of competence, it is useful to use flexible, nonstandardized procedures. Other theorists concur. As Chomsky (1964) put it: "Obviously one can find out about competence only by studying performance, but this study must be carried out in devious and clever ways, if any serious result is to be obtained [p. 36]."

Mixed Cases

The talking aloud procedure and clinical interview methods may be used in various combinations and modifications and may involve elements of naturalistic observation.

A simple variant on the talking aloud procedure involves occasional intervention on the part of the interviewer. Thus, in a study of geometric problem solving, Greeno (1976, p. 482) presents a protocol of a subject solving a geometry problem. For the most part, the inverviewer, after presenting the initial problem, did not intervene except for an occasional "mm-hm." At one point, however, the subject failed to disclose an important part of his reasoning; consequently, the interviewer asked a direct question, "Now, when do you think you could tell that you were going to use angle-side-angle?" This directed intervention was necessary to obtain important data. Aside from the intervention, the protocol is a typical example of the talking aloud procedure.

A more complex example is the following mixed case, which combines the talking aloud and revised clinical interview procedures (Ginsburg, 1977, p. 137).

INTERVIEWER (I): *I'm going to give you another problem. You seem to be doing pretty well adding. Suppose you have 29 again and 4.*
PATTY (P) WROTE

$$29$$
$$+4$$

Before placing the 4 under the 2, her hand hesitated under the 9; apparently she could not decide where to place the 4. Patty then said, "You put the 4 over here . . . that would be . . . that's 9 . . . [she whispered] 2, 3, 4, 5, 6 . . . 69." She had counted to get the sum of 2 and 4. She wrote

$$29$$
$$+4$$
$$\overline{69}$$

This first segment of the protocol involves the interviewer's initial presentation of a problem, the subject's written response, and then a sequence of verbalizations, whispers, and finally a written solution. Part of the sequence has the flavor of a talking aloud protocol: In response to a problem the subject verbalized a complex train of thought, even though she had not been asked specific probing questions to elicit the verbalizations. At the same time, the sequence is not a typical talking aloud protocol since the subject was not explicitly asked to verbalize.

Later in the interview, we have the following interchange (pp. 142–143):

I: *Let's do this . . . there are ten of these* [chips] *and here's one more. How many do think altogether?*
P: *Altogether it would be 11.*
I: *OK. What about 10 **plus** 1, not **altogether** but plus?*
P: *Then you'd have to put 20.*
I: *What if we write down on paper, here's 20, now I write down another 1, and you want to find out how much the 20 and 1 are altogether?*
P: *It's 21.*
I: *Now what would 20 **plus** 1 be?*
P: *20 plus 1?* [Patty wrote   20]
$$\frac{1}{30}$$

In this second segment, the interviewer presented several problems and engaged in flexible questioning, much as in the manner of the revised clinical interview.

In brief, the mixed case involves both the use of the talking aloud procedure to generate "spontaneous" accounts of problem solving and some variant of the clinical interview method to check hypotheses suggested by these accounts.

COMPARISON OF METHODS

Next, we compare the aims of investigators who use the different methods and contrast the methods in terms of certain underlying dimensions.

Comparison of Aims

Investigators employing the talking aloud procedure and the clinical interview method share at least two basic aims: the elicitation of intellectual activities and the specification of the nature and organization of cognitive processes. Both sets of investigators, avoiding a priori definitions, wish to elicit subjects' spontaneous cognitive activities. To accomplish this aim, the clinical interviewer follows the child's thought wherever it leads; the investigator employing the talking aloud procedure intervenes as little as possible when the subject says everything that comes into his or her head. Both sets of investigators also share an interest in the specification of the nature and organization of cognitive processes. To accomplish this aim, the clinical interviewer performs certain critical tests contingent on the subject's prior responses; the investigator using the talking aloud method administers to subjects common problems (e.g., GERALD + ROBERT) and uses the resulting protocols to infer underlying cognitive processes.

The two sets of investigators differ somewhat in their concern with competence. Users of the clinical interview are vitally concerned with the evaluation of competence. The aim is to identify each child's highest level of functioning (given his or her current state of development). Users of the talking aloud procedure seem to *assume* that each individual is performing at his or her typically fairly high level of competence and instead focus on individual differences in expertise, as in the case of expert–novice comparisons. Perhaps the two groups of investigators differ in their concern with competence because of their subjects' characteristics. Investigators generally have used the talking aloud method to investigate problem solving by adults, whereas the clinical interview method was developed for use with children. The "deviousness" of the clinical interview is made necessary by its subjects' lack of verbal skills, poor comprehension, and sometimes minimal cooperation; the unstructured nature of the talking aloud procedure is suited to the sophistication and cooperativeness of its adult subjects.

In brief, the two major protocol methods share similar aims with respect to elicitation and specification of cognitive processes. The differing emphases on competence may reflect differences in the subject populations with which the methods are typically concerned. The general commonality of aims is also reflected in the fact that it is possible to produce a coherent blend of the two procedures; the *mixed method* incorporates both the spontaneous verbalizations of the talking aloud procedure and the focused probes of the clinical interview method.

Evaluation of Other Methods

Both protocol methods imply similar evaluations of other research techniques. Users of both protocol methods share an interest in *eliciting* cognitive activities in an unbiased fashion. To accomplish this, one would like to obtain observations of thought in various natural settings. But such observations are exceedingly difficult to make, partly because much thought is private and partly because the occasions on which thought is public are few and far between. Consequently, for purposes of cognitive study, naturalistic observation is usually not practical as a technique and must be replaced by the protocol methods.

A second possiblity is the *standard method*. This is a procedure by which the same problem or task is presented to all subjects. The standard method includes standarized tests, like achievement tests,[3] and most laboratory procedures, like the reversal shift task, administered to all subjects in the same fashion. Such standard methods are of limited value both for purposes of elicitation and specification of cognitive processes. Consider first elicitation. By necessity, a standardized test—for example, a mathematics achievement test—usually focuses on a predetermined range of subject matter. The tester must ask certain questions and does not enjoy the freedom to vary their form and nature, nor to follow up on interesting leads by asking new questions. The main aim of the standard test is to produce reliable rankings of individuals on some characteristic. But this is not maximally effective for elicitation (nor is it meant to be). The standard test does not offer the kind of flexibility required for extensive exploration, for the immediate pursuit of interesting phenomena.

In the case of specification of process, the standard test may fall short because it usually fails to yield rich data concerning complex cognitive pro-

---

[3]Recently, standardized tests have been elaborated into a more flexible format where tests are "tailored"—constructed contingently based upon examinee characteristics and performance.

cesses. The individual's knowledge base and problem-solving strategies are too rich and subtle to be adequately portrayed by a circumscribed set of questions. In brief, users of both protocol methods concur that, for the purpose of cognitive research, naturalistic observation and standard tests are neither as efficient nor as effective as the talking aloud and clinical interview method.

Underlying Dimensions of Protocol Procedures

The protocol methods may be compared on several underlying dimensions. The dimensions refer both to the interviewer's techinques and tasks and to the subject's responses.

INTERVIEWER DIMENSIONS

The Initial Presentation of a Task.   In all three protocol methods—talking aloud, verbal clinical interview, and revised clinical interview—the interviewer determines the subject's initial task by presenting a problem. In the talking aloud procedure, the initial task is the only one presented by the interviewer. By contrast, in the two clinical procedures, the subject's reaction to the initial task may lead the interviewer to introduce further tasks to clarify various interpretations. Hence, the subject's response influences the selection of all tasks after the initial one, and the tasks frequently change in the course of the interview.

Concreteness.   Tasks may include concrete objects (empirical referents) or problems stated entirely on the verbal level. The concrete–verbal dimension serves mainly to distinguish the verbal clinical method from the revised clinical method. Piaget makes a sharp distinction between the two methods, claiming that the revised method is generally more effective than the verbal one for use with young children. (Of course, there may be cases where the use of concrete problems is not possible, as in Piaget's study of dreams.) The talking aloud procedure may involve tasks of both types.

The Demand for Reflection.   In all three methods, the interviewer asks the subject to verbalize his or her thoughts. Yet the methods differ in the extent to which the interviewer asks the subject to reflect on what he or she has done. In the talking aloud procedure, the interviewer requests only that the subject verbalize ongoing activities. By contrast, in the clinical methods, the interviewer may request reflections on completed activities (Q: *How did you get the answer?* A: *I added on my fingers.*) or on the reasons for a solution (Q: *Why did you add on your fingers?* A: *I always get the right answer that way.*).

Channeling into Areas of Interest.   Sometimes the interviewer directs the subject's remarks into certain areas of interest. Thus, in the clinical interview procedures, the interviewer may desire to discover whether the subject has engaged in a particular kind of thought. To find this out, the interviewer asks a series of questions that push the subject's remarks in a certain direction. In the talking aloud procedure, such channeling does not occur, aside from the initial presentation of a problem.

Contingency.   Sometimes, the interviewer's questions are contingent upon the subject's response. Thus, in the clinical interview methods, the interviewer begins with a particular problem, but the sequencing of questions is determined by the subject's response. In the conservation problem, if the subject believes that two rows are the same in number, then one line of questioning follows; if the subject believes that the lines are not the same in number, then a different line of questioning ensues. If the subject gives one reason for his or her response, then a certain counter-suggestion is offered. The contingency of questioning appears to be the essence of clinical interviewing and sets it apart from the talking aloud method, in which only the initial task is presented.

Standardization of Questions.   In the talking aloud methods, there are no questions aside from the initial one, which is presented to all subjects and hence is standard. By contrast, the clinical interview methods vary in degree of standardization. At one extreme, when the interviewer is attempting an initial investigation of some concept, he or she may not know in advance what the relevant questions are, so that they can hardly be standardized. In this case, the interviewer develops the questions, on the spot, in response to the subject's responses. At another point in the continuum, when an area has been thoroughly investigated (like conservation), the interviewer may employ virtually the same set of questions for all subjects. He or she may ask everyone, "Do these rows have the same number?" and later "Why do you think this row has more?" In a genuine clinical interview this standardized set of questions may not be given in an identical sequence to all subjects, nor may all questions be used. Instead the various "standard" questions will be used contingent on the subject's responses and may be omitted or supplemented as necessary. If the questions are all given to all subjects in the same manner and order, without omissions or additions, then the result is no longer a genuine clinical interview, but a standard test. Contingency defines the clinical interviewing methods.

Use of the Experimental Method.   Sometimes, in the clinical interview procedure, the interviewer may employ the experimental method, in the sense of holding some variables constant while deliberately varying others. Thus, in the interview with Patty, cited previously as a mixed case, the interviewer deliberately varied the problem's language ("plus" versus "altogether") and its concreteness (chips versus numerals). This is essentially a factorial arrangement and, if undertaken with a large enough number of subjects (rather than one) and an appropriate counter-balancing procedure (rather than none), would be an example of traditional experimental design. Sometimes clinical interviews make use of such experiments and sometimes they do not. Of course, the talking aloud procedure does not make use of such experiments since there is no interviewer intervention beyond the presentation of the initial problem.

Use of Naturalistic Observation.   Pure naturalistic observation involves the unobtrusive recording of a subject's behavior occurring "spontaneously"—that is, without the observer's intervention, although the behavior may be caused by any one of several factors—in his or her natural environment (e.g., a playground). Since all of the protocol methods involve the interviewer's deliberate imposition of a problem, they cannot be said to involve naturalistic observation in a strict sense. Thus, the talking aloud procedure is the observation of the subject's provoked (rather than spontaneous) verbalizations elicited by an imposed (rather than natural) problem. Yet sometimes the three methods approach naturalistic observation in certain ways. The talking aloud procedure involves almost no interviewer intervention after the posing of the initial problem and the request to verbalize thoughts. The essential spirit of the talking aloud procedure is to discover the subject's thought processes with as little interference as possible—that is, only enough to get the subject to reveal his or her thoughts. This general aim is quite consonant with the spirit of naturalistic observation. Similarly, in the clinical interview method, the interviewer is often surprised by some of the subject's remarks and behaviors and observes them carefully before deciding what to do next. This frequently occurs in—indeed almost defines—exploratory clinical interviews. Thus, in the mixed case previously described, the interviewer observes Patty solving the problem 29 + 4 in an unanticipated manner, and this almost "naturalistic observation" suggests hypotheses for future investigation. Thus, a kind of naturalistic observation may occur within the clinical interview procedure, especially when exploratory. Of course, very often such observation does not occur, as when the subject gives a constrained response to highly focused questions (e.g., Q: *Does this have more or does this row?* A: *This row does*).

SUBJECT DIMENSIONS

Language and Behavior.   In the talking aloud and verbal clinical interview methods, the main data are the subjects' verbalizations. Occasionally, behaviors like eye movements, facial expressions indicating puzzlement, or hand movements indicating finger counting may be recorded, but attention to such behaviors is usually minimal. By contrast, in the revised clinical procedure, words and behaviors receive heavy emphasis: The interviewer pays careful attention to what the child does with the concrete objects involved in the task.[4]

[4]It seems possible to conduct a kind of clinical interview entirely on a behavioral level. This is essentially what Piaget has done in his study of infancy as reported in the *Origins of Intelligence* (1952). In this case, Piaget often presents the child with a concrete problem and then, depending on the subject's reaction, "interviews" her by presenting new problems especially designed to illuminate certain interpretations.

Multiple Levels of Description.   In the talking aloud method, subjects' verbalizations operate at certain levels of abstraction with respect to cognitive processes and therefore provide some types of information but not others. For example, it is unlikely that a subject would choose to say, *Now I begin to move my fingers to grasp the pencil; I pick it up; then I move my arm and begin to write down numbers and lines.* The subject's verbalizations are unlikely to focus on molecular aspects of the task and instead may describe molar aspects of problem solving like, *First I decided to calculate the sum.* In the talking aloud procedure, the subject seems to select a level of discourse appropriate to the task as he or she perceives it. By contrast, in the clinical methods, the interviewer plays a major role in determining the level of abstraction to be employed in describing the cognitive processes that are the subject of investigation. Thus, the interviewer asks, *How did you know that line was longer?,* rather than *In what sequence did you move your eyes as you compared the rows?* Of course, the interviewer's questions attempt to be sensitive to the child's spontaneous conceptions of his or her own thought processes and may even employ terms suggested by the child (e.g., Q: *How did you know that?* A: *My brain told me.* Q: *Why did your brain tell you?*). Nevertheless, in the clinical methods, the interviewer influences the level of discourse.

The Integral Nature of Verbalizations.   Sometimes, the subject's verbalizations are an integral part of the solution; sometimes they are not. Thus, when the task involves counting a set, the verbalization of the counting words is integral to the task's solution. By contrast, when the subject is asked to remember 6 plus 7, the process of retrieval from long-term memory (assuming that the subject does indeed remember the sum) may not involve verbalization at its core. The dimension of integrality does not seem to differentiate any of the protocol methods from any others; rather it seems to be a dimension of cognitive processes employed on various tasks.

Introspection.   The subjects' ability to report on the processes involved in problem solving influences what kind of information is available both in the talking aloud and clinical interview procedures. Whether or not they are integral to the task, there are certain cognitive processes—for example, the mechanisms of perception—that do not normally lend themselves to introspection and therefore cannot be made available through verbal report. Thus, there is no point in asking subjects how they perceived depth or in expecting a verbal account of the process to appear spontaneously in the talking aloud procedure.

VALIDITY OF VERBAL REPORTS

The previous sections have described the nature, aims, and rationales of protocol methods and have contrasted them along certain basic dimensions. Furthermore, we have shown that users of protocol methods feel that both naturalistic observation and standard methods suffer from serious deficiencies as instruments of cognitive research. Yet the protocol methods themselves must be critically examined on methodological grounds. We have seen that all of the protocol methods make extensive use of verbal reports, and the clinical interview methods rely heavily on nonstandard, contingent procedures and indeed are virtually defined by them. These key characteristics of protocol methods often prompt criticism. Contingency can result in substantial variation from subject to subject in the context of interviews and in the data collected. Such variation obviously raises questions concerning the comparability of information gained by two interviews in the same study or by different interviewers in similar studies. Analogously, introspective self-report methods have a controversial history in psychology; questions concerning their accuracy and relevance naturally arise whenever they are used. The next section discusses the status and vailidity of verbal reports and addresses the following issues of reliability and validity with respect to protocol methods.

History

Central to the protocol methods we have been describing is their dependence on subjects' verbal reports. But reliance on such data is thought by many to be methodologically problematic, if not outright illegitimate. Prejudice against the use of verbal reports, however, is not new. Debates about their significance have been part-and-parcel of the classic controversy over the role of introspection in psychological inquiry. And, as the whole idea of introspective psychology progressively fell into disrepute, the status of verbal data followed suit. Nevertheless, we believe protocol methods can provide fruitful data, although we also allow that much of the early criticism had its point. A brief review of some of the challenges to introspection will thus offer a useful perspective from which to sort out the issues.

Underlying most early introspectionist claims were certain assumptions about the nature of mind as well as certain assumptions about people's abilities to report on their mental life. The two were connected in that the account of verbal reports followed directly from the view of mind presupposed. The model of mind employed was a common one and remains much a part of today's common-sense views of cognition. According to the model, mental life was to be identified with conscious awareness; mental states were conscious states. And what made a particular state the mental state it was were its phenomenally experienced properties. Pains differed from hunger because they felt different, and the thought, belief, or desire that "The Mets will win in 1982" differed from the thought, belief, or desire that "The Pirates will win in 1982" because the ideas consciously entertained were not the same. Thus, the kind and content of a mental state was determined by the characteristics of the experience the person had, and this naturally was something of which the person could be said to be aware.

Given this picture of mind and the mental, the rationale for using verbal reports is straightforward. Minds are transparent to their owners. To have or to be in a mental state is to be in an experiential state to which the mind has direct access. So once we learn an adequate vocabulary, there should be no problem reporting on our mental life. Furthermore, the account of language much in vogue ensured that the vocabulary would be adequate to the job. Acquiring a vocabulary required grasping the meanings of words—the pairing of symbols with their proper ideas. But meanings or ideas were seen as just another kind of mental state. Having once learned the word *red* by associating it with the appropriate experience of a *red-idea,* we could be sure of its suitability for describing future "red" mental states.

Although verbal reports were, in general, assumed to be reliable, some introspective theorists did allow that particular reports might not always be accurate. But the source of difficulty was not traced to the subject's lack of direct access to mental phenomena, nor primarily to unbreachable inadequacies of the reporting language. Rather the difficulties were seen to lie either in the fact that some mental states disappeared very rapidly or in the fact that subjects, conditioned by habit and past experience, failed to limit their reports to what was actually presented to them in experience. They went beyond the "given," embellishing it with interpretation. Whereas it was admitted then that verbal reports might not always be totally accurate, it was nevertheless assumed that the person having or being in the mental state was in a better position to report on it than anyone else. In fact, an even stronger claim was often thought to follow. Since *only* we are aware of our own experiences, our mental life is essentially private. Psychological research *must* rely on verbal reports of the subject, for no one else could *possibly* be in a position to observe them. Psychology thus could not get off the ground without using introspective reports as data.

In the early part of the twentieth century, this intuitively plausible conception of mind came under serious attack. Logical positivists saw in the model a threat to their own commitments to physicalism and the unity of science. They responded by claiming that introspective (or "mentalistic") psychology was methodologically untenable. The model violated basic canons of objectivity, relying as it did on reports and hypotheses that were not amenable to intersubjective testing and scrutiny. If there was to be a *science* of human activity, it, like all science, would have to focus on *publicly* observable events and employ only those concepts definable in terms of test conditions that referred to observable physical attributes. For the positivists, translating mentalistic language into talk of overt patterns of behavior seemed like the only way to go, if psychology was ever to be put on firm scientific footing.

Behaviorists, on the other hand, saw the the positivists' methodological scruples support for their own antimentalistic, antitheoretical tendencies. Behaviorists took these arguments to provide further sanctions for their demands that psychology ground itself in observable S's and R's. Behaviorists also had strong independent reasons for abandoning the introspectionist's enterprise. They felt that the approach had not produced fruitful explanations and that the data gained from introspective reporting were unreliable. Subjects never seemed to agree on what the experienced mental state was really like. What's more, many of the behaviorists' own studies, especially those on animal learning, seemed to be making progress, without taking any account of conscious experience and certainly without using verbal reports. Finally, talk of internal mental states, open only to the introspection of their possessor, seemed unnecessary. Once the principles that linked stimuli with overt responses were discovered, all that was worth accounting for could be explained without going beyond the publicly observable.

Another challenge to the introspectionist's model came from work in the philosophy of mind, especially that derived from Wittgenstein's probing analysis of mentalist language. Philosophers began to question the status of the fundamental tenet of the introspectionist's model of mind—the identity of mental with conscious—for not only did the model seem to lead into metaphysical quagmires, but dependence on conscious qualities to define and give content to mental states was thought at bottom to be an incoherent program. Wittgenstein and others did not deny that persons were conscious organisms; rather they argued that we could not equate the states and dispositions necessary to explain behavior with the ongoing conditions of awareness. Briefly put, their point was that unless a postulated state served to relate and integrate *behaviors* in appropriate ways, the state would not count as a particular mental state, no matter what the conscious experience felt like. You may feel now just as I typically do when I am hungry, but if you are not disposed to pursue and ingest food but instead liquid, your state is not one of hunger but thirst. The supposed similarity or dissimilarity of phenomenal experience goes by the board. Also challenged was the picture of language that supported the introspectionist model. What the earlier view didn't pay enough heed to was the social nature of language learning. The teaching community does not have access to the learner's private experiences, so of necessity must rely on pairing utterances with publicly observable external objects. As long as the learner, for example, applies the word *red* appropriately to physical objects, it will make little difference what the accompanying experiences are "really like." The same holds for the child's acquisition of a vocabulary to describe his or her mental states. The community must key its teaching of words like *anger* and *fear* with observable goings on and not with inaccessibly private feelings. There must then be reliable behavioral patterns associated with mental concepts or there could be no meaningful use of those terms at all.

Another jolt to the transparent mind model came from a seemingly different direction. Relying on his clinical evidence, Freud was led to argue that many of our most important mental states are not consciously available to us. These states play a central role in organizing and determining our behavior, but we are not aware of them and consequently cannot report on them. It is this aspect of Freud's theory—that is, playing down the significance of conscious awareness—that has appealed to behaviorists such as Skinner.

Now although the programs of the positivists, behaviorists, Wittgensteinians, and Freudians have their problems, the challenges raised against the transparent model of mind and the conscious content account of mental states are not without merit. Nevertheless, the rationales underlying some of this criticism have themselves come into question, and the ways in which this has happened are relevant to analyzing current uses of protocol methods. To begin, the basic positivist conception of scientific methodology, which provided an overall context for evaluating psychological theories, is itself no longer accepted. The more liberally minded positivists saw early on that limiting science to observable phenomena and observationally definable concepts was too restrictive. Such limitations would prohibit any rich science from developing. Indeed, the positivists' own paradigm of good science—physics—could not be made to fit this atheoretical mold. More recent work in the philosophy of science has also questioned whether the very distinction between observable and nonobservable phenomena and concepts can be drawn in a methodologically insightful way.

This loosening of positivist scruples has also served to undermine the framework of the behaviorist's position. The behaviorist can no longer argue on purely methodological grounds that it is illegitimate to postulate internal states and mechanisms, or that all legitimate concepts must be reducible to sets of observable attributes, or that a commitment to the behaviorist paradigm is the true mark of doing scientific research. Moreover, the behaviorist's positive program has begun to look less promising. It seems highly unlikely that a science of psychology, limited to describing input–output relations and behavioristically defining away apparent mediating states, will be adequate to explore any rich range of activity.

The purpose of this brief historical review has been to set the stage for examining current controversies concerning the use of verbal reports. The link to these earlier issues is clear. With positivist and behaviorist strictures against talk of theoretically postulated mental states no longer in force, today's *cognitive* psychologist feels free to probe for the underlying knowledge, structures, and mechanisms that are thought to mediate cognition and behavior. A major problem facing the researcher, though, is how to secure data that will be helpful in generating and testing theories about these mental constructs. Many researchers have attempted to embed their investigations in standard experimental methodologies—the usual paradigm of an experiment with a contrasting control group and an analysis of data that links dependent and independent variables (with, perhaps, special emphasis on reaction times). But as the previous section of this chapter indicates, a growing number of investigators have been attempting to expand the data base for doing cognitive psychology. To many, use of protocol methodologies has seemed like a promising means of supplementation. Unfortunately, the contemporary cognitivist can no longer depend on the introspectionist model of mind to provide a simple rationale for the use of these verbal data. Once it is admitted that the mind is not transparent, that mental states cannot be identified with states of conscious awareness, and that language is not essentially geared to describing private mental states, questions concerning the reliability and significance of introspective reports begin to loom large. So the protocol enthusiasts' dilemma is to square methodological needs with a reasonably balanced view of mind and the mental. We believe it can be done. Put simply, our position is that protocol methods do have a place in cognitive research; that there are important limits and constraints on their use; that effective use of protocol data requires paying careful attention to these limits and constraints; and that, when this is done, many of the remaining qualms about using protocol data apply equally to other sorts of data collected by more standard research methods.

Some Uses of Verbal Reports

Even in the heyday of behaviorism, verbal data were not completely ignored. In many cases, verbalizations comprised part of the task itself. The subject is asked to recall the items of a list, or to say which tone sounds louder, or which object appears nearer, or to spell out a solution to a puzzle, or to indicate his or her choice among alternative routes or rewards. In some of these instances, verbal reports could be eliminated in favor of other behavioral indicators. In other cases—for example, list memory or puzzle solutions—it is less apparent how this could be done. Anyway, it is not obvious why anyone should want to employ more elaborate, roundabout procedures for no apparent gain in information or reliability of data.

Another related area, where there seems to be little reason for *overall* skepticism about introspective reports, are subjects' claims about their own propositional attitudes (i.e., their beliefs, desires, thoughts, expectations, etc.) True, we can no longer assume that such reports are the result of direct intuitions of conscious happenings, or that they can not be mistaken, or that, in principle, there can be no other means for determining their existence. And true, we may have to allow that some of our beliefs and desires are not known to us and that attempts to report on their content either may not be possible or may alter their very nature. But with all this said, the fact remains that, over a wide range of conditions and situations, people are reasonably good at telling what they believe, want, and expect. In fact, with some of our more complex propositional attitudes (e.g., having such relatively abstract thoughts, beliefs, or desires as "The price of gold will go up in 1983 if there is a drought in the Midwest") it would be hard to make sense of the claim that such states are independent of an accompanying disposition to articulate them in a language or other symbol system. And although people are frequently unaware of all their motives and the real bases for certain actions, neither do these lapses show that the reasons people do cite for their behavior are generally mistaken. The findings of social psychologists that show that, between two identical objects, people unknowingly tend to choose the one on the right or that people may be unaware that racial or sexual bias influenced a decision should make us cautious of accepting all reports at face value. But proper caution is different from rampant skepticism. In the ordinary course of things, there may be little reason to question a man's claim that he chose the blue tie because it matched his suit better than the brown, or another's claim that she preferred candidate B because he better met the qualifications for the position. And notice too, even in the right–left identity study, subjects' reports that they chose the item on the right because they believed it to be of better quality may be correct. Where they have gone wrong is in not realizing the cause of this inappropriate belief.

Finally, we think it is a reasonable practice to rely on subjects' descriptions of some of their cognitive processing. This point in part depends on how one construes the notion of *processing*. Construed narrowly enough—for example, limited to descriptions of neural processing—introspective reports will be of no help. On the other hand, if stages or steps taken in solving a problem are the aspects of processing that are of interest, verbal reports may be a valuable source of information. For example, in the case of mathematics, it would be gratuitous to deny weight to subjects' claims that they did the column addition-problem starting at the bottom, or that in cryptarithmetic they made use of the fact that adding a number to itself always results in zero or an even number, or that in solving a logic problem they began with the conclusion and worked backward. And although it may be admitted that a person has little or no access to the sources of his or her insight or to the underlying processes responsible for creative synthesis, even here, all needn't be opaque. The mathematician may readily sketch out tried approaches that failed and key realizations that turned the problem around. And the poet, perhaps with the aid of his or her earlier drafts, may very well be able to provide us with an understanding of why one line was dropped, another changed, and a new word chosen to replace the old one.

So we believe that introspective reports can provide useful information and that protocol methods have a place in research. Allowing that introspective reports should be given credence, however, does not mean that we buy the old introspectionist model of mind. Introspective reports are not in general based on direct "inner perceptions" of conscious qualities; such reports may be in error;

and other tests may be used to confirm or refute them. Indeed, the clinical method has resources available to check on subject error. The interviewer can always pose additional questions, challenges, or tasks to test the soundness of any suspect report. The talking aloud method is less suited to take such corrective measures since it attempts to be nondirective. Even so, the talking aloud investigator might give the subject another task and hope the reports it elicits will provide the bases for a better overall evaluation. Alternatively, the talking aloud theorist might temporarily suspend the nondirective stance and resort to clinical interviewing.

Domains

Introspective reports of mental states and processes can be expected to be accurate only within domains to which subjects have access. Now it should be clear that subjects frequently are not in a position to know about various areas of their cognitive activity. Even the heartiest old-time introspectionist allowed that the mechanisms underlying sensory and motor skills were not all available for introspective report; nor was it thought that the mechanisms of association, the nature of the trace, and the relative strengths, weaknesses, and paths of association were things the subject would readily know.

How informative reports will be may therefore depend on the sorts of activities subjects are questioned about. It would be most helpful, of course, for investigators to have at hand a general model of mind and cognition that would tell in advance which processes and states subjects have access to, which they don't, and why. But it is unlikely that anything like this is going to be worked out in detail prior to developing better theories and understanding of the individual domains themselves. So the whole thing must be something of a bootstrap operation. Nevertheless, it seems to be an empirical fact that in some areas—for example, mathematics—subjects can report on various aspects of their activities. Equally, a certain amount of common sense and exploratory inquiry would indicate that in the case, say, of reading it would be foolish to suppose that subjects could report on deep principles of grammatical structure, on how sentence meaning is derived, or on where exactly their points of eye fixation were. What information is accessible to subjects will vary from domain to domain.

Selectivity and Levels of Description

The object in front of you is a pencil, a wooden thing, contains cellulose molecules, is Navy blue, reflects light of predominantly 470 nm wavelengths, is smaller than a breadbox, and is not made in Syracuse. The number of correct descriptions is unbounded; hence a *complete* characterization is impossible. Any attempt at description is always constrained by the need to select aspects and levels of reporting. At the same time, such selection is constrained by the knowledge and data available to the person doing the describing. We may have the knowledge and information to report on some aspects and not on others, and, within an aspect, we may be able to report at one level of detail and analysis and not at another. These same considerations come into play when dealing with verbal reports of mental states and processes. What a subject reports will always involve selectivity and interpretation. Introspective descriptions are not representations of an unconceptualized mental given, but, of necessity, reflect the subject's skills and habits of categorization. There is no unique way to report on our cognitive activity and, within a given domain of description, there will be levels and features of analysis for which we have no information.

When we ask a subject what she did or experienced or why she accomplished a task we must allow that there is no one correct answer. To the question, why she went to the bank today, she might be equally accurate in responding: *Tomorrow is a legal holiday, I wanted to cash a check,* or *John was too busy to do it.* Even within a selected aspect of description, there will always be alternative levels of analysis, some of which the subject may be aware of and others not. The subject may describe her last tennis serve sparsely as, *I adjusted my swing to hit it hard to the backhand,* or in more detail as, *I changed my grip this way and threw the ball higher,* or she may be able to give more specific details about the movements, steps, and adjustments along the way. But, of course, sooner or later the subject will hit a level of characterization—be it neurological or at a finer specification of eye, finger, and arm movements—to which she has no access and cannot accurately report. The same holds for more intellectual tasks. In doing a mental long division problem the subject may be able to explain how she grouped the numbers and in what order she applied subtraction, multiplication, and simple division, but have no idea of the processes underlying the choice of strategies and may be able to say no more about how she multiplied 9×5 than that she knew it equaled 45.

So in seeking to acquire fruitful data on states or processes via protocol methods, we must be careful to assure that the subject has access to descriptions at the level of analysis that interest the invesigator. And where various levels are accessible, steps must be taken to secure reports at the level relevant to the researcher's project. For this purpose the clinical method would seem to have an advantage over the talking aloud approach. To the extent that talking aloud theorists wish to remain noninterventionist, they must rely on context, task structure, or common conversational expectations to key the subject as to type and level of report desired. The clinical interviewer, however, can take a more direct approach by specifically indicating the level and by explicitly questioning subjects about those aspects of interest. In any case, no matter which protocol method is adopted, it must be recognized that self-reports can never be complete or reveal all the causal or processing factors of interest. But, then again, no source of data is ever complete or can provide answers to all of our questions.

Report Interference

That reporting on mental states and processes might interfere with or change the very nature of the mental phenomena was an issue much debated in earlier arguments over introspection. Many theorists who accepted the transparent model of mind still doubted that accurate reports could be given of *ongoing* mental activity. For example, Brentano (1973) argued that although we have inner perception of our mental life and it is the primary source of psychological data, we could hardly ever report on our current inner perceptions without destroying the character of the phenomena being experienced. A case he cites is that of anger. His claim is that were we to stand back and, as calm observers, dispassionately focus attention on the state, we would no longer be in the same state of anger. James (1950) voiced like concerns about simultaneous introspective reports. Both men, however, thought that such obstacles could be overcome by relying on memory reports, as well as other less direct means of self observation. They admitted that delay or deviation from direct inspection would have a cost in accuracy and detail and would leave more room for external interferences, but this could not be helped.

Quite similar doubts about the interfering effects of verbal reporting have been raised against current use of protocol methods. And surely, with regard to various states and tasks, the doubts are justified. It's commonplace that if we attempt to focus on how we are doing some kinds of things, we may not only affect the course of the process, but may no longer be able to complete the task. And even when we can succeed, adopting a reflective attitude may affect such variables as reaction time. Also, there is the possibility that calling attention to the need for reports may make subjects self-conscious and lead them to employ different strategies or means than they might if left on their own. Thus, there often may be real problems involved when requiring verbal reports of ongoing processes. As earlier theorists maintained, under some circumstances at least, it may be more fruitful to ask for reports after the task is completed. Again, there will always be the added danger of memory decay, interference, and the subject's merely making up an account of what he or she thought probably went on. And there is no way to eliminate the influences of past experience, expectation, and habits of categorization and organization. All descriptions, by necessity, are filtered through the reporting subjects' conceptual schemes and systems of interpretation. Such reports are never certain, and in doubtful cases it would be wise to test their correctness by other means. But once we allow for the possibility of error, there seems to be no reason to reject all process reports out of hand. The fact remains that some reports provide useful information and do not appear to be seriously distorted by the factors just described.

With mental states, as opposed to processes, there may be even less reason for wholesale skepticism based on report-interference considerations. We noted earlier that people are reasonably good at describing their propositional attitudes and, over a wide range of conditions, can probably do better than anyone else in describing simple emotional and sensory states. Admittedly, mistakes occur and the demand for introspective reports may trigger the sort of self-reflection that leads to a change of mind. Surprisingly, however, abandoning the introspectionist's mental-conscious identity makes the problem of simultaneous report interference less of a difficulty. In the introspectionist's view, it was the immediately sensed conscious goings-on that constituted the mental state. Given this model, it is easy to understand why pausing cooly to analyze the flow of experience would be thought to change its felt qualities and tones (hence, Brentano's qualms about reporting anger). But once the mental state as experiential state model is abandoned in favor of construing mental states as more theoretical and stable underlying structures and dispositions, there is less reason to expect the act of reporting to interfere. If we see a person in an agitated state of unknown nature, it would seem prima facie reasonable to accept one's claim that he or she is angry because of recent dealings with the telephone company. Stopping to report like this may calm the person down, but it need not remove the anger. He or she may ramain angry at the telephone company for weeks on end, while engaging in many activities of a much more obtrusive nature than introspective reporting. And the fact that we attribute anger to him or her throughout this period is one more reason why shades and tones of ongoing conscious experience cannot in general be taken as essential to being in a particular mental state. All protocol methods must allow for the possibility that introspective reporting may affect the phenomena under study. Since the talking aloud method requires simultaneous reports, it is particularly vulnerable to interference in cases where the introspective activity taps the cognitive resources needed to perform the task at hand. The clinical method must allow for this, too, whenever the interviewer asks for simultaneous reports. But even when the act of reporting is not simultaneous or does not itself disrupt the task, there still remains the possibility that the mere call to self-reflect will alter the very phenomena under examination.

Ambiguity

Ambiguity is another source of variation in verbal reports, and its effects must be taken into account when analyzing protocol data. An interesting case in point are subjects' responses to ''why-questioning.'' We think it important to

distinguish at least three types of anwers that may be run together in the protocols. We label these different senses: (*a*) processing, (*b*) explicatory, and (*c*)
rationales. This classification scheme is neither exhaustive nor precise, but we
believe the sorts of distinctions it points to are of particular relevance to studies in
mathematical cognition.

Suppose a subject is asked to describe how he did a column addition problem, and he reports that he added the rightmost column and carried a 3 to the next
column. If asked why he did this, he may reply that the first column came to 32
and that he was taught a procedure or algorithm that requires putting the 3 in the
nextmost column. He may even be able to spell out the procedure in detail. These
answers then represent the subject's attempt to report on processing activity.
Alternatively, the why-question might be construed as asking for the subject's
explanation for the carrying maneuver, in the sense of "Why does carrying work
to give the correct sum?" Were the subject to understand the question in this
manner, he might cite facts about addition or the modular ten basis of the
numbers. Both sets of verbal reports may be accurate, but they are answers to
different questions. Notice too, the subject may be able to add skillfully, but not
know why the procedure works, or he may be able to provide such an explicatory
account, but have little skill at actually doing computations (e.g., due to memory
or attention failures).

Another example may bring home the contrast more sharply. Suppose a
person is asked to report on doing the division problem $1258 \div 37$. The subject
replies that she played around with the numbers until seeing that 37 goes into 125
at most 3 times, multiplied 3×37 getting 111, and subtracted that from 125.
However, if she were now asked why she subtracted 111 from 125, at least two
sorts of answers would be appropriate. The processing answer may involve
pointing out that subtraction at this stage is part of a method always followed
when doing long division. The explicatory answer would be to explain why
subtraction is an arithmetically sound step at this stage in the computation.
Again, it is possible that the subject may be able to divide and not justify the
procedure, or be able to provide an analysis yet lack long-division skill.

A related, but somewhat different sense in which why-questions may be
answered is for the subject to offer what we might term rationales. Suppose a
person is asked why or how he determined that some object in the room is a
chair. In this context, the subject may respond by offering pertinent data about
chairs or what distinguishes chairs from other items in the room. On reflection,
the subject may answer with a definition, for example, a chair is a portable seat
for one person, and X is such. But this characterization of important chair
properties may play no role in a real process account of how the subject determined that it was a chair and not a table or bench, nor may it justify a step or
stage in such a procedure. Similarly, if asked why he called a mark an instance of
the numeral 4, the subject may, on the spot, make up an explanation citing its

particular angles and connecting lines and contrasting it with other nearby numerals and letters. In this case, we know people are not able to provide definitions or defining properties that determine the application of concepts like *4-shaped* and *the letter A,* but asked to say something, they may cite salient features of the given instance. Furthermore, were the subject to spell out a set of features that characterized all and only 4s, it wouldn't follow that such an analytic features scheme played a role in actual processing or provided an account of why the procedure worked. After all, we can define a triangle as a polygon whose angles sum to 180°, although it is unlikely that any steps of angle measurement and summation precede our determining to label something a triangle. Rationale protocols may thus provide interesting information about the knowledge and beliefs of the subject, but are not accurate reports of real processing. And while they may be offered by the subject to defend or support and, in that sense, justify some action, they are not readily interpreted as providing an explanatory analysis of the particular processes or steps actually taken.

Potential ambiguity of subjects' responses is a feature of both protocol methods. Again, the more directive nature of the clinical interview has an advantage. It allows the experimenter to specify explicitly the type of report wanted or to probe for clarification when the protocols seem ambiguous.

Indirect Reports

In earlier parts of this section, we outlined what might be called the *protocol theorist's dilemma.* Breaking with behaviorism and S-R psychology, the cognitivist wants to talk about internal states and mechanisms. At the same time, abandoning the introspectionist model of mind means that there is no assurance that the subject has access to the features of cognition the theorist is interested in mapping out. In addition, the deeper and more abstract the structures or processes under consideration, the less likely it is that subjects will have access to them. Thus, there are serious limits and constraints on how far it is possible to go with verbal reports of this direct nature. On the other hand, we believe there is no reason to assume that protocol methods can be useful only in cases where the states or processes under investigation are accessible to the subject. Data gleaned from verbal reports may be valuable in generating and testing hypotheses when the reports are not *directly* linked to the particular hypothesis of interest.

We have already touched on such less direct uses of protocol material in our original analysis of the clinical and talking aloud methods and just recently when dealing with the various types of why-questions. When Piaget and others ask subjects to report on their reasons for a decision or answer, the point is not merely to catalog their responses, but to gain insight into the concepts and structures the subject tends to employ in the domain. For example, the subject's particular explanation or justification of a conservation judgment is not important in and of itself. Rather it serves as data for more theoretical hypotheses about the types of logical reasoning a child can or can not engage in at a certain stage of development.

Work in linguistics and psycholinguistics is another area of study where protocol data seem useful, but where the reports are only indrectly linked to the hypothesis of concern. The linguist wants to construct a grammar for English; however, it would be hopeless to expect speakers to be able to verbalize a grammar. Typically, the linguist sits down with informants and secures their judgments of grammaticality, ambiguity, paraphrase, and co-reference. In many instances the linguist doesn't question a subject, but instead relies on his or her own intuitions about the language. Either way, the data obtained are not thought of as direct reports of underlying structures or ongoing processes, but serve instead as touchstones for constructing abstract hypotheses about the nature of grammar and competence. Similarly, if we want to find out about children's comprehension problems, we would not expect children to be able to report directly on their processes of sound-meaning interpretation. But judicious questioning about reference relations, logical connections, and paraphrase may provide rich data for formulating hypotheses about the source of the problem. In both the normal competence and remedial situations, data secured by more standard research methods may also be of use. Nevertheless, in these cases, protocol methods seem at least as good as any alternatives in sight and probably a lot more helpful than naturalistic observation or standardized testing.

Another example of a more indirect use of protocol data can be found in the work of Freud. Although Freud challenged the mental-conscious equation, his method of inquiry—in overall theorizing, as well as in dealing with individual patients—depended heavily on protocol data. Not only does Freud allow that verbal reports may be inaccurate; it is part of the theory that important mental states may be inaccessible to the subject. Indeed, the theory predicts that certain verbal reports will be mistaken. It is the essence of a repressed belief or desire that the subject will deny its presence. Still Freud uses introspective data as a fertile source of material for generating and testing hypotheses about mental states, structures, and processes. With Freudian theory then we have a case where a verbal report denying some hypothesis H is, relative to that theory, taken to support the claim that H truly characterizes the subject.

Data and Theory

The previous Freudian example highlights a point about the function of evidence in general and the function of verbal data in particular. The point is that the use and significance of data is always relative to theory and to those other hypotheses and assumptions held about the domain under consideration. Data serve as evidence only against a set of background beliefs and assumptions. Individual hypotheses almost never entail or are never entailed by data alone. The richer and more theoretical a hypothesis, the more complex and tenuous is the route back from hypothesis to observable data. Verbal reports or data secured by experiments and observations of other kinds cannot serve to *test* hypotheses in isolation. When it comes to *generating* hypotheses, the link between evidence and theory is even looser. At most, data can be suggestive or point the way. Protocols, by themselves, are neither fruitful nor banal; it all depends on what we do with them. This, of course, means that the value or significance of any set of data will be a function of the intelligence and perspicacity of the theorist using it. To evaluate the fruitfulness of verbal data would be to see what its payoff has been or is likely to be. And in the case of research on mathemetical thinking, we believe the payoff has already been significant.

PROBLEMS RELATED TO CONTINGENCY

A defining characteristic of clinical interviewing methodology is its contingent structure. The specific direction an interview takes—the questions to be asked and the answers given—varies as a function of subject characteristics and answers to earlier questions. Both questions asked and answers obtained vary from subject to subject. In talking aloud methods, the experimenter is typically a much less active participant in determining interview direction. Nevertheless, the material covered—the verbal reports obtained—can vary greatly in form and content, as a function of subject knowledge, skills, and general characteristics. Such variability raises obvious questions. In what sense are two clinical interviews in the same study comparable? Can two researchers employing talking aloud methods to investigate the same domain expect to obtain similar results or, for that matter, to compare results? Given the subjective, contingent nature of the raw data in protocol studies, can agreement be obtained on what the data imply for theory? These questions are serious ones, meriting answers, if researchers are to accept the utility of protocol methods as legitimate research methodologies that provide reliable and valid information.

Theoretical Specificity and Research Purpose

Addressing these questions is made more complex because research using protocol methods varies in level of theoretical specificity and in purpose of investigation. As discussed earlier, during initial work in a new domain, the primary function of research activity is discovery. In the study of mathematical cognition, initial work focuses on the identification of interesting phenomena, the definition of the task environment to be explored, and the development of useful categories of mathematical knowledge and cognitive processes. After a period of discovery, a theory specification stage ensues. In this stage of research on mathematical thinking, theoretical ideas on cognitive structures, knowledge, and processes involved in solving mathematical problems become more developed. As successful theory is generated and tested at one level of detail, new theory is needed at the next, more detailed level. Thus, theory and research evolve cyclically, as more and more detailed understanding of mathematical thinking is achieved. The discovery stage is characterized by lack of detailed theoretical knowledge. The specification stage is characterized by the existence of some theoretical ideas that provide a context within which research is carried out.

Regardless of the level of theoretical specificity providing a context for research, any individual study may be oriented toward the purposes of hypothesis generation or hypothesis testing. During the discovery stage of research, initial activities are likely to fall in the hypothesis generation category since very little is known of the research domain. After some initial investigation, however, which identifies some very general hypotheses, a theory-testing phase of research is begun. A typical hypothesis during the discovery stage might be as broad as "Children count on their fingers when first learning to add." Research efforts then collect data bearing on this hypothesis. During the specification stage, research can still be oriented toward hypothesis generation or hypothesis testing, depending upon whether the aim is elaboration of existing theory or testing of theory already constructed. Thus, having determined that, indeed, children learning to add often do count on their fingers, a researcher might then decide to try to construct a more detailed theory of strategy employed, including the role of finger counting. Initial efforts toward this goal would involve hypothesis generation—the construction of a hypothesized strategy for beginning adders. After hypothesis generation, which increased the level of theoretical detail, the researcher would likely decide to test the newly generated theory with earlier theoretical insights providing a context within which theory testing takes place. Thus, hypothesis generation is characterized by lack of *specific* theory at the level of interest, not lack of *general* theory; hypothesis testing is characterized by the presence of specific theory undergoing research scrutiny.

Contingency during Hypothesis Generation

Theoretical specificity and research purpose affect the argument one makes for the use of protocol methods and, indeed, affect how protocol methods are used. In this section, the justification for protocol method use in hypothesis generation activities is developed; the next section presents a different justification appropriate to hypothesis testing efforts.

The hypothesis generation stage of research is, you will recall, characterized by lack of theoretical development at the level of research interest: More precise characterizations of the cognitive structures, knowledge, and processes involved in mathematical thinking are desired. The major criterion for use of a methodology during this stage is utility in fueling the investigator's intuition, in uncovering new phenomena, and in providing rich descriptions for theory induction. Because research activity during this stage is not (completely) theory directed, it is difficult to argue that any particular methodology is most useful. However, if the goal is to investigate the cognition involved in solving mathematical problems, protocol methods are an obvious choice since they provide, in elaborate detail, the subject's own view of the knowledge and cognitive processes involved. Although the subject's view of problem solving cannot be taken as infallibly correct, it surely provides relevant information for hypothesis generation.

Contingency in interviewing results is no special problem during the hypothesis generation stage of research activity. Indeed, contingency is probably useful and to be encouraged. Oftentimes, interviewing uncovers new and interesting phenomena to be explored in whatever level of detail is necessary for one's purpose (discovery or specification). Flexible procedures make it possible to follow up on new insights and leads in the context in which they are uncovered. This is obviously a very desirable characteristic during the hypothesis generation stage of research, where no clear specification or research direction is available. One need not know the answers to research questions to use contingent interviewing procedures, nor would one need to know the questions of interest—the method can be adapted to the identification and pursuit of interesting questions on a subject by subject basis.

Reliability and validity of measurement are not relevant issues during the hypothesis generation stage of research. These psychometric issues become important only when the purpose of measurement has been determined—when the constructs to be measured are theoretically specified. By definition, during the hypothesis generation stage of research, there is a lack of theoretical development; the purpose of the stage is to extend theory to a new level. Without theoretical definition of the constructs to be measured, the psychometric adequacy of the operational definitions of the constructs can hardly be an issue. If research is aimed at discovery of interesting phenomena, criticism based on interviewers behaving differently or obtaining different results is not appropriate. Similarly, if research is aimed at hypothesis generation at a more detailed level of theory, some initial difficulties in construct definition and theory development are to be expected; psychometric precision is simply not a relevant issue until such development has taken place. Consistency of measurement (reliability) must be undefined and unimportant if the purpose and processes of measurement are still being worked out. The accuracy of inferences (validity) drawn from analysis, usually of verbal reports, cannot be a serious problem if the types of inferences to be drawn have not been identified. Verbal reports are, during the hypothesis generation stage of research, analogous to raw, uninterpreted data in traditional testing. Theory dictates how these data are transformed to yield a score of some sort. It also dictates conditions under which that score should change or remain constant and what relationships to other scores should be like. Thus, during the hypothesis generation stage of research, no psychometric criticisms of the clinical interview are appropriate.

It is important to emphasize that, during the hypothesis generation stage, it is a *theory* of the *domain (i.e., mathematical thinking)* that is unavailable at the level of detail desired. A theory of *data* must be available, at least in a rudimentary and, perhaps, implicit form. No investigator begins a study from a totally neutral point of view; minimally, some beginning ideas concerning the general character of the research problem and the classes of data relevant to its investigation are necessary. One of the virtues of flexible protocol methods is that they do not prematurely limit the sorts of data that are admissible as relevant to theory construction: The theory of data can evolve as the researcher gains more experience with the domain of interest.

There is available, in a limited way, a theory of data relevant to the use of verbal reports, as was discussed in the previous section. This theory points out some problems in the accuracy and relevance of introspective, verbal reports to be kept in mind during the hypothesis generation stage of research: In a sense, it provides information for judging the validity of the raw data. If there is reason to believe that subjects have no access to the cognitive components of the area of mathematical thinking under study (e.g., retrieval mechanisms for overlearned number facts), there is little sense in using protocol methods, contingent or not, as a method of investigation. On the other hand, in many instances there is ample reason to believe that subjects can provide at least some information relevant to mathematical cognition, particularly when verbal reports concern macroscopic, integral components of problem solving. Subjects obviously have unique access to the elements of their own cognition, and it is not to be taken lightly in theory development. In such instances, the researcher can safely take verbal reports seriously, though perhaps not literally and infallibly, as a source (often the major source) of information concerning structure, knowledge, and cognitive processes involved in mathematical thinking. Thus, broad guidelines exist that provide information for use of protocol methods.

Although it is difficult to argue that some particular methodology is best during the hypothesis generation stage of research, it is relatively easy to maintain that noncontingent methods, including strict experimental approaches, are probably not maximally effective. The strength of experimental methodology is in the investigation of relationships between theoretical predictions and data, with experimental control over variables extraneous to the investigation. Thus, use of experimental methodology to investigate the cognitive processes involved in mathematical thinking requires a theory, including theoretical predictions, knowledge of relevant extraneous variables and mechanisms for controlling them, and a data collection method that reflects theoretical interests and experimental ends. Use of experimentation is problematic during the hypothesis generation stage of research since many of these components are, by definition, lacking at the theoretical level of interest.

A second difficulty in use of noncontingent methods is associated with the tendency toward rigidity in experimental procedures and measurement techniques. Rigidity implies knowledge of the appropriateness of these procedures and techniques: If no adaptation of method to subject is permitted, the prespecified method must correctly yield the desired measurement information if the research is to be successful. Yet in the hypothesis generation stage of research, firm knowledge of procedures and instrumentation is typically unavailable at the level of detail required for prespecification. Adjustment to individual subject characteristics is particularly problematic, but especially required, as discussed in the next section.

Contingency during the Hypothesis Testing Stage of Research

The argument for use of protocol methods during the hypothesis generation stage of research hinged on the need for rich description of behavior for theory development, with psychometric difficulties associated with contingency irrelevant because of lack of theory. The logic is different for use of protocol methods when theory is available and the predominant research issue is hypothesis testing. In this situation, the justification and rationale for protocol method use must address psychometric issues. Protocol methods can no longer be viewed as *research* methods; rather they are *measurement* methods—ways of acquiring data for theory testing, with the concomitant psychometric requirements of major concern.

IDIOGRAPHIC BARRIERS TO HYPOTHESIS TESTING

In the area of mathematical thinking, idiographic and historical factors pose serious barriers to hypothesis testing. Theories of mathematical thinking typically specify the nature of the cognitive structures, knowledge, and processes involved in solving mathematics problems. Often, some characterization of developmental milestones in learning to think mathematically is also specified, perhaps in terms of common categories of errors. This characterization is typically quite idealized and general, oriented toward competence over a range of problems rather than idiosyncratic performance on a specific problem.

Theoretical notions of competence must, however, make contact with performance factors. Any particular subject, prior to participation in a study, has had extensive mathematical experiences, often specifically in the area under study. *Sesame Street,* teaching by parents, the school curriculum, games and books, and spontaneous learning all may contribute to an individual's knowledge. Thus, the subject's history may obfuscate hypothesis testing: The link between idealized theory and the empirical realization in which it is tested is often quite tenuous since subject history usually varies idiographically. The interview with Patty in the section on mixed cases provides a nice example of idiographic barriers to theory testing that can be surmounted through contingent interviewing procedures. Patty's contact with addition prior to the interview had somehow resulted in a differentiation of the mathematical meanings of *altogether* and *plus. Altogether* may have evolved its meaning through informal, everyday contact with addition, whereas *plus* seems to have been defined more algorithmically, probably in a school setting. Given a research interest in mathematical competence, it is clear that contingent procedures usefully clarified Patty's problem solving capabilities. A conditional set of interviewing procedures, which specify methodological alternatives to be taken when particular circumstances arise, is very useful. Such contingencies in procedure provide adaptations to individual historical differences in mathematical learning.

Contingency in protocol methods in hypothesis testing is present for exactly this reason. Theory provides a fairly detailed framework for analyzing the cognitive structures, knowledge, and processes involved in an individual's mathematical thinking. For example, as a function of idiographic history, a particular subject may adopt one (e.g., finger counting and/or recall of specific number of facts) of a series of plausible methods or a combination of methods to solve problems (e.g., simple addition) posed by the interviewer. These methods will differ individually and in their combination from subject to subject. Errors in one or more components or in the coordination of components may be present. Concurrently, relevant knowledge (e.g., of number facts) may be in error or missing. The result is highly complex behavior for which several alternative explanations are possible. John Seely Brown's work (e.g., Brown & Burton, 1978) on arithmetic ''bugs'' provides many interesting examples of the range of methods and errors that occur in addition, as well as an idealized theory that accounts for the methods and errors. Given a theory of arithmetic behavior like Brown's, contingent interviewing procedures make it possible to break down the complex behavior into discrete, interrelated components and to differentiate alternative explanations of subject behavior, thus investigating whether the theory accounts for the behavior. Without contingent procedures in these cases, it would be very difficult to understand the complexity of behavior and to achieve healthy theory development and testing.

PSYCHOMETRIC CONSIDERATIONS

To the extent that prior pilot testing and methodological development have taken place, no unusual psychometric difficulties arise as a result of contingencies in interviewing. The sequence of experimenter activities varies among subjects, but not in unsystematic ways since the variation in activities is predefined through an explicit linkage between the more general theory and the interviewing methodology used to evaluate it. Similarly, in analysis of the resulting data, procedures are specified for transforming transcripts of the raw (verbal report) data of the interview into the theoretical constructs of interest. Variation in interviewer behavior must be taken into account in these procedures, but, in principle, poses no problems as long as the relationship between interview data and theoretical constructs is well defined.

For example, consider a hypothetical study using protocol methods. Interviewing procedures have been (at least broadly) defined ahead of time, and there are an explicit set of subject attributes for which "scores" are desired. Scores are interpreted quite broadly; for example, a score might be the assignment of a subject to a category as a user of a problem solving method or submethod. Each *subject* is interviewed by several *interviewers*. Each interview involves several *problems,* with interviewing procedure on each problem contingent on the nature of the subject's problem-solving efforts. Each problem yields a set of comparable *scores* (categories) of interest. Interviews are transcribed and several *raters* read each transcript and determine each score on each problem in each interview.

Traditional conceptions of reliability and validity apply in this situation in the usual way. Consider, for example, interrater reliability. For each combination of subject, problem, and interviewer, several raters have scored the interview. Variability in raters' scores associated with a subject × problem × interviewer combination provides an indication of error variance. Variability from subject to subject on these same scores provides an index of total variance in scores. Taken together, these two pieces of information yield an estimate of interrater reliability. Inter-interviewer reliability can be calculated in a similar way. For each subject, several interviewers carried out interviews. Variability in scores given to a subject by different interviewers provides an indication of error variance; variability from subject to subject provides total variance information. Once again a reliability estimate can be calculated, this time for inter-interviewer reliability. By viewing problems as items, we can also derive estimates of internal consistency (homogeneity). Generalizability theory (Cronbach, Gleser, Nanda & Rajaratnam, 1972) provides a single framework for conceptualizing the various reliability coefficients in this situation, just as it does in more traditional (noncontingent) situations.

The psychometric situation is very similar to the one encountered in the scoring of essay examinations. Sometimes, test takers may select the essay question that they answer (i.e., answer five out of eight questions). Oftentimes, different examinees choose to answer the same essay question in different ways. A scoring key identifies criteria for assigning points to answers; the key specifies a contingent relationship between answers provided and scores to be awarded. Talking aloud methods parallel this situation almost perfectly, with essay answers given orally rather than in writing. Clinical interviewing simply takes the contingency procedure one step further by introducing a contingent structure to the "examination" itself. Oral examinations, as often used in professional education, have the exact same psychometric characteristics. Scoring of essay tests and oral exams requires considerable judgment on the part of the scorer: Scores are determined subjectively. Analogously, protocol methods require judgment from interviewers and raters.

Internal consistency of subject performance across problems merits some special consideration. A common finding in research on mathematical thinking is the *lack* of interproblem consistency. Oftentimes, subjects adopt a strategy as a function of problem characteristics that are not immediately obvious to the researcher. Resnick's work (1976) on subtraction provides a nice example. She discovered that elementary school children used predominantly two strategies for subtraction. If the larger number is m and the smaller number is n, the *decrementing strategy* consists of setting a counter to m, decrementing it n times, and reading the counter. The second strategy, the *incrementing strategy*, consists of setting a counter to n, incrementing it until the counter reading matches m, and then reading the number of increments as the answer. However, both these strategies were used by most children, as a function of the relative size of n and $m - n$; children picked the most efficient strategy depending upon problem characteristics. Thus, measures of internal consistency must be interpreted in the light of the subject's perception of the problem-solving environment and other factors. Low internal consistency values—lack of consistency in performance across problems—may be expected on theoretical grounds and do not necessarily indicate flaws in the measurement technique.

The notion of inter-interviewer reliability also deserves some special attention since comparability of results from two studies using flexible methods is a major issue. Results of two studies may differ for several reasons. Obviously, contingency can be a problem if interviewing procedures and measurement constructs are not agreed upon beforehand. Similarly, interviewers can differ in their interviewing skills and judgment. Both these sources give rise to legitimate questions concerning reliability because of the subjectivity involved in use of protocol methods, both in contingency of interviewing procedures and in scoring. Such questions are, however, amenable to psychometric exploration and remediation; the solution is to use skilled interviewers and work out the details of the interviewing procedure, particularly the relationship between contingency in questioning and the measurement constructs of interest. Although some unanticipated variability in subjects' behavior in the interviewing situation often does occur, careful preparation of interviewing and scoring methods minimizes the problems that arise. Along a different dimension, two interviewers may obtain different results because they are interested in different aspects of the same problem; similar procedures leading to different results are perfectly acceptable and expected in such circumstances. No reliability problems arise here. Reliability of a *research* method is not psychometrically defined—only reliability of a measurement method is. Different researchers studying different but related problems should get different results. The analogous problem in experimental methodology arises when two experimenters conduct seemingly similar studies and obtain different or even apparently conflicting results. Such outcomes are very common in psychological research and often are an indication of the need for additional investigation of the differences. Surely they do not call the method into question; however, the *theories* involved are at issue.

Traditional conceptions of validity also apply to the hypothetical research study just outlined. The accuracy of the inferences drawn from the data collection and analysis procedures can be investigated in normal ways. For example, patterns of relationships to other measures of interest (IQ, arithmetic achievement on standardized tests, class performance on problems similar to those used in the study, etc.) can be established as dictated by the theoretical substance of the study. Oftentimes, because verbal protocols can be supplemented by observations of problem solving in the interviewing setting, by eye movement data, and by other process data, rich validation information is available.

The primary purpose of most studies of mathematical thinking is the characterization of the structure, knowledge, and processes involved in the solution of mathematics problems in some domain. Viewed as a measurement activity, this purpose seems somewhat different from the assignment of a quantitative score to an underlying trait, the typical purpose ascribed to traditional standard testing methods. However, one can view method identification as a *classification* activity, with theory specifying the range of components and subcomponents that a method might include. Such a view attributes *nominal* scale properties to the scores which result, rather than interval or ratio scale properties associated with quantitative scores. For example, subjects might be classified as finger counters, rote memorizers, and so on, with additional study then given to other subject characteristics related to those categories. Appropriate psychometric theory is available for nominal scales, the needed reliability and validity indices are already developed.

Throughout the preceding discussion of reliability and validity, it has been maintained that interviewing procedures should be predetermined, at least to some extent. It has been assumed that appropriate relationships among contingent interviewing procedures, subject behavior, and scoring of that behavior have been worked out in advance. It can (and probably should) be argued that very few investigators using protocol methods specify a priori the contingencies used in interviewing; they simply "go with the flow" of the interview, leaving implicit the adaptation of interviewer procedures to subject responses. Analysis procedures may similarly be implicitly related to the theoretical constructs of interest, with interview scores determined "wholistically." This approach may or may not present a problem, depending upon the routine of the interviewer, the complexity of the inferences to be made, the subject matter area, and so on. Hypothesis testing in the discovery stage might proceed quite successfully in this way since the theory to be tested is fairly global—for example, judgments of finger counting use are fairly straightforward. Research in the specification stage is more likely to be adversely affected since the judgments to be made are more subtle. When interrelations among contingency, subject behavior, and scoring are unspecified, reliability and validity of the interview as a measurement tool will probably be adversely affected. This is simply poor use of the methodology, not an indictment of the method itself.

Diagnostic Testing

Thus far, this chapter has addressed the use of protocol methods in research on mathematical thinking. Another, related use of interview-based methods is found in diagnostic testing of mathematics ability and achievement. Diagnostic testing of mathematics skills often employs procedures similar to clinical interviewing. The diagnostician presents a set of (sometimes standardized) problems to a child who attempts to work them. Oftentimes, the diagnostician intervenes in the child's attempts to find out more about the problem-solving process being used. Hypotheses about difficulties are generated and tested as in research efforts. Metaphorically, at least, this process resembles theory development in research on mathematical thinking, but for individual children. Thus, the initial phase of testing is one of discovery; given teacher and parent comments and the curriculum the child is exposed to at school, the diagnostician's first task is to broadly identify the child's problems. Hypotheses about difficulties are generated and tested, perhaps using some fairly standard battery of problems. The diagnostician's next task is one of specification. The broadly identified difficulties the child is experiencing must be defined more specifically. Diagnostic testing in this phase focuses upon difficulties identified earlier and seeks a more specific account of the source of the difficulties. Again, there are both hypothesis generation and testing aspects of the diagnostician's efforts. At this level, there may even be elements of factorial experimentation in hypothesis testing, as depicted previously in the mixed case example, where an interviewer systematically investigated *plus* versus *altogether* effects on problem-solving performance. The specification phase ends when the diagnostician feels that a sufficiently precise "theory" of an individual's mathematical thinking is available to prescribe remediation.

Diagnostic testing using such interviewing procedures has been termed *impressionistic* testing by Cronbach (1960), in contrast to the traditional *psychometric* philosophy of testing. Psychometric testing seeks to obtain (usually numeric) estimates of a limited number of aspects of a person's performance. These aspects are the measurement constructs of interest. Impressionistic testing, on the other hand, views the tester as a sensitive observer who uses a broad range of methods to obtain multiple cues, which are integrated into an overall impression, often a verbal description of subject characteristics. Interviewing methods fall naturally into the battery of techniques used by the impressionistic tester. Formal scoring plays a major role in psychometric testing, but almost no role in impressionistic testing.

In a sense, the protocol methods advocated in this chapter seek to draw on both testing philosophies. During the hypothesis generation phase of research, impressionistic testing is used to gain familiarity with the domain of interest and to bootstrap theory and measurement techinques. During hypothesis testing, what appears to be an impressionistic testing procedure is actually closer to a psychometric procedure because the contingencies in interviewing and scoring are worked out in advance, along with the specific measurement constructs to which scores are to be assigned.

Recent developments in measurement theory in the area of tailored testing (e.g., Lord, 1980) further reinforce this view of protocol methods as a flexible variant of standard psychometric methods. Tailored testing involves contingent selection and presentation of items to examinees using the best available estimates of examinee status on the traits being tested. For example, a tailored test of arithmetic achievement would contain an initial set of items based upon school background, teacher ratings, or prior test performance. Performance on the intitial set of items would provide a revised estimate of achievement, which would influence selection of the next item(s). Estimates are continually updated, in light of performance on new items, until a final estimate (with an appropriately small error of measurement) is determined.

Such tests are often computer-based, with sophisticated procedures for estimating scores and selecting items. Obviously, there are also analogs that are not machine-administered. There is a broad range of individual intelligence tests (e.g., the Binet) that uses contingent procedures: The particular items administered are a function of success or failure on prior items. Other examples are also found in tests of growth and development (the Denver Developmental), and of achievement (Lord, 1980). Psychometric theory for estimating reliability and validity of tests and inferences based upon them is well developed for these versions of tailored testing.

Diagnostic testing in mathematics, because of the state of the theoretical and measurement art in research on mathematical thinking, cannot really be viewed as tailored psychometric testing; it falls almost exclusively in the impressionistic camp of testing. There is little agreement on a taxonomy of general mathematical disabilities. Any particular child is likely to have a mixture of conceptual and procedural difficulties contributing to math learning problems, as well as more general learning problems and emotional difficulties (Chapter 8 by Allardice and Ginsburg). Thus, diagnostic testing takes on (and should take on) an artful, impressionistic philosophy in which each child is viewed to have a relatively unique set of problems to be understood and remedied.

The use of psychometric, rather than impressionistic testing methods in diagnosis of math learning difficulties awaits additional research on mathematical thinking. There is simply too little known about the nature of mathematical thinking and the nature of mathematical learning difficulties to work out the interviewing contingencies required for a true tailored testing approach. Our studies of mathematical learning are only beginning to make the transition from hypothesis generation to hypothesis testing at a level of theoretical detail relevant to the needs of practitioners. We believe that protocol methods have a large role to play in advancing theory to meet the practical challenge.

REFERENCES

Brentano, F. *Psychology from an empirical standpoint.* New York: Humanities Press, 1973.

Brown, J. S., & Burton, R. B. Diagnostic models for procedural bugs in basic mathematical skills. *Cognitive Science,* 1978, *2,* 155–192.

Chomsky, N. "Discussion" in U. Bellugi & R. Brown (Eds.), The acquisition of language. *Monographs of the Society for Research in Child Development,* 1964, *29* (Serial No. 92), 35–39.

Cronbach, L. *Essentials of psychological testing* (2nd ed.). New York: Harper and Row, 1960.

Cronbach, L., Gleser, G., Nanda, H., & Rajaratnam, N. *The dependability of behavioral measurements.* New York: Wiley, 1972.

Gelman, R. What young children know about mathematics. *Educational Psychologist,* 1980, *15,* 54–68.

Ginsburg, H. P. *Children's arithmetic: The learning process.* New York: D. Van Nostrand, 1977.

Greeno, J. Indefinite goals in well-structured problems. *Psychological Reivew,* 1976, *83,* 479–491.

Groen, G. J., & Parkman, J. M. A chronometric analysis of simple addition. *Psychological Review,* 1972, *79,* 329–343.

James, W. *The principles of psychology.* New York: Dover Publications, 1950.

Lave, J. What's special about experiments as contexts for thinking? *The Quarterly Newsletter of the Laboratory of Comparative Human Cognition,* 1980, *2,* 86–91.

Lord, F. *Applications of item response theory to practical testing problems.* Hillsdale, N.J.: Lawrence Erlbaum Associates, 1980.

Newell, A., & Simon, H. *Human problem solving*. Englewood Cliffs, N.J.: Prentice-Hall, 1972.

Piaget, J. *The child's conception of the world*. New York: Harcourt, Brace & World, 1929.

Piaget, J., & Szeminska, A. *The child's conception of number*. London: Routledge & Kegan Paul, 1952.

Resnick, L. Task analysis in instructional design: Some cases from mathematics. In D. Klahr (Ed.), *Cognition and instruction*. Hillsdale, N.J.: Lawrence Erlbaum Associates, 1976.

Scribner, S., & Cole, M. Études des variations sub-culturelles de la mémoire sémantique: les implications de la reserche inter-culturelle. *Bulletin de Psychologie*, Spécial Annuel, 1976, 380–390.

KAREN C. FUSON
JAMES W. HALL

CHAPTER **2**

The Acquisition of Early Number Word Meanings: A Conceptual Analysis and Review

The normal adult in our society is conversant with a wide variety of meanings and uses of the number words. These include knowledge of the conventional sequence of words, their use in counting, their corresponding numerical symbols, their use in describing the numerosity of sets (cardinal meaning), their use in measuring with arbitrary units (measure meanings), their use in indicating relative (ordinal) position, and even their use as a convenient coding or categorizing device. The adult can shift meanings rapidly and with minimal confusion. The statement that ''Despite a seventy-eight yard run by number thirty-four the Bears lost by two touchdowns and dropped into sixth place'' may bring consternation, but not confusion, at least to anyone familiar with American football.

Our concern in this chapter is with the young child who is faced with the acquisition of this variety of meanings. We assume that number words come to attain meaning by their use in particular contexts. Different kinds of number words are used in contexts that have different structural characteristics. The English language differentiates some, but not all, of these structurally different number contexts. For those contexts in which the same number word is used, a developmental sequence is hypothesized in which the young child first learns the number word as several different context-dependent words. Later these different meanings of the word become interrelated, resulting in a mature closely connected set of meanings for that word. The different number contexts treated in

THE DEVELOPMENT OF MATHEMATICAL THINKING

our analysis are sequence, counting, cardinal, measure, ordinal, and nonnumerical contexts. In a sequence context, number words occur in their conventional sequence and no external entities are referred to by the users of such words. In a counting context, each word produced is attached to one entity in a well-defined set of discrete objects or events. In a cardinal context, the number word describes the numerousness of a well-defined set of discrete objects or events (i.e., it tells how many objects or events there are). In a measure context, the number word describes the numerousness of the units along some continuous dimension into which some entity has been divided (i.e., it tells how many units there are in some entity). In an ordinal context, the number word describes the relative position of an entity within a well-defined totally ordered set in which the ordering relation has a specified initial point. In a nonnumerical context, number words are used for convenience to differentiate and identify particular entities or are used as nonnumerical codes.

Our intention is to review the distinctions among these contexts and to describe skills and knowledge that must be acquired by children during the preschool and early school years within each meaning category or context. The present state of research with respect to many aspects of early number is minimal or nonexistent. Where such a body of research exists, it will be briefly summarized. We use terms such as *cardinal number word* to mean a number word used in a context having a cardinal structure. We do not imply that a child (or an adult) has a mature understanding of cardinal number just because he uses a cardinal number word. We are also using terms such as *cardinal number* and *set* in their ordinary usage, rather than in their technical mathematical usage.

SEQUENCE WORDS

Sequence number words are those produced in the conventional sequence order (one, two, three, four, five . . . .) when no entities are being counted. The English language does not differentiate between such contexts and those in which the sequence is used to count entities: Both are often called *counting*. We will make this differentiation by the terms *sequence words* and *counting words*. Meanings accrue to the number words from their sequence context (i.e., from their embeddedness within the sequence and from their relationships to other words in the sequence). Fuson, Richards, and Briars (1982) described developmental levels in these sequence relational meanings and in sequence skills that involve more complex productions of parts of the sequence (see Table 2.1). Initially the number word sequence is a relatively meaningless sequence produced with some effort. At the higher (later) levels the words are produced with little effort and have become the objects of thought, and the sequence itself can be used as a representational tool in various numerical contexts. These sequence

Table 2.1
Sequence Production Levels

Sequence levels	Forward sequence skills	Backward sequence skills	Relations	Counting, cardinal, ordinal, and measure context uses
String •——→ 1	Produce word sequence from one; words may be undifferentiated.			Count: No intentional one-to-one correspondences can be established.
Unbreakable chain •-o-o-o-o-► 1	Produce word sequence from one; words are differentiated.			Count: Intentional one-to-one correspondences can be established.
				Card: Cardinality rule can be acquired. (Can count to find out *How many?*)
				Ord: Ordinality rule can be acquired. (Can count to find out *What position?*)
				Meas: Measure rule can be acquired. (Can count to find out *How many units?*)
				Card op: Simple addition problems if objects for the sum just need to be counted.
	Count-up from one to *a*.		And then, and then (before): The sequence can be used to find these relations. Comes after, comes before:	Card: Make a set of numerosity *a*.

Ord: Find the "*a*th" entity. |

(*continued*)

Table 2.1 (*continued*)

Sequence levels	Forward sequence skills	Backward sequence skills	Relations	Counting, cardinal, ordinal, and measure context uses
			The sequence can be used to find these relations.	Meas: Make (find) a quantity made up of *n* units. Card op: Count-all and count-part procedures for addition and subtraction.
Breakable chain o-o-o ●-o-o-o➤ a	Start counting-up from *a*.		And then, and then (before): These relations can be produced immediately. Comes after, comes before: These relations can be produced immediately. Between: Partially correct solutions may be given by using ''and then'' or ''and then (before)'' relations.	Card op: Count-on without keeping track (addition).
	Count-up from *a* to *b*.		Between: Can produce all words between *a* and *b* going forward.	Card op: Count-on from *a* to *b* without keeping track (subtraction).
		Count-down from *b*.		Card op: Count-back from *b* without keeping track (subtraction).

(continued)

skills also come to be used in other number word contexts (especially counting and cardinal contexts), leading to the construction of links between sequence meanings and other meanings. Fuson *et al.* presented their own data on these various sequence skills and reviewed findings of other investigators. Our summary here will be brief and selective.

Table 2.1 (*continued*)

Sequence levels	Forward sequence skills	Backward sequence skills	Relations	Counting, cardinal, ordinal, and measure context uses
		Count-down from *b* to *a*.	Between: Can produce all words between *a* and *b* going backward.	Card op: Count-back from *b* to *a* without keeping track (subtraction).
Numerable chain ┉ ●┉┉┉▶ a	Count-up *n* from *a;* give *b* as answer. 1. *n* = 1 ("and then") 2. *n* = 2,3 (4) 3. *n* > 4			Card op: Count-on *n* keeping track (addition)
	Count-up from *a* to *b* keeping track; give *n* as answer.			Card op: Count-on from *a* to *b* keeping track (subtraction or missing addend problems).
		Count-down *n* from *b;* give *a* as answer.		Card op: Count-back *n* keeping track (subtraction).
		Count-down from *b* to *a;* give *n* as answer.		Card op: Count-back from *b* to *a* keeping track (subtraction).
Bidirectional chain ●┉┉┉▶ a	Can count-up or down quickly from any word; can shift directions easily.			

Sequence number words may be produced in a variety of situations. They may be produced spontaneously or in response to a command and alone or as a part of group recitation. They also may be produced for various purposes—for sheer practice, for timekeeping (e.g., saying the words to 20 in Hide-and-Go-Seek), for showing-off (*I can count higher than you can*), as well as for the various numerical contexts in Table 2.1. Each such sequence production should contribute to the acquisition of the sequence and to its eventual fluent production but probably does not contribute substantially to any further knowledge regarding the number words. Few data exist at present on the extent or frequency of the various circumstances in which the sequence number words are produced.

Middle-class children's acquisition of the conventional number word sequence up to 100 covers the period from age 2 to about age 6. Gelman and Gallistel (1978) reported that 15 of their 16 2-year-old subjects produced a number word sequence beginning with "one, two, three" and continuing in various ways. Fuson and Richards (1979), Fuson and Mierkiewicz (1980) and Fuson, Richards and Briars (1982) reported mean lengths of accurate number word sequences produced in nonobject contexts by a combined urban and suburban sample to increase from 13 at age 3½–4 years to 51 at 5½–6 years, with high variability within each age group. Word sequences produced by children aged 4½–6 years indicated that they knew the repeating one to nine pattern in the decades (e.g., thirty, thirty-one, . . ., thirty-nine) but they had not yet solved the "decade problem" (i.e., they did not yet produce the decades in the right order). Subsequent observations by Siegler and Robinson (in press) were consistent with those findings. None of these authors found evidence that children understood the teens structure of the number words at the time they were acquiring those words. The number word sequence thus seems to be for young children an unstructured list until twenty (or perhaps twenty-nine) and then the decade structure is evident.

Fuson, Richards and Briars (1982) reported that during the period of acquisition the form of the number word sequence seems to be the following: a first "conventional" portion that is some beginning part of the conventional sequence (e.g., "one, two, three, four, five"), a second stable nonconventional portion that is different from the conventional sequence but produced with some consistency by a given child (e.g., "five, eight, nine, eleven"), and a final nonstable portion with little consistent pattern over repeated productions. The stable nonconventional portion varies from child to child and usually consists of words from the conventional sequence with some omissions. The final nonstable portion consists of single unrelated words, runs of two to four words in the correct order with no omissions, and/or runs with omissions. For some but not all children, this final portion contains only words not appearing in the first two portions. The stable nonconventional portion may continue to be produced without change over a considerable period of time (they reported examples of 5 months without change).

Ginsburg and Russell (1981) reported in their Study I mean lengths of accurate number word sequences for samples of middle-class and poor inner-city 4-year-olds in preschool or day-care centers to be 19.89 and 15.52, respectively, with only two children having sequences below 10. In study II, these means were 18.36 and 30.97, and every child was correct to 5. However, Wang, Resnick, and Boozer (1971) reported that 13% of their low SES *kindergarten* children did not produce a sequence to 5 and 32% did not produce a sequence to 10. This apparent discrepancy may be due in part to the steps taken by Ginsburg and Russel to ensure subject interest and attention, to greater preschool experience in the Ginsburg-Russell sample, or perhaps even to general cultural changes occurring in the several years intervening between those studies. A recent study employing an object counting task with entering first-graders from a wide range of socioeconomic and residential backgrounds found that over 80% of the children could produce correct number word sequences above 30, and 95% could produce sequences correctly to 18 (Hendrickson, 1979). Bell and Burns (1981) found somewhat lower performance in a heterogeneous sample from a small city bordering Chicago. At the beginning of the year, about half of their kindergarten children could not produce an accurate sequence to 30, about a fourth had correct sequences between 30 and 70, and about a fourth could go from 196 to 201 accurately. For first, second, and third graders, these figures moved to 14%, 7%, and 48%; 0%, 8%, and 31%; and 0%, 0%, and 25%, respectively. In addition, about 60% of the second graders and 70% of the third graders could produce sequences above 296 accurately. The age at which the whole sequence to 100 is acquired seems heavily dependent upon the practices of individual teachers as well as on subject variables. Our observations in suburban schools suggest that this can be accomplished by the end of kindergarten for most middle-class children if teachers provide moderate amounts of sequence production activities.

COUNTING WORDS

Counting is the successive assignment of sequence number words to items. In a counting context some procedure is used to assign number words to the items in a set of well-defined "countables" (i.e., items that exist in space or in time and constitute a well-defined set). In accurate counting, each countable is paired with one and only one sequence word. Thus, in any particular counting act, a counting word has a referent—the countable to which it is attached by the counter.

This pairing of words and countable items is often accomplished by a pointing action. Fuson and Mierkiewicz (1980) reported that with fixed objects almost all of the 96 3- to 5-year-olds in their sample spontaneously pointed at objects as they counted. Gelman and Gallistel (1978) reported the use of pointing by children in this age range, but also found that some 2-year-olds failed to use pointing on some of their counting trials. Children's use of pointing in the counting of objects implies the involvement of three sets of correspondences: (a) a pairing *in time* between a word and a pointing action, (b) a pairing *in space* between that pointing action and an item, and (c) the resulting pairing of the word and the item (Fuson and Mierkiewicz, 1980). The pointing act (which exists in space and in time) creates a space–time unit, which connects the entity existing in space and the word existing in time. Failures thus can occur in a word–point correspondence, in a point–object correspondence, or in both. Fuson and Mierkiewicz (1980) reported differences in these kinds of errors as a function of age and of item characteristics. Findings by Briars and Seigler (1981) indicate that pointing is an important part of children's conception of counting and that a developmental progression may exist in which children first consider that correct counting requires a three-way one-to-one correspondence among a word, an entity, and a point, and only later understand that it is the derived word–entity correspondence that in fact is crucial to a correct count.

An internalization of the pointing act also seems to occur with age. In the Fuson and Mierkiewicz study, 3-year-olds usually touched the objects as they counted whereas 4- and 5-year-olds often just pointed. Some of the 5-year-olds also counted to themselves without pointing, a procedure also reported for some kindergarten children by Ginsburg and Russell (1981). In a later study using the same objects, college students often used gaze fixation rather than any external pointing gesture.

With objects that are not fixed in position, an indicating act other than pointing may be used. Counting creates successive partitions of the set of counta-bles into two sets, the counted and the uncounted (Beckwith & Restle, 1966). As an entity is counted, it can be physically moved from the pile of uncounted to the pile of counted entities. Age differences in the accurate use of this type of indicating gesture and accuracy differences between this type of indicating act and points have not yet been systematically studied; nor are there systematic data for children's counting of items presented auditorially (e.g., clock chimes), where no indicating act is required, but wherein word and entity are paired by continguity in time.

The aforementioned research has been primarily descriptive of the external behavior of the child. Steffe, Richards, and von Glasersfeld (1981) have been more concerned with the internal representations that are involved in such behav-ior. Counting, in their view, consists of the production of a counting word and of a counting unit item, where that unit item is a mental construction. That is, some act or internal representation constitutes the unit that is counted. In the simplest case (perceptual), the counter produces unit items from physically present ob-jects. Other entities that serve as units for counting are figural unit items (con-structed from mental representations of entities), motoric unit items (from body movements), verbal unit items (from sequence or counting words), and abstract unit items. A concrete-to abstract developmental progression is proposed in the nature of countables from which unit items are produced. Counting words thus are paired with increasingly abstracted entities.

Gelman and Gallistel (1978) identified five "principles" which they argued underlie accurate counting. Two (the cardinal principle and the order irrelevance principle) concern cardinal contexts and will be discussed in the next section.

The other three are the abstraction principle (any collection of entities can be a set of countables), the stable order principle (the words used in counting must be produced in a fashion that is stable from trial to trial), and the one–one principle (each item in an array must receive one and only one counting word). Reviewing the research on the types of objects preschool children are able to count, Gelman and Gallistel argued convincingly that the studies reporting children's failure to consider heterogeneous objects as countables are methodologically flawed. Thus, at the time of school entry, children's knowledge of the abstraction principle seems to extend to most kinds of groups of objects physically present in space. We do not at present know the extent to which children also consider other items (e.g., entities in time such as clock chimes, mental representations of physical objects, imaginary entities) to be countables, nor do we know the extent to which children can count these latter types of entities.

The evidence cited by Gelman and Gallistel for the child's understanding of the stable order principle was the fact that any given child tended to produce the same idiosyncratic list on each counting occasion; the list was idiosyncratic in that it was only partially correct, and it varied from child to child. Fuson, Richards, and Briars (1982) and Fuson and Mierkiewicz (1980) noted a similar tendency and observed that the errors seem to be ones of omission, as in "one, two, three, five, six, eight" (see first section on sequence words). The interpretation of such productions in terms of a principle possessed or understood by a child seems unnecessary. A simpler characterization is that an imperfectly learned serial list is elicited reliably by requests to count; that is, the child uses his or her sequence of number words in whatever imperfect state of acquisition it exists at the time. The use of the term *principle* seems to imply knowledge on the part of the child (e.g., understanding *why* the same sequence must always be used) for which there is at present no evidence.

Currently, little evidence exists concerning children's state of knowledge of the one–one principle (that each item in an array must receive one and only one counting word). Gelman and Gallistel offered no direct evidence concerning children's knowledge of this attribute of correct counting. They simply described children's ability to produce correct correspondences in counting. Similarly, the Fuson and Mierkiewicz data concerned only children's performance. Gelman and Gallistel argued that children's knowledge about the counting act precedes totally accurate counting (i.e., that counting errors are primarily errors of execution rather than of understanding). This of course at some time must be true; we certainly would characterize adult errors in this way. In practice, though, it is difficult to sort out execution from knowledge and even more difficult to discover the nature of the knowledge or understanding.

Mierkiewicz and Siegler (1980) and Briars and Siegler (1981) addressed this latter question by asking 3- through 5-year-old children to identify errors and correct counting from puppet demonstrations of counting. The errors chosen were based upon the analysis of counting errors presented in Fuson and Mierkiewicz (1980). The ability to detect errors increased with age and at each age level varied with the nature of the error. Consistent with Gelman and Gallistel's (1978) position, errors in which objects are skipped appear mainly to be execution errors. That is, such errors were readily detected even by most 3-year-olds, whereas such errors are among the most frequent made by the preschool children studied by Fuson and Mierkiewicz (1980). The opposite pattern was found for certain other errors, leading Briars and Siegler to conclude that correct execution often precedes awareness that a component is necessary for correct counting. Further evidence of young counters' incomplete awareness regarding counting was indicated by the relatively high frequency with which children classified as "mistakes" certain procedural variations that, in fact, had no effect on the accuracy of the observed counting.

Thanks to relatively analytical research over the past few years, our knowledge of children's counting competence is substantial. Knowledge about the cognitive bases and correlates of counting is more difficult to obtain, and only a beginning has been made in that respect.

CARDINAL WORDS

A cardinal context is one in which a number word describes the numerousness or numerosity (the "manyness") of a well-defined set of discrete objects or events. Our language, as well as many others, has special cardinal words for particular kinds of things: duet, trio, quartet, etc.; twins, triplets, quadruplets, etc. In may languages the first three or four cardinal words are adjectives, whereas, the larger words are noun forms (Menninger, 1958/1969). Thus, for small numbers, the concept of *the number of things* was originally very closely tied to the things themselves. *Two shoes* are different from *one shoe,* but they are also different from *two oxen*—thus, a *pair of shoes* and a *brace of oxen.* It was a large step forward from these object context-specific number words to the abstract *two,* which could apply to *any* object.

Young children in our culture now learn to use the word *two* in cardinal contexts as early as their second year (Walters and Wagner, 1981). However, these various experiences in which the child uses the word *two* may initially be for him or her somewhat different, and these various operational meanings of *two* may at some point need to become connected to form a more general concept of *twoness.* Some possible such early meanings of *two* are *two cookies* (one in each hand), *two sticks of chalk* (one for me and one for my friend), *two shoes* (no bare foot), *two kitties* (my kitty and another kitty). There may not be a single developmental route to the abstraction of *two* or of other particular numerosities. Alternate paths may exist, such as kinaesthetic (*one in each hand*), visual pattern (*that one and that one*), or sensorimotor (*one and then another*) paths. In fact, each of these experiences contains kinaesthetic, visual, and sensorimotor aspects, though in differing amounts, and individuals may differ in their emphasis on these aspects. Von Glaserfeld (1981) has proposed that what is common about each of these experiences is the pattern of attention required; each requires two moments (or foci) of attention. *Twoness* in his view consists of the abstraction of this particular attentional pattern. No doubt there are other plausible bases for numerosity generalizations. The age at which and the way in which this more general conception of *two* or of other numerosity words comes to exist is not known at present.

How young children become able to use cardinal words for particular cardinal contexts is at present a source of controversy. Adults can accurately and very quickly produce the cardinal word for small sets of objects (one to four, at least) without counting. This phenomenon has been labeled *subitizing*. There is evidence that children aged five and up also can subitize sets of one, two, and three, though not as rapidly as adults (Chi & Klahr, 1975; Klahr & Wallace, 1973, 1976). Klahr and Wallace (1973, 1976) proposed that subitizing predates counting as a means of determining numerosity and that counting initially takes on quantitative meaning by being used in the subitizing range. Gelman and Gallistel (1978) argued the opposite position: For small numerosities counting precedes subitizing.

No completely definitive data exist on this issue at present, as methodological difficulties limit the interpretation of much of the present work. However, the case for subitizing seems a bit stronger. Walters and Wagner (1981) reported much more frequent use of the *two* in home situations where counting did not occur than where it did occur and described some examples in which it seemed quite unlikely that counting had occurred. For *three*, equal numbers of occasions were observed in which entities were counted and subitized, and for *four*, more counting occasions were observed. Schaeffer, Eggleston and Scott (1974) found that almost all of their least skilled group of children (mean age 3–8, range 2–0 to 5–0) correctly recognized and labeled displays of two, and half did so for displays of three and four. Only one child from this group was able to count arrays of two, three, and four correctly. Thus, most of these children were capable of subitizing, but not of counting small numerosities. Some recent evidence also indicates that preverbal human infants are able to differentiate and categorize groups of items on the basis of numerosity if the groups are very small and that nonhuman primates can perform discriminations based on numerosity. Young infants habituated to a particular group of two or of three items (i.e., no longer visually attending to them) will gaze significantly longer at a new group of items if the numerosity differs than if it remains the same (Starkey, Spelke, & Gelman, 1980). Thomas, Fowlkes, and Vickery (1980) reported training squirrel monkeys to select from a pair of cards having differing numbers (from two through seven) of various sized dots the card with the fewer number of dots. Both of these results appear to be independent of other attributes of the displays such as size, length (or area), density, and arrangement of the items. These performances could not depend upon counting, so at least some important aspect of the ability to abstract numerosity from small displays is independent of counting or of specific number words.

On the other side of the argument, Gelman and Gallistel (1978) reported data indicating that children counted when they used the specific numerosity words *two* and *three*. However, it is not clear from the report of their data whether some of the counting might not have been functioning as a justification or explanation to the the experimenter rather than being the solution method. Silverman and Rose (1980) reported for 3-year-old children equivalent levels of accuracy for subitizing and for counting for displays of two and three and considerably greater accuracy for counting than for subitizing for displays of four items. The finding of equal and good counting and subitizing abilities for numerosities two and three at this age implies that we will need to look at 2-year-olds to settle the issue of counting and/or subitizing as the earliest source of numerosity. The authors also reported more intrusions of counting into subitizing trials than vice versa and concluded that children of this age range have a preference for counting over subitizing. However, several design problems (items spaced further apart on subitizing than on counting trials: the presence of a count-inducing practice card in the subitizing task; instructions not to count or point on subitizing trials, which might have suggested such procedures to some children; use of a correspondence rather than a correct final cardinal word criterion for accurate counting) render this conclusion problematical.

There may in fact not exist one identical developmental route through subitizing and then counting (or vice versa) for all children. The developmental relationship between these two may be messier than that sought for or postulated by most researchers, and it may depend considerably upon the experiences of individual children. A 2-year-old whose parents play cards often may receive a great deal of practice at labeling patterned displays with feedback and may develop a preference for and accuracy at subitizing, which is extended to other types of numerosity displays in the environment. Another 2-year-old whose parents or siblings are interested in counting may instead receive considerable experience in counting small displays. Thus, resolving the subitizing–counting controversy may prove quite complex and may require reformulation to account for different experience, especially for the numerosity *three* and perhaps for *four*.

The nature of the subitizing procedure is not clear at this point. Mandler and Shebo (1980) found that adults quickly learned to respond to canonical patterns (e.g., a rectangular pattern of dots for four) for numerosities from one through ten (except subjects confused the particular patterns chosen for ''seven'' and ''ten''). For numerosities of one, two, or three, the nature of the display pattern (canonical versus random) did not influence response speeds, whereas with larger arrays, response speeds were slower with random arrays. Mandler and Shebo concluded that subitizing involves the use of canonical patterns for two (a line) and for three (a triangle). For school-aged children, Maertens, Jones and Waite (1977) provided data suggesting that children also are able to make quite accurate numerosity judgments with sets larger than four if the items are arranged in canonical patterns (e.g., as spots are on dice) and some practice is given. However, the judgments in both of these studies may not involve genuine assessments of numerosity, but only verbal responses paired to the various visual patterns.

Consideration of numerosity generally has been confined to objects and events in space. That seems to be a quite arbitrary and undesirable restriction, particularly since there exists a body of research with quite different origins (in verbal learning) concerning numerosity in time (i.e., event frequency). It seems clear from that research that information about event frequency is encoded with considerable fidelity (for reviews, see Hintzman, 1976, and Howell, 1973). The encoding of frequency information is widely thought to occur automatically with the perception of the event (Hasher and Zacks, 1979), although the necessary and sufficient conditions for such encoding are not yet completely understood. For small numbers (one, two, and three), even children as young as 5 years are impressively accurate, although probably not as accurate as somewhat older children (e.g., Lund, Hall, Humphreys, & Wilson, in press) in their judgments of the frequency of recent events within a particular context (situational frequency). This is true both for absolute (how many times?) and relative (which occurred more often?) judgments and occurs whether or not the children are instructed to attend to event frequency. Hasher and Zacks (1979) have proposed that our species is "genetically prepared" for the automatic encoding of event frequency. In light of the numerosity research and of the wide utility of frequency information (e.g., Estes, 1976), such a proposal seems plausible.

The capacity to process numerosity information for small arrays does not mean that the child is aware that numerosity is a property of all sets. When and how that knowledge is acquired is unknown, but experience with the application of the number word sequence to large as well as small sets (i.e., counting) probably is instrumental here and will be discussed in more detail in later sections. First we will discuss numerical estimation and approximation.

Numerical Estimation and Approximation

Much of the numerical information that we possess is not exact. Approximate numerosity is sufficient in many cases and, in fact, may be even more desirable under some circumstances. Ordinarily we do not want to know that the population of a medium-sized city is 95,126 rather than 95,000 or even "about 100,000." The excess information not only is unnecessary, but may actually get in our way. Although exactness may not be required, very often a close approximation is necessary. Skill in the estimation of perceived numerosity, therefore, is desirable.

Relatively little research has been devoted to numerical estimation, especially on children's acquisition and use of estimation skills. One distinction that may prove fruitful is between what we shall term *direct and indirect estimates*. Direct estimates do not involve any source of numerosity information other than the entities to be estimated; indirect estimates do. First, to what extent and by what means is numerosity information estimated directly from sets of objects? Mandler and Shebo (1980) have concluded from reaction time measures that if a display of dots is flashed on for a brief time (200 msec), adults count a mental representation (icon) of the displays for sets of four, five, and six, but use an estimation procedure for larger arrays. The mental counting is less accurate than subitizing, but much more accurate than estimating; direct estimates fall from an accuracy of about 30% for seven to less than 5% for eleven on up. Direct estimation also may involve numerical information that is derived from information about area and density. This may be particularly true for relative numerosity estimates. For example, Pringle (1980) has proposed that when relative numerosity judgements are required for sets of identical objects that vary both in total space occupied and in density, a "normalizing process" occurs. That is, one of the sets is imaged as transformed to match the other set on one of the two dimensions; then the sets are compared on the other dimension. The extent to which humans can extract from large sets numerical information that is not derived from area and density attributes evidently is not known at present.

In the ordinary course of events, indirect estimation probably is much more common than direct estimation. It involves some source of numerosity information (generally in memory) other than the entities to be estimated. Siegel, Goldsmith, and Madson (1980) have distinguished between two such indirect procedures. One involves the use of a bench mark (a well-known object): the numerosity of a set is judged by its similarity to some well-known set (the benchmark). Thus, if one is teaching in a new classroom, one might quickly generate an accurate estimate of the number of seats in it based on experience with (memory for) a well-known classrom of similar size. Benchmark estimation is used commonly for attributes such as length and area, where one estimates, for example, the length of a room by comparing it with similar rooms whose length is known. This would seem to be an extremely common procedure for numerical estimation. It would seem also that the extent of use of this procedure would depend primarily on the child's memory for specific numerosity benchmarks.

The second indirect estimation procedure described by Siegel *et al.* was termed *decomposition–recomposition*. This procedure involves breaking down the set into subsets for which numerosity either can be estimated by a simple benchmark procedure or determined by other means (e.g., subitizing), and then reassembling the subsets. An example would be to estimate the number of seats in a large auditorium by first estimating how many classrooms of known numerosity would fit within it (decomposition), then multiplying that number by the number of seats in the known classroom (recomposition). Klahr and Wallace (1976) have discussed this estimation procedure and point out that it is essentially a measurement procedure. That is, a unit of some known numerosity is selected, and the estimator attempts to determine how many of those units are required to account for the total array. This procedure obviously is more complicated than the simple benchmark procedure.

The research of Siegel *et al.* indicates that estimations of large sets (much above 100) generally are difficult even for adults, and children, expecially those of elementary school age, tend to give up or to generate wild guesses or extremely "loose" estimates (e.g., *Somewhere between 100 and 1,000,000*). This is true even when some reasonable approximation through decomposition–recomposition seems possible (to the experimenter at least). With small sets the simple benchmark estimations sometimes are used by children as young as 8 years (the youngest in Siegel's sample).

Estimation of numerosities obviously seems to require experiences of various kinds. Young children may lack many or most of these experiences and thus their estimating abilities may be quite weak. They may particularly be lacking in stored, labeled representations of numerosity. The knowledge and skills involved in estimation may be acquired during the middle and late elementary school years and probably would continue to be acquired throughout adulthood as new experiences with numerosity occur.

Cardinal Meanings and Counting Meanings

A crucial step in a child's learning to apply the counting procedure in a cardinal context is learning that the last number word said in counting a set of items gives the numerosity (the cardinal word) for that set. This has been called *the cardinality rule* by Schaeffer, Eggleston, and Scott (1974) and *the cardinality principle* by Gelman and Gallistel (1978). Schaeffer *et al.* covered a set of objects after a child had counted them and asked the child to say how many objects were in the set. A correct response was taken to mean that the child had acquired the cardinality rule. Gelman and Gallistel inferred the acquisition of the cardinality principle from any of four behaviors: (*a*) ability to respond immediately with the correct cardinal word to a *How many?* question about a set, (*b*) emphasis (louder or slower pronounciation) on the last word produced in counting, (c) repetition of the last word in counting, and (d) stating without counting the correct cardinal word after that same set had been counted on an earlier trial.

A common response of children judged not to possess this rule is to count the set again if asked *How many objects are there?* For example, in the Fuson and Mierkiewicz (1980) study, some children recounted sets as many as seven times in response to each repeated question of *How many blocks are there?* rather than giving the final word from the count. In such cases the *How many?* question seems to function as a request for the counting act rather than as a request for information gained from the counting act. The actual response required to a *How many?* question (at least for sets too large to subitize) has two successive stages, the first being counting and the second being the reporting of the final count word as a cardinal word. The child who recounts to a repeated *How many?* question then may be viewed as having mastered only the first of these two stages. However, in one limited sense, the child's counting response may be viewed as an entirely satisfactory answer to the *How many?* question. That is, the adult is actually fully informed of the cardinality of the set simply by hearing the child's count.

Markman (1979) reported a verbal manipulation that increased the number of children who regarded a given context as a cardinal as well as a count context, (i.e., it increased the number of children showing the cardinality rule). Performance was compared when class terms (*pigs, nusery school children, animals, blocks*) and collection terms (*pig family nursery school class, animal party, pile*) were used. Children hearing collection labels responded to the *How many?* question with their last counting word a mean of 3.46 (out of 4) times, whereas those hearing the class labels so responded a significantly lower mean of 1.85 times. Markman argued that collective nouns focused the child's attention upon the set as a whole rather than upon the individual objects within it and thus facilitated the appropriate use of the cardinal word as a referent to the whole set. In the terms of the present analysis, the collection manipulation increased the number of children who identified that context as a cardinal as well as a counting context. This research indicates that rather subtle differences in the way the cardinality question is posed may lead to quite different inferences about the child's possession of the cardinality rule.

Another factor that might, in some instances, lead to an apparent absence of the cardinality rule is forgetting. For example, in the Schaeffer *et al.* procedure the *How many?* question apparently was posed after counting was completed, and a child's failure to respond with the correct counting word may on occasion be due to a failure to remember what that word was rather than to a lack of understanding that the last counting word can also convey a cardinality meaning.

The main body of evidence to date indicates that middle-class children are able to apply the cardinality rule by the age of four and that inner city children may be somewhat dealyed in this task. Schaeffer, Eggleston, and Scott (1974) identified two groups of children who did not display the cardinality rule for sets of five to seven. The mean ages of these groups were 3–8 years and 3–5 years, with the ranges being 2–0 to 5–0 years and 2–9 to 4–6 years, respectively. Ginsburg and Russell (1981) reported in Study I that most of their middle-class pre-schoolers (mean age 4–3 years) displayed the cardinality rule for sets of three, five, eight, and eleven, whereas less than half of their inner-city age-mates (mean age 4.5 years) did so. In their Study II, virtually all of their somewhat older pre-school children (mean age 4–8) of both social classes demonstrated considerable understanding of the cardinality rule. Markman (1979) reported good cardinality performance by her 13 subjects given collection terms (aged 3–2 to 4–9 years, mean 4–0 years) and considerably lower performance by the 13 subjects given class terms (same age). Gelman and Gallistel (1978) reported almost unanimous demonstration of the cardinality rule by 2-, 3-, and 4-year-olds in their magic experiments on very small numerosities (numerosities of two and three for the youngest children and two, three, and five for the two older groups). Pergament (1982) found with 48 middle to upper-middle-class children the same age as Markman's sample that eleven of the twelve children who did not demonstrate the cardinality rule were 3-year-olds (mean age 3–7). The remaining 36 children did demonstrate the cardinality rule on sets of size four through 14. Cardinality rule performance was very much "all or nothing" phenomenon and was not affected by the size of the sets (4, 5, and 6 versus 9, 12, and 14), by the consistency (homogeneity versus heterogeneity) of the objects in the sets, or by the verbal label (class or collection terms) used in the *How many?* question. No effect of the Markman class/collection manipulation appeared in either a replication analyses using only the set sizes and exact words Markman used or in the larger study of 12 trials per subject. Two-thirds of the twelve children who did not display the cardinality rule were in the collection condition. The reason for the lack of replication of the verbal label effect is not clear at present.

An exception to the above general pattern of results was reported by Gelman and Gallistel (1978) in their videotape study of counting. They reported a much lower level of cardinality rule performance than in their magic study (and than in the studies above) and a very strong dependency of cardinality rule performance on set size. For example, the performance of 3-year-olds dropped from 68% for sets of three to 25% for sets of 11; the figures for 5-year-olds are 100% and 48%, respectively. Gelman and Gallistel suggested that this pattern indicated that children, to some extent, monitor their own skill at counting and therefore stop using the cardinality principal when their counting performance becomes inaccurate. However, with respect to the effect of accuracy of counting, Pergament did not find the cardinality rule to depend upon accurate counting (nor on set size), and Ginsburg and Russell also made the same observation about their sample. Therefore, the anomolous results reported by Gelman and Gallistel may be due (as they observed with respect to other aspects of their cardinality rule videotape data) to the considerable number of repeated trials for each set size in the videotape experiment or to the fact that their design necessitated inferring the use of the cardinality rule for some children from their counting performance on previous trials of the same size.

There are several levels of knowledge concerning relationships between counting and cardinality. Because in English the counting and cardinal words are the same and because in adults our cardinal and counting meanings have become so closely related, it is useful in considering these levels to imagine that our counting words are the alphabet because no cardinal meanings are attached to the alphabet (only sequence meanings are). Therefore, for example, when we count some toy animals (e.g., A, B, C, D, E, F, G, H, I, J, K) and then say *There are K toy animals,* we shift from a counting meaning of the K as we point to the last animal to a cardinal meaning of K when we report the numerosity result. This is what has just been discussed as the cardinality rule. We term this *the count–cardinal transition,* for it requires a transition from a count meaning to a cardinal meaning. This transition requires that the user understand the cardinal (numerosity) meaning of K. It seems possible that a more primitive form of this rule exists in which the child has simply acquired a chain of responses (counting, followed by the repetition of the final count word) that has become regularly elicited by the *How many are there?* question. In this earliest case, the child has no clear idea that the repeated word (K) refers to the numerosity of the counted group, but only that this response is the one desired by the adult. When a child makes a count–cardinal transition in the fashion just described, a more complete understanding is implied. Children who met the Schaeffer *et al.* criterion for the cardinality rule would seem to be making a count–cardinal transition, whereas such understanding would not be necessary, at least in principle, to meet all of the Gelman and Gallistel criteria. Determining just what children do and do not understand about cardinality from their success or failure in obeying the cardinality rule is difficult at best and obviously depends on the criteria for that rule.

A slightly later level of understanding is the opposite transition, *the cardinal–count transition,* discussed by Fuson (1982). This transition is required in the addition counting-on procedure. It involves a transition from a given cardinal meaning of a number word (the first addend in an addition problem) to the counting meaning of that word (and then the subsequent counting of the second addend continuing on from the first addend).

A further level of understanding of count–cardinal relationships is involved in Saxe's definition (1982) of counting as *the progressive summation of correspondences,* that is, in our terminology, that each count word also is a cardinal word stating the cardinality of the set of correspondences established between the sequence–counting words and the entities counted thus far. Saxe considers the *five* said in counting a larger set (e.g., eight) to be a notation for the *summation*

of "the five correspondence relations" involved in counting the first five entities. This view of counting is that as each counting word is produced, it carries
with it a cardinal meaning. This clearly is not true for children who have not yet
acquired the cardinality rule. Even for those who have done so, however, it is not
clear that as each counting word is produced a cardinality meaning is also
accessed. It would seem inefficient to be accessing unnecessary meanings.
Rather what seems likely is that if an adult or older child were stopped in
counting and asked how many entities they had counted so far, they could then
make a count–cardinal transition and produce the desired information. Thus, it
seems more sensible to view the progressive summation cardinal meaning of
each count word as an advanced level of understanding that is acquired, but
which is not necessarily used each time the counting words are produced. However, it also seems likely that the first notion of progressive summation a child
acquires is that of the numerosity of the set of *objects* he or she has just counted
(i.e., of the first five objects in the sample above) rather than of the five *correspondence relations* involved in counting. Piaget (1941, 1965) discussed this
object summation aspect of cardinal number. The correspondence summation
discussed by Saxe would seem to be quite a bit more abstract. However, the
correspondence summation notion provides one important way of understanding
why the count–cardinality rule works. Evidence is needed concerning when
children in fact do acquire any notion (*a*) of counting as the progressive summation of entities counted, and (*b*) of counting (as in Saxe's definition) as the
progressive summation of the correspondences in the counting act.

Gelman and Gallistel (1978) have termed the understanding that entities can
be counted in any order *the order-irrelevance principle*. On successive counts of
the same set they asked children to make different specific objects in linear sets
of four or five objects the one, that is, to begin counting at different points on
succeeding trials. Two-thirds of the 5-year-olds and almost half of the 4-year-
olds were able to do this. Some were also able to explain why an object could be
given a different number word on successive counts. Ginsburg and Russell
(1981) reported that their 4- and 5-year-old subjects were able to carry out
different orders of counting better than they were able to anticipate that such
counting would result in the same cardinality. One fifth of the 4-year-olds and
two-fifths of the 5-year-olds could not anticipate this effect. Results of
Mierkiewicz and Siegler (1980) showed that children's understanding of the
irrelevance of the order in which a set is counted is limited to certain types of
counting orders. Although many of their 3-, 4-, and 5-year-old subjects indicated
that counting a row of objects by starting at the opposite end to that used by the
child was accurate counting, from half to two-thirds of the sample identified two
other kinds of nonlinear counting as errors: (*a*) starting in the middle of a row,
completing it, and then counting the beginning of the row, and (*b*) in a row
composed of blocks of alternating colors, counting all blue blocks moving to the

right and then counting all green blocks moving back to the left. A second study (Briars & Siegler, 1981) demonstrated that these children could imitate these types of counting, so their judgements of these as errors were not based on a failure to encode the counting adequately. However, the possibility remains that the results for some children indicate a confusion of "wrong" or ("a mistake") and "not usual counting." This problem would seem to be a difficult one in attempting to ascertain what children *know* about correct counting. Future research on the irrelevance principle might examine not only effects of different orders of counting, but also whether children understand the limits of this principle, for example, that it is true of counting in cardinal contexts, but not true of counting in ordinal contexts; in ordinal contexts the order of counting is not irrelevant. Furthermore, there would again seem to be at least two levels of understanding of this principle: an understanding that the order in which objects are counted does not matter and an understanding of why this is so. The former might be arrived at by empirical verification, especially for small groups of objects, whereas the latter requires a higher conceptual level of understanding concerning why the count–cardinal transition works.

How does the child become aware of the absolute reliability of correctly executed counting as a means to assess cardinality? One important aspect of this knowledge is knowing that the last word in counting is the same for repeated counts of the same group of entities. In the Fuson and Mierkiewicz study, when 3- and 4-year-old children ended their counting on a different word on two successive counts of the same group of entities, they usually announced each last word as the cardinal word for that group, giving no indication that they noticed or were disturbed by the two different answers. Some 5-year-olds and all high school students (later run on part of the tasks) would spontaneously recount a third time if they got two different answers. When the experimenter pointed out to the younger children the two different answers that had been given and asked which answer was correct, children always responded that the answer they had just given was correct. It is not clear whether this reflects the belief that the cardinality now was the second word but that earlier it had been the first word (i.e., that both answers were correct at the time they were given) or rather a belief that the first answer was wrong (i.e., the result of an incorrect counting act) and the second was right. These alternatives might be examined in future research.

Three alternative sources seem possible for the specific knowledge that the counting result is maintained over repeated counting acts and for the more general knowledge that correctly executed counting always produces the correct cardinal word. All three relate in some way to the amount and nature of the child's counting experience. The first is direct successful experience with counting for cardinality. That is, if a child has consistently obtained correct cardinality information through counting, then a sense of the absolute reliability of the procedure, including the production of the final count word as the cardinal value, may be induced. Two conditions are required. The child must be correct, and the child must beware of the correctness. Saxe (1977) explored the relationship between counting accuracy and counting use by 3-, 4-, and 7-year-old children in tasks such as making a display of the same numerosity as another display. He reported an existing but somewhat "loose" relationship between these for the numerosity nine. Experience with very small arrays are likely to provide strong support for the reliability of counting for cardinality because of the possibility of use of subitizing for such arrays. As Klahr and Wallace (1976) point out, with small arrays the reliability of counting for determining numerosity can be confirmed by consistent agreement between counting and subitizing. However, with arrays larger than three or four, and especially during the early stages of learning the counting procedure, the child's counting often is not correct. Thus, when asked to determine how many blocks there are, the child may make a procedural error such that the final count word does not correctly indicate the numerosity of the set. He or she may be told to count again or simply that the answer was not right, in which case there may be no reason for the child to assume incorrectness of his or her execution rather than an inherent unreliability of counting. Furthermore, even when counting is consistently executed correctly, there may not be sufficient practice with feedback for the child to induce the reliability of the procedure very rapidly. Thus, if consistently successful experience with feedback is the principal basis for the child's awareness of the reliability of counting for cardinality, it is not surprising that such awareness might be a long time in coming.

A second main potential source of information on the reliability of counting for cardinality is direct instruction to that effect. After all, young children tend to believe what adults (especially their parents) tell them. However, we suspect that there is less of such instruction than might be assumed and that it is a rather unusual parent who patiently describes (and redescribes) the relationship involved, even at the simplest level. Because parents are so convinced of the reliability of counting, it may not even occur to them that the child might not know that counting reliabily yields an accurate cardinality. Thus, the parents may never remark on this. Some indirect instruction probably is common, of course, Children are asked how many marbles (or whatever) they have and, if they do not know, often will be told to "count and see". No doubt these are important experiences, but they may not be sufficient to produce the reliability knowledge in question. They also may have less effect than one might consider them to have because there are many other counting experiences in which the final cardinality is not important. Much counting practice given by parents and others and much self-practice is done to ensure proper execution, not because the outcome matters. Put another way, much early counting is done for the sake of counting and not for the purpose of ascertaining some cardinality.

The third potential source of the conviction that counting is a reliable means of determining cardinality is a conceptual grasp of the relevant concepts and relationships. From knowledge of why the count–cardinal transition and the one–one principle work, one can derive the validity of counting for the determination of cardinality. This probably is not an important source in practice because the required level of conceptual understanding is likely to have been reached long after the validity of counting has been accepted. At present little is known about when or how more advanced knowledge about counting and cardinality is acquired.

Relative Numerosity: Relations on Cardinal Contexts

Cardinal contexts can be considered in isolation, in which case an absolute numerosity is involved or two cardinal contexts (or numerosities) can be compared. The outcome of such a comparison process can be described by one of three relations: an equivalence relation (is equal to) or two order relations (is greater than, is less than). Tasks requiring the use or the establishment of these equivalence or order relations on two cardinal contexts may be verbal or nonverbal. Verbal tasks require the use of words for the equivalence and order relations. Siegel (1978) reviewed a large number of studies that indicate difficulties young children have with the words for these relations. Children confuse relational words for different dimensions (e.g., bigger, more, longer), and they confuse the words for the two opposite order relations on the same dimension (e.g., less and more). They also seem to understand the word for the marked order relation (Clark, 1969) before understanding the word for the unmarked relation (e.g., Marschark, 1977). Siegel described evidence indicating that in a sizable percentage of 3- and 4-year-olds, concepts of order and equivalence relations on cardinal contexts exist before comprehension of the words for these relations, that use of the words for these relations does facilitate the performance of some 4-year-olds and of many older children, and that production of the correct relational words lags behind comprehension of these words (i.e., ability to use verbal instructions in a relational task). Thus, children must learn not only the concepts of equivalence and order relations on cardinal contests, but they must also learn to comprehend and to produce the correct words for these relations.

In the remainder of this section we consider three common bases for judgments about the relative numerosity of two groups of entities: degree of perceptual similarity of the groups, physical one-to-one correspondence between entities in each group, and the use of counting and the identity–nonidentity of the final count number words for each group. Relative numerosity judgments when only number words or symbols are presented are discussed later in this chapter.

PERCEPTUAL SIMILARITY

Perceptual similarity is a common basis for relative numerosity judgments by children. Children may focus on the length, area or density in each display. This is not a senseless strategy, for when other things are equal, each of these attributes is directly related to numerosity. For example, for a linear array and holding density and object size constant, array numerosity is a positive function of array length. This relationship between numerosity and other measure attributes complicates tasks involving numerosity judgments. The young child's tendency to attend to single dimensions of stimuli (Piaget's "centering"), and the stronger perceptual salience of other measure attributes (particularly length), lead the young child to incorrect judgments of relative numerosity when perceptual cues are misleading. Piaget (1941/1965) reported this finding in conservation of number tasks, and many studies have replicated this result (see Brainerd, 1979, and Siegel & Brainerd, 1978, for reviews and Modgil & Modgil, 1976, for a compilation of abstracts). A similar erroneous focus on array length has been reported for young children in relative numerosity situations other than conservation (i.e., transformation) ones. Comiti, Bessot, and Pariselle (1980) found that gross perceptual comparisons (rather than matching or counting) were used to compare sets of 37 objects by one-third of a sample of French children midway through their first year of school. Siegel (1974) found that length was used by many 3- and 4-year-olds in a two-choice numerosity match-to-sample task. Brainerd (1979) reported the use of array length (when counting but not matching was disallowed) by many children, even in the later elementary school years and even when length was a misleading cue and such judgments were incorrect.

When two numerosities are the same, length and density are inversely related; that is, if the length of one array is greater than that of the other array, its density will be less. Children eventually become able to focus upon *both* length and density and to appreciate this inverse relationship. This is the basis for the Piagetian compensation justiciation used by some concrete operational responders: *They are still the same because this one sticks out more here, but the toys are farther apart.*

Perceptual comparison procedures for disorganized nonlinear numerosity arrays evidently can be used by nonhuman primates. Thomas, Fowlkes, and Vickery (1980) reported that squirrel monkeys were successfully trained to choose the lesser of two dot displays having numerosities ranging from two to nine; and Thomas and Chase (1980) reported successful performance in response to cue lights that indicated a choice of the least, the greatest, and the middle numerosity of three dot displays. Nonnumerosity cues seemed to have been controlled in these studies. Thus humans may also be able to learn to use perceptual estimates for judging order relations on numerosity when misleading cues such as length are not present.

The use of perceptual comparison strategies by children may initially arise not only from the actual correlation between numerosity and other perceptual attributes, but may also be an effect of early subitizing experiences. As mentioned before, subitizing is observed with very young children, and the basis of subitizing is thought to be in the distinctive and easily recognizable perceptual patterns formed by very small arrays. Thus, the young child finds that for such numerosities accurate and approved answers to numerosity (*How many?*) questions can be derived reliably from the perceptual characteristics of the arrays. It follows, then, that the young child's earliest notion of numerosity and of the meaning of number words may be that these refer to perceptual pattern-like qualities of arrays. Small wonder, then , that the somewhat older child is "drawn" toward inaccurate judgements of equivalence by differences in the perceptual qualities of arrays. That is, because of the early and successful subitizing experiences, the child must later "unlearn" his or her original concept of cardinality and abandon "rapid perceptual impressions" as a means of deriving numerosity information for arrays larger than numerosities of three. These perceptual impression strategies gradually become replaced with age by the next two procedures to be discussed: matching and counting.

CORRESPONDENCE AND MATCHING

One-to-one correspondence between the elements of the two sets can be used to establish the relative numerosity of two sets. If no entities are leftover after the correspondence is established, the two sets are equivalent in numerosity. If one or more entities in one set are not used in the correspondence, that set is greater than the other set. We shall use *correspondence* to mean the mental relation established between elements of two sets and *matching* to refer to the empirical procedures by which correspondence between sets is established.

The likelihood of implementing a matching procedure in a relative numerosity situation would seem to vary with the effort required by such a procedure, the availability of other procedures to test equivalence, and the child's desire for accuracy. Unfortunately, the influence of these variables has not been examined systematically. This unsystematic variation may account for apparent conflicts among results across studies, and it greatly limits our ability to map out the true state of affairs regarding children's matching and use of correspondence information.

Apparently a matching procedure can be readily implemented by many children by about age 6 even for large numbers. In a sample of French children midway through their first year of school, over 50% of the children used matching to make a set of blue buttons that was equivalent to a set of 37 red buttons (Comiti *et al*, 1980). Comiti *et al*. reported the use of three main one-to-one correspondence procedures. In the first instance, children made a row of the red buttons and then made a row of blue buttons of the same length. In the second instance, children made red and blue pairs by either placing a blue button partially or completely on top of a red button, by inserting a blue button next to each red button, or by pulling one red button at a time out of the red pile and associating it with a blue button. In the third instance, children pointed simultaneously at a red and a blue button from the red and blue piles and then removed them from the red and blue piles. All of these procedures involved moving objects around.

Other studies have involved stimuli that cannot be moved. Gullen (1978) presented children (grades K, 1, and 2) with cards depicting two rows of circles and with lines drawn between corresponding objects. Number and length cues were varied. The children were permitted to count, but the lines permitted rapid and reliable determination of correspondence. The most striking result was the very small reliance on correspondence by the youngest group. Fewer than 10% of the judgements by kindergartners were based on correspondence (length was by far the most prevalent basis for them). In contrast, by grade 1, correspondence was the single most common basis; and by grade 2, it was the basis for more than 60% of the judgments.

Brainerd (1979) has presented data on correspondence that at first seem quite remarkable. Relative numerosity judgements were required for sets of closely spaced circles, counting was not permitted, and array numerosity and length were varied independently (i.e., length was not consistently correlated with numerosity). The remarkable aspect of the results was the very low use of correspondence information. Only 48% of 11-year-olds established correspondences as a means for making judgments, even though that was the only reliable way to make such judgments. Upon closer examination of the task, the results appear somewhat less surprising. The establishment of correspondence required a difficult matching procedure (circles were small and closely spaced), which may have reduced the number of children willing or able to carry it out. It also may be that many of these children could have used matching if they had thought of doing so.

This interpretation suggests that it is useful to distinguish between conditions in which matching is suggested or demonstrated and those in which it is not. Fuson, Secada, and Hall (in press) found that in a conservation task 80% of children aged 4½–5½ who were helped to match by placing connecting strings between the pairs and who were then asked to show the corresponding object for three such objects did use this correspondence information to judge the two sets so matched to be equivalent after a lengthening transformation of one set; this was significantly more than the 14% giving correct judgments when matching was not induced, and these spontaneous judgments were not based on matching. Thus, children seem to be able to use correspondence information to judge numerical equivalence considerably before they spontaneously obtain such information. In a different task, Hudson (in press) compared two conditions in which 4- through 6-year-olds were asked to report the number of objects by which one set (e.g., birds) exceeded another (e.g., worms). Prototypical of one condition was the question: *Suppose the birds race over, and each one tries to get a worm: will every bird get a worm? . . .How many birds won't get a worm?* In the second condition, the corresponding question was the more typical school and research phrasing: *How many more birds than worms are there?* The former condition exceeded the latter by a large degree both in the accuracy of responses and in the extent to which a matching procedure was employed. Hudson concluded that young children can use matching more effectively and more readily than often has been suggested. He also concluded that the low performance of the second condition just described was attributable to the linguistic complexity of the question. However, the form of the final question here was confounded with the extent to which the question directly suggests the use of a particular solution procedure, namely, matching. The "Won't Get" question form contained quite specific action directives about matching one set to the other. Other work on *more* and *less* has indicated that young children do not spontaneously quantify the difference in two sets and they thus can be led to make mistakes in judging which set has more entities (Blevins, Mace, Cooper, Starkey, & Leitner, 1981; Brush, 1978). It therefore seems quite likely that some part of the superior performance in Hudson's study was due to the creation of cues to use a matching procedure.

Thus, the Brainerd results and many usual conservation results seem to reflect a failure of children to use matching spontaneously rather than a failure to understand the relationship between correspondence and equivalence or a failure to be able to carry out a matching procedure effectively. These latter two each are understandings that children must of course acquire. Our point here is that we cannot infer the lack of either of them simply from failures to use matching spontaneously. It is true, however, that such failures do have practical consequences, for they limit the ability of children to obtain adequate numerical information from their world. Future research in this area will be more useful if it attempts to determine all of these various capabilities: when and under what conditions (*a*) children can use matching procedures correctly, (*b*) children can understand and use correspondence information obtained from matching, and (*c*) children will spontaneously obtain such information.

COUNTING

Cardinal words can also be used to determine the relative numerosity of two sets. If the cardinal words are the same, the two sets are equivalent. If they are not, the order relation on the words involved will dictate and reflect the order relation on the two sets. Determining the order relation on two cardinal words is a very complex issue. It potentially involves knowledge about relations on the number sequence words and about relations on measure words as well as about relations on cardinal words. This issue of order relations on cardinal words is discussed on p. 93. This section will deal only with the relationship between counting and order or equivalence relations on cardinal words. Counting may be used in a static situation or in a conservation of number situation in which a transformation of one of the sets is made so that the lengths (or some other spatial attribute) of the sets are in the relation opposite to that on the numerosities (e.g., non-equivalent if the numerosities are equivalent). Both static and nonstatic situations will be discussed.

In order to use counting to determine the relation between two cardinal contexts, children must decide to turn these contexts with unspecified numerosities into contexts with specified numerosities, they must be able to do this by counting accurately, and they must be able to use the specific numerosity information derived from counting to determine the equivalence or the nonequivalence of the two sets. We have already discussed counting accuracy and will therefore treat only the first and third of these requirements.

Children must learn a "same counting word implies same numerosity (cardinality)" rule for cardinal equivalence. That is, they must learn that if they count two sets and get the same final counting word for each set, then the two sets have the same number of objects. This seems trivial to adults (and to older children) who may find it difficult to imagine any other conclusion. However, such a view reflects a pre-Piagetian innocence. Almost all readers of this chapter will have seen or read of a child like the child described in Flavell (1963, pp. 360–361) who will "count two sets and state that there are 'seven' here and 'seven' there . . . ,but he may nonetheless argue that there are more here than there, if the arrangement is perceptually compelling in favor of inequality." This finding has been replicated by many researchers since Flavell reported it from the Genevan research, and there is now little doubt that a "same counting word implies same cardinality" rule is one aspect of cardinal words that young children must learn.

However, in our view, the striking (and almost unbelievable) nature of this primitiveness in young children's notions of cardinal equivalence has led in the years since this phenomenon was first reported to an emphasis on the children who do not use such a rule and to a resulting neglect in answering when children can use such a rule. Consequently, relatively little information is available concerning the latter compared to the large number of conservation of number studies emphasizing the former. One study which is available on this point indicates that children can use specific numerosity information to make judgments of cardinal equivalence in conservation situations before they spontaneously obtain such information for themselves and before they conserve in an operational sense. Fuson, Secada, and Hall (in press) reported that when in a conservation of equivalence situation, 4½ to 5½-year-old children were induced to count both sets (of numerosity seven) after the transformation but before the question, 69% gave correct equivalence judgments, i.e., they demonstrated the "same counting word implies same cardinality" rule. Only 14% of these children's age-mates gave correct equivalence judgments when specific numerosities were not introduced by the experimenters (i.e., when the usual conservation situation was given). Two other studies indicate that children can use specific numerosity information to infer static equivalence before they do spontaneously obtain and use it in a conservation situation. Saxe (1979) reported that half of the 4- and 5-year-olds and almost all of the 6-year-olds did use counting to ascertain cardinal equivalence in a static task (with numerosity nine) while many of these children did not conserve. Moreover, all but one of the children who did make correct equivalence judgments in the conservation of number task did use counting information in the static task. However, the static-conservation comparison is not as clean as one might have wished, for children who did not spontaneously count in the static task were urged to do so but no counting directions were given in the conservation task. Similar evidence with the similar limitation comes from a study by Pennington, Wallach, and Wallach (1980) who reported that when 45 disadvantaged third-graders who failed a conservation of numerosity task on two successive days were induced to count circles in two static rows arranged as in the Brainerd task (length and number cues conflicting), 34 made perfect equivalence judgments and most of the remaining 11 only failed to do so on one trial.

The focus in most conservation studies on operationally-based equivalence judgments has frequently led to a prohibition on counting or to a lack of reporting of solution procedures other than perceptual or operational. This has resulted, in our view, in an underemphasis on the number of children who do spontaneously count in the conservation of number task, i.e., on those who do obtain information about specific numerosities. Fuson, Secada, and Hall (in press) recently found 81% of the 5-year-olds in their sample who gave other than consistent perceptually-based responses in a conservation of number task spontaneously counted on at least one trial. Counting-based equivalence judgments were correct on 89% of the occasions on which they were used. Half of the wrong counting judgments resulted from counting errors, and half from the use of perceptual reasons (e.g., length) rather than the spontaneously obtained count information. Siegler (1981) also reported spontaneous counting in a conservation of number task, and LaPointe and O'Donnell (1974) reported a considerable amount of counting by children giving correct equivalence judgments. Fuson et al. discussed several roles such counting might play in the acquisition of operational conservation responses.

Several writers have earlier discussed such spontaneous uses of specific numerosities. Klahr and Wallace (1976) argued that children in a conservation setting will turn a situation with unspecified numerosities into one with specified numerosities and that such an act is crucial in the attainment of conservation. Gelman and Gallistel (1978) proposed that in general (though not necessarily in the conservation situation) young children come to reason with specific numerosities first. Murray (1979) discussed the use by children of what he termed a "testing action" in conservation tasks (obtaining specific numerosities is one such action).

Children also evidently use counting to establish nonequivalence (order) relations and to find a part of one set equivalent to a smaller set. Some children evidently can do this very early if the numbers are very small. Bullock and Gelman (1977) found that children 2½-years-old trained in a non-verbal task to select the greater (or lesser) of sets of one or two objects transferred the relationship when presented with sets of three and four objects if the original pair of sets was kept in sight. The basis for these judgments is not quite clear. Bullock and Gelman found a relationship between correct responding and the spontaneous verbalization of counting words. They concluded that children counted and then represented the sets as specific numerosities, and these specific numerosities were used to make the judgments. However, in a second experiment with the same twelve 2½-year-olds, only nine of the eleven correct responders could separately give accurate numerosities for all four sets. Thus, at least some of the children seem likely to have been making the judgment on some noncounting basis. Hudson (1981) showed kindergarten children sets of drawings each of which contained two numerosities between two and nine whose numerical difference was one, two, or three. Children were told *Suppose the birds all race over, and each one tried to get a worm. How many birds won't get a worm?* Eighty percent of the responses were accurate, and more than half of the children overtly displayed a "count to equivalence" solution procedure: they counted one or both of the sets, then counted a subset of the larger set that was equivalent to the smaller set, and then announced the difference (evidently they did not count to determine these differences of one, two, and three). For some (27%) of these trials, children evidently first determined which was the larger set by counting both of the sets, but for most they simply counted the small set or counted only the subset. Children also used a count-all strategy on 12% of the overt solution procedures; here they counted all of each set and then announced the difference. Thus, they must have known the difference between the two numerosities they announced. Children also used matching on 22% of the overt solution procedures. The comparatively heavier use of counting procedures when the verbal cues to a solution procedure emphasized matching may have resulted partly from the arrangement of the items in the two sets: they consisted of one vertical and one horizontal line, making matching somewhat difficult to accomplish. In Hudson (in press) a similar procedure was used, but the stimuli were arranged randomly and the numbers were smaller (this study was discussed in the previous section). Although exact amounts of matching and counting solutions were not reported in the earlier study, the conclusion that young children could carry out a matching procedure correctly implied that the predominant solution procedures in that study were matching ones. The conclusion from these two studies by Hudson seems to be that many children aged 4, 5, and 6 can effectively use matching and counting procedures to find exactly how much larger one set is than another if making a correspondence is verbally suggested in some way.

In summary, children must learn the "same word implies same cardinality" aspect of cardinal words, and they later can and do use this property to change a task with unspecified numerosities into one in which counting and the resulting number words are used to establish the existence of an equivalence relation. They also use counting and the resulting specific numerosities to establish order relations. Because counting does not require the simultaneous and contiguous presence of two sets as does matching, it is an especially useful means of determining equivalence or of nonequivalence.

Developmental Differences in
Equivalence and Nonequivalence Performance

There has been some interest in developmental differences between establishing the numerical equivalence and establishing the nonequivalence of two sets, i.e., in the developmental precedence of order relations and the equivalence relation in cardinal contexts. Brainerd (1979) argued that nonequivalence precedes equivalence developmentally. He regarded equivalence relations as requiring a one-to-one correspondence and order (nonequivalence) relations as requiring many-to-one or one-to-many correspondences. He proposed that many-to-one and one-to-many correspondences are understood before one-to-one correspondences. However, this definition of order relations is inaccurate. Mathematically, a one-to-many correspondence is one which takes one element in one set to many elements in the second set and does so for each element in the first set: $\sqrt{x}$ (for x a positive integer) is an example of a one-to-many correspondence, for each positive integer has a positive and a negative square root. A many-to-one correspondence consistently takes many elements in one set to one element in the second set (e.g., x^2 which take $+x$ and $-x$ to the same element in the second set). Neither of these kinds of correspondences is used to establish numerical equivalence or nonequivalence (the order relations). As just discussed, these are established by the same type of correspondence, a one-to-one correspondence. For order (nonequivalence) relations, one or more elements of one of the sets is left over after the one-to-one correspondence has been completed; no such leftover element exists for equivalent sets. Young children apparently have some intuitive understanding of this mathematical definition, for they employ just such a leftover strategy in their use of perceptual (length or density) strategies in conservation situations. Apparently it is the strong salience of the leftover part (e.g., *See, there are some extra ones out here*) that makes them infer that an order relation exists and that contributes to their failure to give equal attention and care to the nature of the (incorrect) matching within the corresponding parts of the two linear sets.

In order to examine developmental differences in the understanding of equivalence and nonequivalence relations, Carpenter (1971) distinguished between two types of nonequivalence situations in conservation settings: (*a*) those in which nonequivalent sets are transformed from appearing nonequivalent in a dominant dimension (e.g., in length) to appearing equivalent in that dimension and (*b*) those in which the transformation is from nonequivalence along some perceptual dimension to nonequivalence in the opposite direction (e.g., the larger–longer set is pushed together to look shorter than the other set). Carpenter reviewed studies on nonequivalence relations and found some evidence of earlier performance on the first type of nonequivalence tasks. In his own conservation of numerical equivalence studies, however, no clear evidence of such a difference was found. He did find such evidence in liquid measure tasks.

Interpretation of equivalence–nonequivalence comparisons may be clouded by a methodological problem noted by Brainerd (1979): The inequality of the sets to be compared may make perceptual strategies relatively more successful for the nonequivalence comparisons. It is easier to tell that a set of nine has more than a set of seven than to tell that a set of seven has the same as a different set of seven. Brainerd found that 23 of 189 kindergarten and first-grade children who were asked to make equivalence judgments on stimuli arranged in the conservation end-state (two rows of circles of unequal length) correctly answered the nonequivalence problems but failed the equivalence problems. Brainerd concluded from the justifications children gave for their nonequivalence choices that they in fact were using perceptual (density) strategies rather than establishing one-to-one correspondences. Thus, although there may be developmental differences between performance on equivalence and nonequivalence tasks, especially on some types of nonequivalence tasks, the literature to date is not clear on this point.

MEASURE WORDS

In a measure context, the number word describes the numerousness of the units into which some continuous dimension of an entity has been divided; that is, it tells "how many" units there are in some entity. A child first must attend to the particular entity and then isolate from other attributes the dimension along which the entity is to be measured (e.g., length, distance, weight, volume, time).

Many variables would seem to affect the ease with which this isolation takes place, but very little is presently known about such effects. A dimensional salience procedure such as used by Odom, Astor, and Cunningham (1975) would seem to be useful in exploring this issue.

The child must then learn to identify (or create) the units into which this isolated dimension is broken. This unit division process may involve filling units (e.g., filling unit cups for liquid or food solids—lard, peanut butter, flour) or covering the to-be-measured entity with units (e.g., an area with centimeter squares, a length with foot-long sticks). With units that must be filled, the child must learn that all of the entity to be measured must be used up and that the measuring unit must be completely filled. Evidently many first-grade American and Soviet children do know the former, but not the latter; they fail to fill the measuring cup completely when measuring out quantities (Carpenter, 1971, 1975; Gal'perin & Georgiev, 1969). When covering with length units, American first-graders evidently complete successfully both requirements of a correct measuring procedure. That is, they cover the whole length, and they do not leave spaces between the units (Carpenter, 1971; Hiebert, 1979). After making the filled units or covering with units, the child must then count the units and make a *count-measure transition,* i.e., shift from the count meaning to the measure meaning of the number word (e.g., four cups or six centimeters).

If only one copy of the relevant measuring unit is available (e.g., a single 1-centimeter strip), the covering measuring procedure must be replaced by one involving the iterative use of the single copy. Instead of a final count of the number used, a running count must be kept with each use of the copy. This requirement increases the information-processing demands of the measuring task in that the count now must be remembered while the unit is being placed each time. This may be difficult for young children who are just learning the unit placement task, and even skilled users of this method may lose count if they are distracted during the procedure. Some data indicate that the use of the iterative process in a length context introduces additional new errors. Some children who successfully carry out the multiple unit covering process put their finger down to mark the end point of the unit measurer in such a way as to include a whole finger width between successive placements (Hiebert, 1979). This seems to be a conceptual error for some children (they assume that the whole length must be covered, but not that it must be covered by units) and an error resulting from limited procedural knowledge for others (the underestimation of the measure is at least vaguely understood, but no better method is known).

More sophisticated and efficient measuring procedures use a scale (usually a length scale) of units marked with numbers. Such scales may be direct (a metric length scale for measuring length) or indirect (a length scale on a thermometer for measuring temperature or on a bathroom scale for measuring weight). Children must learn to use various such scales: measuring tapes, measuring cups, thermometers, weight scales, and clocks. Particular aspects of the use of each scale must be learned. For example, the measuring tape must be placed at one end of what is to be measured and the measure number read off the other end; food in a measuring cup must be level; the thermometer must have been in the environment to be measured for a little while; one's entire body must be on the scales. An enormous amount of quite specific information must be acquired. Problems evidently exist even with the simplest of such scales. In the 1977–1978 National Assessment of Educational Progress in mathematics, 77% of the 9-year-olds and 40% of the 13-year-olds responded that a line segment starting at the "1" on a length scale and extending to the "5" was 5 (rather than 4) units long (Carpenter, Corbitt, Kepner, Lindquist, & Reys, 1981). These responses indicated that they did not use the underlying notion of measure as the number of units comprising the segment. To further complicate the use of most such scales, very few entities to be measured fall exactly on the major units marked on the scale. Thus, the numerical value for each subinterval must be learned. Little research presently exists on these matters.

Some dimensions are quite different to measure directly (e.g., temperature, volume). Measure words for these dimensions are obtained by the use of derived scales or by formulas that use measure words for one dimension to derive measure words for another dimension (e.g., $L \times W \times H = V$). Children must learn how to use these formulas, and hopefully they will also learn at least some intuitive basis for why such formulas work.

Finally, estimation can be used to produce a measure word in a given situation. In fact, estimation may play a more important role in real life *measure* situations than in cardinal situations. For many purposes, estimation is sufficient (e.g., approximate wall area to decide how many gallons of paint to purchase), and in some situations *Will this leftover spaghetti sauce fit into this container?*, estimation may be the only practical procedure. Siegel, Goldsmith, and Madson (1980) have begun to examine children's estimation abilities in measure contexts and have found the same direct and indirect kinds of procedures described earlier in the numerosity estimation section.

Additional skills may be involved in the comprehension (as opposed to the selection) of a given measure word. In all of the direct measuring procedures just described, a representation of the selected unit was physically present. When instead a given measure word is reported and must be comprehended (e.g., 10 inches, 15 centimeters, 5 pounds, 150 meters, 4 years), some representation of the unit must be available and some way of generating a representation of the specified number of units must be devised. These processes probably involve a good deal of estimation, and there likely are wide individual differences in the ability even of adults to understand (generate a "reasonably good" representation of) the given measure word. Again considerable specific experience would seem to be required to develop good comprehension of measure words, and thus young children may be quite poor at it.

Given the quite marked perceptual differences for different measure dimensions and the amount of information specific to particular types of measure dimensions, it seems likely that the measure concepts of children consist of a scattering of relatively isolated fragments, with little overall generality. The generalization of measurement procedures and concepts would seem to be facilitated by the use of the first measure procedure (the use of multiple copies of a single unit) across different types of measure dimensions. Even adults (in-service and preservice teachers) have reported (informally) increased understanding of measurement resulting from the use of such direct measuring experiences (e.g., as described in Bell, Fuson, & Lesh, 1976).

Relations on Measure Contexts

All of the Piagetian conservation tasks other than numerical equivalence concern equivalence and order relations on measure word contexts, but in most such studies the measures of the quantities are not specified by words. Much of the conservation research since the original Piagetian studies has focused upon relationships among conservation on various tasks and on possible sources of the Piagetian explanations of conservation (see Beilin, 1971, 1977; Modgil & Modgil, 1976; Siegel & Brainerd, 1978, for summaries and reviews). In such conservation tasks, children attempting to learn to determine order or equivalence relations on measure contexts also have all of the difficulties with the equivalence and order relational words (e.g., big, tall, high, fat, same).

A certain group of conservation studies has focused upon the role of measurement operations (i.e., the division of the continuous quantities into units and the matching or counting of units) upon the acquisition of conservation of continuous (measure) quantities. Several studies have reported success with such approaches (e.g., Bearison, 1969; Carpenter, 1975; Carpenter & Lewis, 1976; Hiebert, 1979; Inhelder, Sinclair, & Bovet, 1974; Kingsley & Hall, 1967). Such studies have indicated that American kindergarten, first-, and second-graders, when given the measure words for two entities, can evidently use the words correctly to determine whether the entities are equivalent. Hatano and Ito (1965) reported similar findings with Japanese children. Furthermore, the provision of measure words leads to considerably and significantly more correct responses than in control situations without such words, and training with matching or with counting and the use of measure words evidently leads to significant, lasting, and generalizable conservation judgements.

However, some studies indicate that in fact children may have been treating the measure words only as cardinal words (i.e., they ignored the type of unit accompanying each number word). When an initial equivalence or order relation was established by the usual perceptual means and the two quantities were then measured by units of different size, children tended to give responses fitting the final measure words rather than the initially established relation (Carpenter, 1975; Hiebert, 1979). For example, when glasses of water (which children agreed were equivalent) were measured with different units so that one glass had a measure of five and the other a measure of three, the child would respond that the first glass had more water in it. Similar results were obtained for equivalence and for nonequivalence relations. Here is, therefore, a new problem with the very complex word *same*: Two measure words have the same measure meaning only if they refer to the same units. Thus, for measure contexts, children need to learn a more complex rule when establishing equivalence and order relations on measure contexts: "same unit *and* same measure words imply same measure." Furthermore, children must learn that an inverse relation exists between the size of the unit and the number of units in some entity. Carpenter and Lewis (1976) reported that first- and second-graders do recognize this inverse relationship, but are not able to use it when the perceptual characteristics of the entities contradict the measure results.

Results for relations on measure contexts seem to be similar to those for numerosity: the child's use of specific measure words for quantities does facilitate judgments of equivalence and nonequivalence and the ability to use measure words effectively in this way occurs considerably before children spontaneously use them. Children may acquire their eventual understanding of conservation (i.e., of the lack of effect of spatial transformations on quantity) in part through the use of such empirical measuring procedures in the ordinary course of events. Alternatively, the understanding of conservation may be based on the use of general reasoning involving some conception of units rather than explicit empirical actions, or through the generalization of rules devised in conservation of numerosity situations (see also Siegler, 1981). It seems quite possible, as suggested by Brainerd (1979) and Fuson, Secada, and Hall (in press), that quite different routes to an understanding of conservation may coexist and that the use of these means may differ considerably from child to child.

MEASURE WORDS AND CARDINAL WORDS

Measure number words and cardinal number words are similar in that they each describe the numerousness of the units in some set. The unit for cardinal words is a special one—a singular individual entity. American preschool and elementary school curricula begin with a heavy emphasis on cardinal words and only later and in a minor way move to a consideration of measure words. There is a considerable emphasis on counting individual objects, and computation is introduced as occurring with sets of discrete objects. Even when a measure model such as a number line is introduced, it is almost always used in a "cardinalized" fashion. For example, four is four discrete hops of the frog rather than a length of four units (or a Cuisenaire rod of length four units). Thus, for American children a measure number word seems likely to be (and the Carpenter and Hiebert results reported in the last section indicate that it is) psychologically only a "special" cardinal number word that has some measure unit (e.g., foot, liter, cups) stuck after it. In contrast to this approach, several Soviet educational psychologists (Davydov, 1975, 1982; Gal'perin & Georgiev, 1969; Minskaya, 1975) have argued that children ought to learn the most general meaning of number (that based on quantity and real numbers) from the earliest time, rather than learn number only as a natural number based on discrete objects, i.e., they focus on measure number words rather than on cardinal number words. The latter approach requires a subsequent extension of the limited concept of number to fractions and decimals. Gal'perin and Georgiev provided from training experiments some evidence that an early curriculum in which young children measure all sorts of entities with a variety of nonstandard as well as standard units led to an avoidance of common errors in understanding measurement and provided a good foundation for understanding relations and operations on cardinal numbers—the bulk of the elementary school curriculum. In particular, a very general concept of unit was induced so that a unit was considered as an aggregate (e.g., two red beakers or three buttons), rather than only as an individual discrete object. The rationale for this approach was described by Gal'perin and Georgiev as follows: "Orientation by individuality leads to many other mistakes. Individuality is a universal property. If it is taken as a feature of a unit, the child develops a natural and justifiable indifference to the size of the unit measure and to its fullness [p. 191]." As noted earlier, Carpenter (1971, 1975) and Hiebert (1979) found just such neglect of the fullness of the measure unit and of the size of the unit by American first- and second-graders. Montgomery (1973) found, however, that 9 days of teaching area measurement by stressing different units was sufficent to make second- and third-graders aware of the importance of the measuring unit; they spontaneously recorded units with each measure number, whereas a control group did not.

Davydov (1975, 1982) has also described a curriculum for slightly older children that stresses the algebraic recording of manipulations with objects and with measuring units. According to Davydov, that curriculum led to a more general concept of number and of relations and operations on number than does more typical instruction. From the beginning these children operate on general, unspecified (a,b) numerosities and measures rather than on specified ones (e.g., 3, 6, 8).

The very considerable difficulty primary school children experience in learning to use our base ten system of numeration may reflect a more general difficulty in the understanding of units. Our base ten system of numeration (and other systems of numeration) is in fact a measure system: The words (and symbols) used in them refer to a certain number of different sized units. The units attached to the larger number *words* are explicitly named (e.g., four *thousand,* two *hundred,* six-*ty* (six *tens*) eight), but the units in our system of *numerals* (4268) are implicit.

Comprehension of this measure basis of our system of numeration is very difficult for children. Moving from the idea of ten ones as a set of individual entities to the ten ones as a single unit (i.e., as a *one zero* or a 10) is quite difficult (Underwood, 1977). At least through second grade and often later, words up to one hundred seem to elicit primarily counting, sequence, or cardinal meanings, rather than base ten measure meanings. This tendency to consider sequence or cardinal rather than base ten measure meanings may remain strong even in adults. Hinrichs, Yurko, and Hu (1981) reported that reaction time responses of adults asked to tell whether two-place symbols were greater or less than 55 showed an effect of distance from 55 *within* as well as across decades; for example, it took longer to decide that 61 was greater than 55 than it did to decide that 68 was greater than 55. If subjects had been considering the symbols only as measure words (e.g., 68 as six *tens,* eight ones), they should have been able to focus only upon the tens word (all "sixty" words are greater than 55, all "forty" words are less than 55), and reaction times should have been a step function (i.e., equal for all "forty" words and for all "sixty" words). Thus, in spite of the fact that focusing only upon the tens word of a given symbol would have been fully reliable and very simple, adults seemed to persist in considering these double digit symbols as whole sequence or cardinal words.

ORDINAL WORDS

An ordinal number word describes the relative magnitude or the relative position of a discrete entity within a well-defined, totally ordered set of entities in which the ordering relation has a specified initial point. Children learn to under-stand and to use such phrases as *third fastest runner on the block, fourth child, sixth in line.* Few data exist on children's acquisition of the ordinal words. It is doubtful that the sequence of these words receives anything like the amount of practice (rehearsal) that the standard number words do, and they may not really be learned as a sequence at all. It seems likely rather that the first several words (first, second, third, and maybe fourth) are learned from their isolated use in ordinal contexts and that much of the total ordinal word sequence subsequently is derived largely from the standard number word sequence. Beilin (1975) reported that only one of his sample of 54 2- and 3-year-olds could produce a sequence of ordinal words up through *fifth.* However, 57% of the 5-year-olds, 91% of the 6-year-olds, and 98% of the 7-year-olds could do so, although only 2%, 10%, and

46% of these age groups could produce the ordinal sequence up through *twenty-third*. The comparative figures for these three oldest age groups for the sequence words up to *twenty-third* were 81%, 90%, and 100%, respectively; so production of the ordinal word sequence lags quite far behind that of the conventional counting sequence.

The words *ordinality, ordinal number, ordinal aspect of number,* and *ordinal concepts* have been used by different writers to mean different things. Some of these different uses have arisen because of different needs or foci of particular fields of the writer—for example, mathematical foundations, psychological measurement, developmental psychology, and history of mathematics. Most of these uses of terms require only the existence of some kind of linear or total ordering (e.g., *is bigger than, comes later than, is tastier than*) and two or more entities on which this order relation is defined. Many different kinds of entities and many different kinds of orderings can be used in an ordinal word context. The entities can even be in sequence, cardinal, or measure contexts with their usual order relations (for example, the cardinal (numerosity) words ordered by their magnitude (one, two, three), and the sequence number words ordered by their conventional order). Such an ordering only becomes an *ordinal word context,* however, when the point at issue is the relative magnitude or relative position of *one* such entity within the ordering. We believe that a great deal of confusion would be eliminated if, in the future, writers used *order* or *ordered* or *ordering for those circumstances in which only an ordering was involved and restricted the use of the word ordinal* to contexts in which an ordinal word could potentially be used. Otherwise we find ourselves continually finding cardinal, sequence, and measure contexts referred to as *ordinal* as soon as any reference to an ordering has been made. An ordering thus is necessary but not sufficient to regard some context as an ordinal context.

To understand the development of children's use and understanding of ordinal number words, we must know about children's acquisition of the ordinal words, their ability to understand and to construct various types of orderings and their ability to comprehend and to apply an ordinal word to a given entity within an ordered context. We have described the present lack of knowledge about the first of these. We shall now turn to the other requirements.

Children's Understanding of Orderings

Piaget's work on seriation (1941/1965) focused on children's use of physical measure dimensions to construct a total ordering on a set of objects. Sticks of various lengths, balls of different diameters, and dolls of various heights were among the materials used, Piaget concluded that young children below the age of 6 or so can make successful pair-by-pair comparisons, but they are not able to coordinate several ordered pairs into a coherent ordered set of objects or to insert a missing object into an already ordered set. Piaget argued that the latter failures occur because the children cannot simultaneously consider an object as less than a given object and as greater than another object. They begin to be able to do so near the beginning of the concrete operational period, usually somewhere from ages 6–8. These findings about children's construction of a perceptually ordered set of objects have been replicated by other researchers (see, e.g., Flavell, 1970).

Several researchers have reported seriation or ordering abilities in preoperational children. Brainerd (1979) identified an early ordering ability, which he termed *perceived ordinality:* the perception by 3-year-olds of the ordering relation inherent in some physical progression containing at least three terms (e.g., three circles of increasing size). Given the task involved and our earlier discussion of the distinction between ordering and ordinality, this ability might better be termed *perceived ordering.* Blackstock and King (1973) found that 4- and 5-year-olds could select the set of cylinders seriated in height before they could reconstruct such a set. Using bars of varying height with 3- through 9-year-olds, Siegel (1972) examined in a nonverbal task what she called *the concept of seriation.* The bars were presented in an unseriated row, and the child had to select the larger or smaller of two bars, the smallest or middle-sized of three bars, or the largest or next-to-smallest of four bars. The 3-, 4-, and 5-year-olds all found the smaller, larger, smallest, and largest within 20 trials. The 3-year-olds failed on the other two tasks, and the 4-and 5-year-olds eventually did middle-sized, but very few ever did next-to-smallest. Performance over ages 4–8 showed steady improvement on the latter two tasks. Several other studies pursued these findings. In these studies (Gollin, Moody, & Schadler, 1974; Griep & Gollins, 1978; Marschark, 1977) the critical condition in which seriation apparently was displayed was one in which the child was forced (through some manipulation) to identify the littlest (or biggest) object in a nonseriated array of four entities varying in height, and then asked to identify the next littlest (or next biggest) member of the array. Griep and Gollin (1978) found a similar result for stimuli varying in brightness and also found significant transfer of the effect of the manipulation from brightness to height and vice versa. Whether or not accurate performance under the manipulated conditions indicates seriation ability in the Piagetian sense seems questionable to us. That is, it seems possible that once a child was led to identify the littlest entity (as the children could do quite readily), the child might then exclude that entity from further consideration and search again for the littlest (of those remaining under consideration). In contrast, the essence of seriation in Piaget's sense is that an item be compared with both longer and shorter items to arrive at its correct position in an array. Thus, these data clearly indicate young children's ability to find the starting point of an ordering, but do not unequivocally imply more advanced ordering abilities.

Very little is known about children's ability to construct or to understand sets ordered in ways other than by such physical dimensions as length or to use a spatial ordering to represent some other ordering. For example, teams are often listed vertically from first to last place, and children must learn to make and interpret such representations. In our pilot work several 4-year-old children confused the desired correct *spatial* waiting-in-line order of the given set with the *time* order in which the experimenter asked the ordinal word identification question. That is, in response to the first question (*Which animal is fourth?*), these children would point at the first rather than the fourth animal. With each additional question, regardless of the ordinal word used, they would continue to point to successive animals in the line. Thus, there may be unanticipated difficulties in children's using nonmagnitude orderings or in using one type of ordering to represent a different type of ordering.

Three- and four-year-old children evidently are interested in constructing orderings, as indicated by their spontaneous and enthusiastic use of many of the Montessori seriation materials (cylinders of graduated height, color tablets of a hue from light to dark, bells of graduated pitch). Some of these materials have self-corrective features (e.g., cylinders stick up if they are placed in the wrong slots) that enable the seriation procedure to be completed by trial and error before the concrete operational stage has been reached. These materials provide useful examples of ways in which one might provide an environment rich in practice in constructing orderings.

Labeling Ordinal Contexts

A specific number word can be assigned to an ordinal context through the same means used for cardinal contexts: counting, subitizing, estimating, and being told. However, such uses in ordinal contexts may differ from the cardinal uses. For example, the use of counting to determine ordinal position is subject to constraints not present in cardinal contexts. First, the counting procedure in the ordinal context must begin at the specified starting point for the ordering on the set, and the procedure must follow the order on that context until the entity in question is reached. Thus, the order irrelevance principle of counting identified by Gelman and Gallistel (1978) does apply to cardinal contexts but does *not* apply to ordinal contexts. Children also must learn that when counting in an ordinal context, one does not count every entity (also in contrast to a cardinal context). Rather, counting must stop at the object whose ordinal word is desired. So far as we know, there has been little systematic study relative to such issues.

It is possible to use the list of ordinal words (*First, second, third, fourth,* etc.) to count in an ordinal situation. However, in various ordinal tasks with children aged 3–6, we more often have observed the use of the already known and overlearned list of counting words followed by a transition to the appropriate ordinal word (e.g., *One, two, three, four. Fourth. This one is the fourth.*). Such a *counting word–ordinal word transition* obviously is dependent upon knowing ordinal words, even though less use is made of them.

Some research on ordinal word contexts has used words other than ordinal words. Holmes (1963) gave young children an ordered set of 10 dolls of increasing height and asked them to locate particular dolls. The identification question did not involve an ordinal word, but rather the child was told to *touch Doll Six.* The mean percentages of correct identification of dolls was 66% for kindergarten children and 16% for preschool children. Because Holmes gave neither identification tasks involving ordinal words nor tasks labeling particular ordinal positions, the relationship between performance with nonordinal words and performance with ordinal words is not clear. In our own research on conservation tasks with ordinal number words, quite a few 5-year-olds did not understand questions such as *Is this car still third in the yellow race?,* but did understand and respond correctly if asked *Is this car still the Number Three car in the yellow race?* However, it may be that when expressed in this latter way, a counting meaning rather than an ordinal meaning is conveyed to children. That is, the *Number Three* car may be so because it is the car pointed to when the child says *three.* Thus, studies in which such terminology is used are inconclusive with respect to the child's understanding of ordinal position.

Beilin (1975) has reported a considerable difference between children's ability to use the words *second* and *third.* Percentages of 5-, 6-, and 7-year-olds who were able to label the second and third stick from a row of differently colored sticks of the same length were 33% and 8%, 63% and 33%, and 82% and 58%, respectively. In the same sample 57%, 91%, and 98% of those children were able to produce the ordinal word list through *fifth,* so these figures indicate both that the use of *third* is more difficult than the use of *second* and that a sizable gap exists between being able to produce *second* or *third* in the ordinal word sequence and using them as labels for a referent in an ordered set. Additional data are needed to indicate whether the absolute levels and the pattern of relationships in these figures for nonmagnitude orderings (e.g., an ordering by color) will hold for magnitude orderings. Beilin did not report how the children in his study selected their ordinal words *second* and *third.* The considerable gap in performance on these two items might stem from the use by many children of subitizing for *second* but not for *third* and the failure of these children to count and then to use the count–ordinal transition for *third.*

An interesting aspect of ordinal word contexts that has been little explored is that the numerosity of the set of entities that follows a given ordinal position does not alter the item's ordinal position, but it may alter the psychological interpretation of that ordinal position. It may be quite different psychologically to be fourth in a school-side spelling bee (e.g., fourth out of 300 children) and to be fourth in a reading group spelling bee (e.g., fourth out of 6 children). In ordinary affairs the full meaning of ordinal position information may depend in part on the numerosity of the set up through that position *compared to* the numerosity of the total ordered set. If so, this may interfere with the child's acquisition of the mathematical notion of ordinal position as dependent only upon the numerosity of the set up to and including the particular entity.

Relations on Ordinal Contexts

As with cardinal and measure contexts, two ordinal positions can be accurately compared by the use of one of three relations: One of the ordinal positions will be less than, the same as, or greater than the other position. These relations can be determined by the same means used in cardinal and measure relational contexts: perceptual comparison, matching, or counting.

Very few data exist about any of these matters. Research on double seriation does provide data on children's ability to establish equivalence relations on ordinal contexts. In Piaget's original experiments, two different ordered sets were constructed by the child or by the experimenter. The construction of the orderings on these sets involved order relations on measure contexts: all of the sets (e.g., dolls and walking sticks) were ordered by height (or overall size). The double seriation questions asked a child to find an unspecified ordinal position in one set that was equivalent to an unspecified ordinal position in the other set (e.g., *Which walking stick goes with this doll?*). Piaget found that children below about 6 years confused relative position and absolute position, for if the objects in one set (e.g., the sticks) were spread out, the youngest children would choose the stick in the place across from the specified doll (absolute position) rather than the stick in the same relative position as that doll (e.g., the fourth stick). In addition, Piaget reported that somewhat older children confused the cardinality of the set in front of the specified object with the ordinal position of its corresponding object. For example, if the fifth doll was pointed to, a child would count the four objects in front of that doll and would then count *one, two, three, four* walking sticks and select the last counted (fourth) stick as belonging to the fifth doll. By age 8 or so children had coordinated these ordinal and cardinal aspects of number.

To our knowledge no one has reported data on conservation in ordinal contexts, that is, concerning children's knowledge of transformations on the equivalence of the ordinal positions (or ordinal words) of two specified entities. This area is interesting because transformations that preserve numerosity (cardinality) do not necessarily preserve relative position (or the ordinal word) and vice versa. For example, if an object is added to a set, the cardinality of the set is changed, but adding an object to an ordered set only changes the relative position of an object if the added object is placed "in front of" (in a relative position lower than) the specified object. Any transformation in which objects are moved around does not affect cardinality, but such moves can affect ordinal position if the movements result in a change in the size of the set in front of the specified object. Thus, the crucial aspect of transformations that conserve relative position (ordinal word) is that the cardinality of the set of objects up to the specified object must be preserved. Because this is so, conservation of ordinal position would seem to be logically dependent upon conservation of cardinality. In fact, however, children may use empirical strategies such as matching or counting to ascertain relative position and may not use this relationship between cardinal and ordinal conservation at all.

CARDINAL WORDS AND ORDINAL WORDS

In an extensive series of experiments, Piaget examined several relationships between cardinal and ordinal concepts (1941–1965). He reported finding the same three stages (at roughly the same ages) in the development of cardinal and ordinal concepts. He argued that these similarities arose because children dealt with the concepts in the same manner at each stage. A second cardinal–ordinal relationship discussed by Piaget was that described in the last section between an ordinal position and the cardinality of the set of objects in all the ordinal positions preceding that ordinal position. Finally, Piaget argued that a mature conception of number requires the simultaneous conception of both the equivalence (cardinal) and the nonequivalence (ordinal) aspects of number. His argument is complex, particularly with respect to the way in which counting is related to these cardinal and ordinal aspects of number, and we do not have sufficient space here to treat it further.

Brainerd (1979) described logical theories of ordinal and cardinal number and proposed an ordinal theory of the development of number concepts. This theory postulated that ordinal number concepts are learned before cardinal concepts and that the acquisition of cardinal concepts extends through the elementary school years even into the high school years. We do not have space here to discuss Brainerd's treatment of the logical bases for ordinal and cardinal number and can only briefly summarize his empirical results. Brainerd selected transitivity as the most basic construct of ordinality. Children were shown that $a < b$ and $b < c$ for two sticks (or balls of clay) and were asked about the order relation "longer than" (or "heavier than") on the objects a and c. Spontaneous use of one-to-one correspondence to evaluate the equivalence of two rows of circles varying in length and density was selected as the criterion for cardinality. Brainerd found in several different experiments that children consistently were able to do the ordinality–transitivity task before the cardinality–equivalence task. Brainerd also administered different tasks designed to tap different aspects of the cardinal concept and found that performance on these tasks improved even into the high school years.

Brainerd's work added to a field previously restricted mainly to considerations of cardinal number and numerosity the important ideas of order and ordinal number. However, his choice of transitivity of an order relation as the sine qua non of ordinal number is not acceptable. Presumably this choice was made because order (asymmetric) relations are transitive. However, equivalence relations are also transitive. The property that does distinguish order relations from equivalence relations is that of being *asymmetric:* Equivalence relations are *symmetric* (if $4 + 1 = 3 + 2$, then $3 + 2 = 4 + 1$) and order relations are *asymmetric* (if $6 > 4$, then it is not true that $4 > 6$). Furthermore, Brainerd actually examined the existence of transitivity of an order relation in measure, not in ordinal, contexts. Similarly, Brainerd's criterion for cardinality is not a feature limited to cardinality: One-to-one correspondence can be used to determine the equivalence of two *ordinal* positions. To summarize our position, Brainerd's data are instructive concerning children's knowledge about certain aspects of cardinal and ordinal number but do not settle the issue of the developmental procedence of these, nor were the criteria chosen for cardinal and ordinal features of number appropriate: Transitivity and one-to-one correspondence can each be used in both cardinal and ordinal contexts.

Brainerd (1974) used another type of experiment in comparing ordinal and cardinal performance. Cardinal and ordinal responses to the number symbols from one to five were trained. In that experiment most task facets were controlled and were equivalent for the cardinal and ordinal conditions. An exception was that in the ordinal condition E always pointed to the ordinal positions in order, so a child had merely to learn to find (or to produce orally) the number words in their sequence or counting order. In the cardinal condition the sequence of number words could not be used to find the answers. The conditions would have been comparable if the experimenter had pointed to the ordinal positions in the same random order (e.g., three, one, four, two, five) used in the cardinal context. Therefore the finding here of earlier attainment of ordinal concepts seems likely to have been greatly affected by this task difference.

Finally, even if transitivity *were* a distinguishing feature of ordinal word contexts, Brainerd's inordinate emphasis on it as *the* central feature of the mature understanding of ordinal number would be inappropriate. There are many aspects of both ordinal and cardinal concepts and several corresponding procedures for each which the child must master. A number of possibilities have been described in this chapter. Brainerd (1979) actually examined several such different tasks for cardinal number and found an extended period of acquisition for these. The very question of the developmental priority of ordinal or cardinal concepts is thus a misleadingly simple one, and future work would do better to focus upon the rich and varied aspects of each of these.

At the moment, we do not know very much about how the acquisition of knowledge about parallel features of ordinal word contexts and cardinal word contexts relate developmentally. In such research in the future, it will be important to distinguish between studying *order relations* (which may be defined on cardinal, measure, sequence, and ordinal contexts) and studying *ordinal word contexts*. This distinction has been discussed by us above and by Nesher (1972). Furthermore, in our language and in most others, except for the first three words, ordinal words are constructed from the sequence/counting/cardinal words. This derivation would seem to argue for the historically later construction of most of the ordinal words. As discussed earlier, young children also acquire the ordinal word vocabulary considerably after the sequence/counting/cardinal vocabulary. Whether this delay in the acquisition of the ordinal word vocabulary is due to the additional facets that must be understood in an ordinal context (the ordering and its starting point) or to a relative lack of experience with everyday ordinal situations is not clear. These language and experimental factors will need to be considered in future research.

NONNUMERICAL NUMBER WORD CONTEXTS

The final number word context considered here is the nonnumerical context. In these contexts number words (or symbols) are used simply as identification "names" or codes for particular objects. Telephone numbers, driver's license numbers, football jersey numbers, and dorm food ticket numbers are a few of the many such applications. Some such uses may be a combination of numerical and nonnumerical meanings (e.g., telephone area codes use small numbers for the largest cities).

Nonnumerical contexts are fairly ubiquitous and may be a source of confusion to young children. Sinclair (1980) reported that young Genevan children try to interpret such nonnumerical contexts as cardinal or ordinal contexts (e.g., *The #2 bus is called that because it goes on two different streets.*). It seems possible that the nonnumerical use of numbers might therefore undermine the child's progress toward understanding the numerical meanings. At present there simply is not much information on these points.

CARDINAL, SEQUENCE, AND COUNTING WORDS

Relationships among Their Order Relations

Bodies of research literature exist that relate to order relations on sequence number words (the linear order and alphabet comparison literature) and to order relations on cardinal words (magnitude comparison and digit comparison literature). At the moment considerable controversy exists in these literatures. The issues in each will be described briefly, then developmental relationships between order relations on sequence and those on cardinal words will be discussed.

The sequence number words are learned in a total (or linear) order. The production by a child of this conventional list of number words does not imply that the child has any comprehension that the words are ordered in a certain way (any more than he or she may be aware that the words in a nursery rhyme are ordered in a certain way). Later, however, the sequence becomes an object of reflection, and order relations on pairs or words from the sequence can be derived from the conventional order in the list (see Fuson, Richards, & Briars, 1981, and the Breakable Chain Level in Table 2.1). A child can then understand and produce such statements as *seven comes after five* or *six comes before nine*. These order relations on sequence words seem quite similar to those that can be made on letters of the alphabet. Alphabetic order judgement studies exist and concern the processes by which adults make judgments concerning the order in the alphabet of two given letters (Hamilton & Sanford, 1978; Hovancik, 1975; Lovelace, Powell, & Brooks, 1973; Lovelace & Snodgrass, 1971; Lovelace & Spence, 1972; Parkman, 1971). Some studies showed distance effects in the speed of responses; that is, responses were faster for letters farther apart than for letters close together. Judgments were faster for the beginning than for the end of the alphabet, and this difference apparently was not due to weaker associative bonds at the end of the alphabet, but rather to slower entry into the alphabet in its later portions. Most of the investigators implicitly or explictly suggested that some sort of serial production of some part of the alphabet was involved in these order tasks.

A recent article (Hamilton & Sanford, 1978) has provided evidence for the use of two alternative approaches to this task: (*a*) the use of directly available order information about a given pair (e.g., knowing without further processing whether the pair S, G are in alphabetic order), and (*b*) the use of a subvocal run-through of some part of the alphabet. Reaction times using run-throughs were slower than those for the direct knowledge accessing. Reaction times for the latter did not vary with a pair's distance apart in the alphabet, and length of run-through did not vary with the distance apart of the two letters. There was a tendency for the run-throughs to begin with stable subject-defined groupings within the alphabet. The factor that did account for the distance effect was that run-throughs, though not longer for shorter distances, were more frequent at the shorter letter separations. Because run-throughs were slower, the greater number of trials on which run-throughs occurred for the shorter separations led to an apparent distance effect.

The so-called linear order literature also is relevant here. In fairly typical linear order studies (e.g., Carroll & Kammann, 1977; Holyoak, 1977; Moyer, 1973; Moyer & Bayer, 1976; Polich & Potts, 1977; Potts, 1972, 1974a, 1974b;

Trabasso, 1975; Woocher, Glass, & Holyoak, 1978), subjects are given pairs of words related by a binary relation. The set of words is well-ordered under that relation, and subjects are then asked questions about the relation upon pairs of words not originally given. For example, subjects may be given A < B, B < C, C < D, and D < E (where A, B, etc. might be people and < is "is smarter than") and then asked about the relation on B and D, about which they were given no direct information. The evidence at this point from all these studies seems quite clear that subjects as young as 4 years of age do not use the pairs they have been given to make transitive (logical) inferences about the new pair. Instead they construct a total linear ordering of the items (e.g., A, B, C, D, E) and somehow derive the order relation on the new pair from this linear order. Reaction times from these studies typically increase as the distance along the linear order continuum decreases, that is, "close" elements are less quickly discriminable than are "far" elements. Thus, most authors argue that some comparison of analog representations of the elements in the linear order is the process by which these order judgments are made. However, it seems possible that in some of these linear order situations the distance effects might be resulting from the same combination of two processes found by Hamilton and Sanford (1978) in their alphabetic study described earlier. So far as we know, this possiblity has not been fully explored.

The literatures that relate to order relations (greater than, less than) on cardinal contexts involve magnitude comparison and digit comparison. Reaction times on relational questions for both digit and nondigit magnitudes typically show a distance effect such that reaction time increases as a pair of compared entities become closer on the magnitude dimension (e.g., Banks, 1977; Banks, Fujii, and Kayra-Stuart, 1976; Moyer and Dumais, 1978; Moyer and Landauer, 1967, 1973; Parkman, 1971; Sekuler & Mierkiewicz, 1977). Also typically found (e.g., Holyoak & Mah, 1981; Marschark & Paivio, 1979, 1981; Paivio, 1978) is a semantic congruity effect such that when both entities match the comparative used in the relational question, responses are faster than when they do not (e.g., *two is smaller than three* would be answered faster than *eight is smaller than nine* because two and three are small). These areas are currently the focus of a considerable amount of research, with several competing models of the mental processes involved and competing explanations for the semantic congruity effect. In the digit comparison literature, these models include a "mental counter" which searches very quickly (twenty times faster than the Hamilton & Shapard subvocal alphabet run-through) through the ordering of digits from one through nine until it reaches one of the presented digit pair (Parkman, 1971), a random walk model (Buckley & Gillman, 1974), some general magnitude comparison process similar to what one might use to compare two rows of dots equal in density (Hu & Heinrichs, 1978, Sekuler & Mierkiewicz, 1977), and semantic coding models in which certain groups of digits are labeled "small," "large," and, perhaps, "medium" (Banks, Fujii, & Kayra-Stuart, 1976; Siegler & Robinson, in press). To further complicate matters, Katz (1980) found that the choice of the larger of two digits was faster and more accurate when the digits were presented to the left visual field (right hemisphere processing) than to the right visual field (left hemispheric processing) and that the two hemispheres seemed to be solving the task in different ways. Katz discussed other research (Besner, Grimsell, & Dairs, in press) that reported the opposite finding of left hemispheric superiority for digit comparison and argued that the different findings resulted from differences in the size of the stimuli and in the rates of presentation of the digit pair.

An assumption has been made in all of this literature that the meaning accessed by adults for digits is a cardinal one, that is, that the task is a numerosity/magnitude one rather than a sequence one. If a given author implicates in the processing a serial list for the digits, this list is assumed to be an ordering of the digits according to their cardinal (numerosity) size. For adults this assumption would seem to be valid, for otherwise similar patterns of results would have been reported for digit and for alphabetic comparisons. That is, if adults were treating digits by their sequence meanings only, their digit responses should have fit their response patterns for alphabetic comparisons and they did not. However, some recent evidence indicates that adults do seem to process order relations on single digits and those on two-digit numbers differently. The choice of the larger or smaller of two-digit numbers (11–29 and 81–99) for pairs that differed by one, two, or three did not show the usual distance effect (Marschark, 1982). When subjects did not know the range of the digit pairs, no significant effect of distance was found. When subjects were informed about the possible range, pairs differing by two were faster than pairs differing by one or by three. No digit comparison task with single digits has found this result. The smallest number in each pair was the best predictor of reaction time for these two-digit stimuli, as one might expect if adults were using the sequence words between these two-digit numbers to make the comparison. Further work with these two-digit numbers obviously will be necessary. Such work might consider the extent to which adults are reacting to these stimuli as members of the counting sequence (i.e., to their sequence meanings) rather than to any cardinal meanings. It is difficult to see how adults could in fact have specific magnitude representations for each of the cardinal numbers between 10 and 100 and use them to make order decisions about which set of entities (e.g., 83 or 84) is larger. Marschak was convinced also that his subjects were not reacting simply to the second digit, but were treating these two digit numbers as wholes. The pattern of reaction times different from that for single digits supports his contention.

It is evident that we do not fully understand the mature process toward which children are developing in learning to make such order judgments on digits. Moreover, there exist few data from studies with children, and these have not yielded unequivocal results. Studies by Hu and Heinrichs (1978) and Sekuler and Mierkiewicz (1977) involved children from grades K, 2, 4, or 6, as well as adults responding to simultaneous visual presentations of digits. The reaction times showed distance effects, and psychophysical magnitude comparison models were proposed. Siegler and Robinson (in press) asked 3-, 4-, and 5-year-olds to make order judgments about pairs of words (from one to nine) presented auditorially (and thus sequentially). Multiple scaling and hierarchical clustering analyses on these data, together with evidence from a training study, supported a semantic category model in which children categorized words as small, medium and large, and made order judgments on the basis of these categories. It is not clear whether the conflicting results from these studies with children are due to age, presentation differences, or some other factors.

Unpublished data collected in 1979 by Fuson also are relevant here. Because sequence number word contexts and cardinal contexts are separate ones for young children, it seemed possible that young children process order relations on sequence words differently from those on cardinal words. Perhaps children might address sequence word order questions by the use of an overt or covert run-through procedure (as subsequently reported for adults by Hamilton & Sanford, 1978) but approach cardinal word questions by the use of a magnitude comparison process similar to that used for rows of dots (as reported by Hu & Heinrichs, 1978). Another possibility is that comparisons of one type of word might be accomplished at an earlier age than the other type. If so, children might use the first type to respond to questions of the other type (e.g., use a cardinal word comparison process to answer a question about sequence words or vice versa). Mierkiewicz, in personal communication, reported to us that some kindergarten children in the Sekuler and Mierkiewicz (1977) study on cardinal numbers had done just that: they had run through the sequence words on some pairs before responding with the larger (or smaller) word. That is, they apparently had used an order relation on sequence words to solve an order relational question on cardinal words. Another possibility examined was whether differences exist in processing words below and above ten. It seemed questionable whether children would possess individual magnitude representations for number words above ten and quite likely that for such items they would use the sequence words to solve relational questions.

Children within three age levels from 4½ to 7½ were randomly assigned into either a sequence or a cardinal condition. Presentation of the words was auditory. Questions in the sequence condition were of the form, *Which comes later (comes earlier), five or nine?*, and questions in the cardinal condition were of the form, *Which is bigger (smaller), five or nine?* Within each condition, half of the pairs consisted of adjacent items (e.g., five, six) and half of the pairs consisted of the same initial item paired with the item four larger (e.g., five, nine). Half of the comparisons were below ten and half used words between ten and twenty. Various counterbalancing procedures were used to avoid possible order artifacts.

The data indicated differences in performance on these tasks. The children below 6½ responded more accurately in the cardinal condition than in the sequence condition for comparisons between one and ten, but response levels for the two conditions were equivalent for comparisons between ten and twenty. Sequence responses were at similar levels for both sizes of number words. A ceiling effect for the older children prevented clear interpretation of their data. In short, the pattern of the results is consistent with the interpretation that children below 6½ use a single process for deriving order relations on sequence words over the whole range of words one to twenty, and that this same sequence process is used for the cardinal relations for words between ten and twenty. Lower cardinal items appeared to have been processed differently, perhaps by a magnitude comparison process, as suggested earlier. The results, however, should be regarded as only suggestive and clearly require confirmation. A particular problem in exploring these issues is that differential familiarity may exist between the sequence relational terms *comes later than (earlier than)* and the cardinal relational terms *bigger (smaller) than*. However, the issue of connections between relations on cardinal number words and relations on sequence number words definitely seems worth pursuing.

Arithmetic Operations

The focus of this chapter is upon the early acquisition of various meanings of number words and upon ways in which these meanings become related to each other. Since early school number learning is discussed in Chapter 3, we will restrict ourselves to some selective comments about ways in which number word meanings become related during the early school years and to references in which these additional relationships are discussed more fully.

The sequence of number words becomes a representational tool that is used for solving operations (addition, subtraction, multiplication, division) in cardinal contexts. Fuson, Richards and Briars (1982) described and presented data about children's acquisition of advanced sequence skills that are used in addition and subtraction problems. These sequence skills involve counting-up and counting-down from and to specified words (e.g., *How many words from "eight" to "fourteen"?*). When these skills are applied in cardinal problem contexts, they become solution procedures that have been termed *counting-on* for addition and *counting-back x* or *counting-back to x* for subtraction. These solution procedures are discussed in more detail in Carpenter, Hiebert, and Moser (1979), Carpenter and Moser (1982), Fuson (1982), Resnick (Chapter 3), and Steffe, Thompson, and Richards (1982).

Fuson, Richards, and Briars (1982) also discussed more advanced sequence skills that can be used in addition and subtraction of two-digit numbers and in multiplication and division. The former are counting-up or counting-down by tens and by ones (this use of the sequence is discussed in Chapter 3). Bell and Burns (1981) reported data concerning elementary school children's abilities on these tasks. The multiplication and division skills described by Fuson, Richards, and Briars are skip counting: repeatedly counting-up or down by a given number. For example, counting-up by 8 will generate an ordered list of the multiples of 8: 8, 16, 24, 32, and so on. Such lists can be used to facilitate the memorization of multiplication facts or to generate one fact from another known fact (e.g., 8 times 5 is 40, plus 8 is 48—that's 8 times 6).

Fuson, Richards, and Briars (1982) and Steffe, Richards, and von Glasersfeld (1981) each suggested that the use of the sequence of number words for the solution of addition and subtraction problems permits an even more advanced understanding to be acquired—the understanding that addition and subtraction are inverse operations. This understanding contributes considerable flexibility to children's solution of addition and subtraction problems, for they can then choose procedures for convenience or efficiency, rather than because they directly model some basic meaning of a given problem.

The use of physical manipulative materials (embodiments) in learning the underlying meaning bases for the conventional algorithms for solving symbolic problems is discussed in Chapter 3. The particular embodiments discussed there are measure embodiments: They are based on measure meanings of numbers (e.g., one tens block is ten units long). Bell, Fuson, and Lesh (1976) made a basic distinction between measure embodiments and count embodiments that can be used for teaching basic meanings of whole numbers, integers, and rational numbers (fractions), and they described how to use count (actually, numerosity) and measure embodiments for whole numbers (small and large base ten numbers), integers, and rational numbers, and for operations on these numbers. Some of the embodiments proposed were developed by others (e.g., Cuisenaire rods—a measure embodiment) and some were invented by these authors. They reported that in their experiences in using count and measure models with adults, individual differences seemed to exist in the ability to understand each type of model, that is, some adults preferred working with and seemed to benefit from the use of a measure model more than a count model and vice versa. This finding based upon observation and adult self-report has not been explored by other research means, nor has it been extended to children. Evidence discussed earlier in this chapter indicated that certain aspects of measure concepts are more difficult to understand than are counted numerosity concepts. Whether these differences might affect children's comprehension of numerosity and of measure embodiments might be explored in future research.

The basic assumption underlying the use of physical embodiments in teaching number concepts or symbolic algorithms is that the concepts or symbols will take on meanings attached to the embodiments. Based on extensive experience in the use of such embodiments in teaching these concepts to teachers, Bell *et al.* (1976) recommended the use with children and teachers of the column by column recording method of relating the symbolic algorithm to the embodiments that Resnick reports in Chapter 3 as being successful with her 10 child subjects. For other embodiments, ways of closely relating the sumbols for numbers and for operations and the meanings as embodied by the physical materials (and by the physical operations on these materials) were also proposed by Bell *et al.* Through the use of such embodiments in close conjunction with number words and number symbols, particular numerosity, counting, sequence, and measure meanings described in this chapter become combined into the more advanced and complex meanings of base ten numbers (large whole numbers and decimals), integers, and rational numbers.

We have chosen in this chapter not to treat the number word-number symbol link at all. Some young children do learn the written symbols for number words before entering school. However, evidence concerning the extent and source of this knowledge is spotty, and the theoretical and mathematical issues involved in symbols above nine are more advanced than treated in this chapter (though some of them were briefly touched upon in this section and in the section on measure words and cardinal words). Bell, Fuson, and Lesh (1976) discuss some of these issues with respect to the structural differences between the number word and the number symbol systems.

CONCLUSIONS AND COMMENTS

Early in this century psychologists were actively involved in the analyses of various bodies of academic knowledge and skills. Elementary concepts and skills relating to mathematics received considerable attention from experimental–theoretical psychologists of the day. Toward the middle of the century, psychology had much less to say to those interested in early mathematical (or mathematics-related) skills and concepts. Pedagogical research in mathematics education continued, but was increasingly removed from research and theory in psychology. Mathematics educators, generally speaking, received heavy training in mathematics and in education, but relatively little in psychology. Psychologists (with a few exceptions) seemed neither to know or to care much about issues in mathematics learning and teaching, and the nature of much psychological theory during that period seemed irrelevant to an understanding of mathematics education issues. In our view the results of this separation, and an inevitable result, was a severe limitation in the progress made during this period.

The influence of psychology on mathematics education increased sharply during the 1960s. The increasing visibility in the United States of Piaget's work and the rapid influence of that work on developmental psychology were especially important for education. Gagne, who emphasized both the logical and the psychological analyses of school learning tasks (especially arithmetic tasks), also was a notable influence, as were Bruner and Ausubel. These (and a few other) psychologists had a significant influence on some of the most productive continuing research programs in early mathematics learning and instruction undertaken since then by mathematics educators.

On the other hand, quite recently there has been a virtual explosion of research activity by developmental and cognitive psychologists regarding preschool and early elementary school children's understanding of mathematical and premathematical concepts and their acquisition of mathematics-related skills. Much of this recent work has been stimulated in some way by Piaget's work, and very recently some work has been influenced heavily by recent advances in psychological theory in memory and cognition, especially by information-processing notions. However, some of this work suffers from an inadequate knowledge of mathematics and/or of research and practices in mathematics education.

As we see it, research and theory development in mathematics learning has entered a rapid development period similar to that which characterized the reading field in the 1970's. As should be clear from our review, although there have been many rather recent advances, the state of knowledge regarding the many and complex issues remains sketchy and the quality of research designs and techniques is uneven. The psychologists involved are generally less sophisticated in mathematics and in their understanding of educational practices and issues than would be ideal. Many from the mathematics education tradition remain more removed from current psychological theory and experimental paradigms than would be ideal. Neither has yet developed the full range of research strategies and tools and that are sufficiently sophisticated for the complexity of the issues being addressed. Increased collaboration of psychologists and mathematics education researchers seems advisable, as do joint doctoral and postdoctoral programs. The present collaborative authorship is of this sort, and we believe that the issues we have addressed, and the research results we have summarized have been sharpened as a result. There are signs of increases in such collaboration and exchange so a research review in this area 10 years from now may show substantial progress on issues of basic theoretical and instructional importance.

ACKNOWLEDGMENTS

Thanks to Art Baroody, Max Bell, Charles Brainerd, Paul Cobb, Tom Hudson, John Richards, Don Saari, Walter Secada, and, of course, Herb Ginsburg for helpful comments on the earlier drafts of this chapter. Conversations with many different colleagues over the past 3 years have informed our thinking in this area, and we are grateful for these interactions. This material is based upon work supported by the National Science Foundation and the National Institute of Education under grant SED 78-22048 and by the Spencer Foundation under a National Academy of Education Spencer Fellow Award to the first author. Any opinions, findings, and conclusions or recommendations expressed in this chapter are those of the authors and do not necessarily reflect the views of the National Science Foundation, the National Institute of Education, or the Spencer Foundation.

REFERENCES

Banks, W. P. Encoding and processing of symbolic information in comparative judgments. In G. H. Bower (Ed.), *The psychology of learning and motivation* (Vol. 11). New York: Academic Press, 1977.

Banks, P., Fujii, M., & Kayra-Stuart, F. Semantic congruity effects in comparative judgments of magnitudes of digits. *Journal of Experimental Psychology: Human Perception and Performance,* 1976, *2,* 435–447.

Bearison, D. J. Role of measurement operations in the acquisition of conservation. *Developmental Psychology,* 1969, *1,* 653–660.

Beckwith, M., & Restle, F. Processes of enumeration. *Psychological Review,* 1966, *73,* 437–444.

Beilin, H. The training and acquisition of logical operations. In M. F. Rosskopf, L. P. Steffe, & S. Taback (Eds.), *Piagetian cognitive development research and mathematical education.* Reston, Va.: National Council of Teachers of Mathematics, 1971.

Beilin, H. *Studies in the cognitive basis of language development.* New York: Academic Press, 1975.

Beilin, H. Inducing conservation through training. In G. Steiner (Ed.), *Psychology of 20th century, Piaget and beyond,* Vol. 7. Bern: Kinder, 1977.

Bell, M. S., & Burns, J. *Counting, numeration, and arithmetic capabilities of primary school children.* Proposal submitted to National Science Foundation, April, 1981.

Bell, M. S., Fuson, K., & Lesh, R. A. *Algebraic and arithmetic structures.* New York: The Free Press, 1976.

Besner, D., Grimsell, D., & Dairs, R. The mind's eye and the comparative judgment of number. *Neuropsychologia,* in press.

Blackstock, E. G., & King, W. L. Recognition and reconstruction memory for seriation in four- and five-year-olds. *Developmental Psychology,* 1973, *9,* 255–259.

Blevins, B., Mace, P. G., Cooper, R. G., Starkey, P., & Leitner, E. *What do children know about addition and subtraction?* Paper presented at the Biennial Meeting of the Society for Research in Child Development, Boston, April 1981.

Brainerd, C. J. Inducing ordinal and cardinal representations of the first five natural numbers. *Journal of Experimental Child Psychology,* 1974, *18,* 520–534.

Brainerd, C. J. *The origins of the number concept.* New York: Praeger, 1979.

Briars, D., & Siegler, R. S. *Preschoolers' abilities to recognize counting errors.* Paper presented at the Biennial Meeting of the Society for Research in Child Development, Boston, April 1981.

Brush, L. Preschool children's knowledge of addition and subtraction. *Journal for Research in Mathematics Education,* 1978, *9,* 44–54.

Buckley, P. B., & Gillman, C. B. Comparisons of digits and dot patterns. *Journal of Experimental Psychology,* 1974, *103,* 1131–1136.

Bullock, M., & Gelman, R. Numerical reasoning in young children: The ordering principle. *Child Development,* 1977, *48,* 427–434.

Carpenter, T. P. *The role of equivalence and order relations in the development and coordination of the concepts of unit size and number of units in selected conservation type measurement problems.* Report from the Project on Analysis of Mathematics Instruction (Tech. Rep. No. 178). Madison: The University of Wisconsin, Wisconsin Research and Development Center for Cognitive Learning, August 1971.

Carpenter, T. P. Measurement concepts of first- and second-grade students. *Journal for Research in Mathematics Education,* 1975, *6,* 3–14.

Carpenter, T. P., Corbitt, M. K., Kepner, H. S., Lindquist, M. M., & Reys, R. E. National assessment. In E. Fennema (Ed.), *Mathematics education research: Implications for the 80's.* Alexandria, Va.: Association for Supervision and Curriculum Development, 1981.

Carpenter, T. P., Hiebert, J., & Moser, J. The effect of problem structure on first grader's initial solution procedures for simple addition and subtraction problems. *Journal for Research in Mathematics Education*, 1981, *12*, 27–39.

Carpenter, T. P., & Lewis R. The development of the concept of a standard unit of measure in young children. *Journal for Research in Mathematics Education*, 1976, *7*, 53–64.

Carpenter, T. P., & Moser, F. The development of addition and subtraction problem-solving skills. In T. P. Carpenter, J. M. Moser & T. A. Romberg (Eds.), *Addition and subtraction: A developmental perspective.* Hillsdale, N. J.: Lawrence Erlbaum Associates, 1982.

Carroll, M., & Kammann, R. The dependency of schema formation on type of verbal material: Linear orderings and set inclusions. *Memory and Cognition*, 1977, *5*, 29–36.

Chi, M. T. H., & Klahr, D. Span and rate of apprehension in children and adults. *Journal of Experimental Child Psychology*, 1975, *19*, 434–439.

Clark, H. Linguistic processes in deductive reasoning. *Psychological Review*, 1969, *76*, 387–404.

Comiti, C., Bessot, A., & Pariselle, C. Analyse de comportements d'eleves du cours preparatoire confrontes a une tache de construction d'un ensemble a un ensemble donne. *Reserches en didactique des mathematiques*, 1980, *1.2*, 171–217.

Davydov, V. V. The psychological characteristics of the "prenumerical" period of mathematics instruction (A. Bigelow, trans.). In L. P. Steffe (Ed.), *Soviet studies in the psychology of learning and teaching mathematics* (Vol. 7). Chicago: The University of Chicago, 1975. (From *Learning capacity and age level: Primary grades.* D. B. El'konin & V. V. Davydov (Eds.), Moscow: Prosveschenie, 1966.)

Davydov, V. V. The psychological characteristics of the formation of elementary mathematical operations in children. In T. P. Carpenter, J. M. Moser, & T. A. Romberg (Eds.), *Addition and subtraction: A developmental perspective.* Hillsdale, N. J.: Lawrence Erlbaum Associates, 1982.

Estes, W. K. The cognitive side of probability learning. *Psychological Review*, 1976, *83*, 37–64.

Flavell, J. H. *The developmental psychology of Jean Piaget.* New York: D. Van Nostrand, 1963.

Flavell, J. H. Concept development. In P. H. Mussen (Ed.), *Carmichael's manual of child psychology.* New York: Wiley, 1970.

Fuson, K. C. An analysis of the counting-on solution procedure in addition. In T. P. Carpenter, J. M. Moser & T. A. Romberg (Eds.), *Addition and subtraction: A developmental perspective.* Hillsdale, N. J.: Lawrence Erlbaum Associates, 1982.

Fuson, K. C., & Mierkiewicz, D. *A detailed analysis of the act of counting.* Paper presented at the Annual meeting of the American Educational Research Association, Boston, April 1980.

Fuson, K. C., & Richards, J. *Children's construction of the counting numbers: From a spew to a bidirectional chain.* Unpublished paper, Northwestern University, Evanston, Ill., 1979, and paper presented at the Annual Meeting of the American Educational Research Association, Boston, April 1981.

Fuson K. C., Richards, J., & Briars, D. J. The acquisition and elaboration of the number word sequence. In C. Brainerd (Ed.), *Progress in cognitive development, children's logical and mathematical cognition* (Vol. 1). New York: Springer-Verlag, 1982.

Fuson, K. C., Secada, W., & Hall, J. W. *Matching, counting, conservation of numerical equivalence. Child Development*, in press.

Gal'perin, P. Ya., & Georgiev, L. S. The formation of elementary mathematical notions (D. A. Henderson, trans.). In J. Kilpatrick & I. Wirszup (Eds.), *Soviet studies in the psychology of learning and teaching mathematics* (Vol. 1). Chicago: The University of Chicago, 1969. (From a series of four articles published in Reports of the Academy of Pedagogical Sciences of the USSR, Vol. 1, 1960.)

Gelman, R., & Gallistel, C. R. *The child's understanding of number.* Cambridge, Mass.: Harvard University Press, 1978.

Ginsburg, H. P., & Russell, R. L. Social-class and racial influences on early mathematical thinking. *Monographs of the Society for Research in Child Development, 46,* 1981.

Gollin, E. S., Moody, M., & Schadler, M. Relational learning of a size concept. *Developmental Psychology,* 1974, *10,* 101–108.

Griep, C., & Gollin, E. S. Interdimensional transfer of an ordinal solution strategy. *Developmental Psychology,* 1978, *14,* 437–438.

Gullen, G. E. Set comparison tactics and strategies of children in kindergarten, first grade, and second grade. *Journal for Research in Mathematics Education,* 1978, *91,* 349–360.

Hamilton, J. M. E., & Sanford, A. J. The symbolic distance effect for alphabetic order judgments: A subjective report and reaction time analysis. *Quarterly Journal of Experimental Psychology,* 1978, *30,* 33–43.

Hasher, L., & Zacks, R. T. Automatic and effortful processes. *Journal of Experimental Psychology: General,* 1979, *108,* 356–388.

Hatano, G., & Ito, Y. Development of length measuring behavior. *Japanese Journal of Psychology,* 1965, *36,* 184–196.

Heinrichs, J., Yurko, D. S., & Hu, J. M. Two-digit number comparison: Use of place information. *Journal of Experimental Psychology: Human Perception and Performance,* 1981, *7,* 890–901.

Hendrickson, A. D. Inventory of mathematical thinking done by incoming first-grade children. *Journal for Research in Mathematics Education,* 1979, *10,* 7–23.

Hiebert, J. *The effect of cognitive development on first-grade children's ability to learn linear measurement concepts* (Tech. Rep. No. 506) Madison: Wisconsin Research and Development Center for Individualized Schooling, 1979.

Hintzman, D. L. Repetition and memory. In G. H. Bower (Ed.), *The psychology of learning and motivation* (Vol. 10). New York: Academic Press, 1976.

Holmes, E. E. What do pre-first-grade children know about number? *Elementary School Journal,* 1963, *63,* 397–403.

Holyoak, K. J. The form of analog size information in memory. *Cognitive Psychology,* 1977, *9,* 31–57.

Holyoak, K. J., & Mah, W. A. Semantic congruity in symbolic comparisons: Evidence against an expectancy hypothesis. *Memory and Cognition,* 1981, *9,* 197–204.

Hovancik, J. R. Reaction times for naming the first next and second next letters of the alphabet. *American Journal of Psychology,* 1975, *88,* 643–647.

Howell, W. C. Representation of frequency in memory. *Psychological Bulletin,* 1973, *80,* 44–53.

Hu, J. M., & Hinrichs, J. V. *Comparison of number and numerosity: A development study of psychophysical judgment.* Paper presented at the Midwestern Psychological Association, Chicago, May 1978.

Hudson, T. *Children's use of counting in establishing correspondences between disjoint sets.* Paper presented at the biennial meeting of the Society for Research in Child Development, Boston, April 1981.

Hudson, T. Correspondences and numerical differences between disjoint sets. *Child Development,* in press.

Inhelder, B., Sinclair, H., & Bovet, M. *Learning and the development of cognition.* Cambridge, Mass.: Harvard University Press, 1974.

Katz, A. N. Cognitive arithmetic: Evidence for right hemispheric mediation in an elementary component stage. *Quarterly Journal of Experimental Psychology,* 1980, *32,* 69–84.

Kingsley, R., & Hall, V. Training conservation through the use of learning sets. *Child Development,* 1967, *38,* 1111–1126.

Klahr, D., & Wallace, J. G. The role of quantification operators in the development of conservation of quantity. *Cognitive Psychology,* 1973, *4,* 301–327.

Klahr, D., & Wallace, J. G. *Cognitive development: An information-processing view.* New York: Erlbaum, 1976.

LaPointe, K., & O'Donnell, J. P. Number conservation in children below age six: Its relationship to age, perceptual dimensions and language comprehension. *Developmental Psychology,* 1974, *10,* 422–428.

Lovelace, E. A., Powell, C. M., & Brooks, R. J. Alphabetic position effects in covert and overt alphabetic recitation times. *Journal of Experimental Psychology,* 1973, *99,* 405–408.

Lovelace, E. A., & Snodgrass, R. D. Decision times for alphabetic order of letter pairs, *Journal of Experimental Psychology,* 1971, *88,* 258–264.

Lovelace, E. A., & Spence, W. A. Reaction times for naming successive letters of the alphabet. *Journal of Experimental Psychology,* 1972, *94,* 231–233.

Lund, A., Hall, J. W. Humphreys, M. S. & Wilson, K. P. *Frequency judgment accuracy as a function of age and school achievement (LD vs. Non-LD) patterns. Journal of Experimental Child Psychology,* in press.

Maertens, N. W. Jones, R. C., & Waite, A. Elemental groupings help children perceive cardinality: A two-phase research study, *Journal for Research in Mathematics Education,* 1977, *8,* 181–193.

Mandler, G., & Shebo, B. J. *Subitizing: An analysis of its component processes.* Paper presented at the Psychonomic Society, St. Louis, Mo., November, 1980.

Markman, E. M. Classes and collections: Conceptual organization and numerical abilities. *Cognitive Psychology,* 1979, *1,* 395–411.

Marschark, M. Lexical marking and the acquisition of relational size concepts. *Child Development,* 1977, *48,* 1049–1051.

Marschark, M. *Comparative magnitude judgments with digits and other number stimuli.* Unpublished manuscript, University of North Carolina at Greensboro, 1981.

Marschark, M. Expectancy, equilibration and memory. In J. Yuill (Ed.), *Imagery, memory, and cognition: Essays in tribute to Allan Paivio,* New York: Erlbaum, 1982.

Marschark, M., & Paivio, A. Semantic conguity and lexical marking in symbolic comparison: An expectancy hypothesis. *Memory and Cognition,* 1979, *7,* 175–184.

Marschark, M., & Paivio, A. Congruity and the perceptual comparison task. *Journal of Experimental Psychology: Human Perception and Performance,* 1981, *7,* 290–308.

Mierkiewicz, D., & Siegler, R. S. *Preschoolers' abilities to recognize counting errors.* Paper presented at the Fourth International Conference for the Psychology of Mathematics Education, Berkeley, Calif.: August 1980.

Minskaya, G. D. Developing the concept of number by means of the relationship of quantities (A. Bigelow, trans.) In Steffe, L. P. (Ed.), *Soviet studies in the psychology of learning and teaching mathematics* (Vol, 7). Chicago: The University of Chicago, 1975. (From *Learning capacity and age level: Primary grades.* D. B. El'konin & V. V. Davydov (Eds.), Moscow: Proveschenie, 1966.)

Menninger, K. *Number words and number symbols.* (Revised German ed. published 1958 as *Zahlword Und Ziffer* by Vandenhoeck & Ruprecht.) Cambridge, Mass.: The M.I.T. Press, 1969.

Modgil, S., & Modgil, C. *Piagetian research: Compilation and commentary* (Vols. 2, 5, 7, 8, 12). London: NEFR Publishing, 1976.

Montgomery, M. E. The interaction of three levels of aptitude determined by a teach-test procedure with two treatments related to area. *Journal for Research in Mathematics Education,* 1973, *4,* 271–278.

Moyer, R. S. Comparing objects in memory: Evidence suggesting an internal psychophysics. *Perception and Psychophysics,* 1973, *13,* 180–184.

Moyer, R. S. & Bayer, R. H. Mental comparison and the symbolic distance effect. *Cognitive psychology*, 1976, *8*, 228–246.

Moyer, R. S. & Dumais, S. T. Mental comparison, In G. H. Bower (Ed.), *The psychology of learning and motivation* (Vol. 12). New York: Academic Press, 1978.

Moyer, R. S., & Landauer, T. K. Time required for judgments of numerical inequality. *Nature*, 1967, *215*, 1519–1520.

Moyer, R. S., & Landauer, T. Determinants of reaction time for digit inequality judgements. *Bulletin of the Psychonomic Society*, 1973, *1*, 167–168.

Murray, F. B. The conservation paradigm: The conservation of conservation research. In I. Sigel, R. Golinkoff, & D. Brodzinsky (Eds.), *New directions and application for Piaget's theory*. Hillsdale, N.J.: Lawrence Erlbaum Associates, 1979.

Nesher, P. *From ordinary language to arithmetical language in the primary grades (What does it mean to teach "2 + 3 = 5?")* (Doctoral dissertation, Harvard University, 1972). *Dissertation Abstracts International*, 1976, *36*, 7918a–7919a. (University Microfilms No. 76-10, 525).

Odom, R. D., Astor, E. C., & Cunningham, J. G. Effects of perceptual salience on the matrix task performance of four-and six-year-old children. *Child Development*, 1975, *46*, 758–762.

Paivio, A. Comparisons of mental clocks. *Journal of Experimental Psychology: Human Perception and Performance*, 1978, *4*, 61–71.

Parkman, J. M. Temporal aspects of digit and letter inequality judgments. *Journal of Experimental Psychology*, 1971, *91*, 191–205.

Pennington, B. F. Wallach, L., & Wallach, M. A. Non-conservers' use and understanding of number and arithmetic. *Genetic Psychology Monograph*, 1980, *101*, 231–243.

Pergament, G. G. *Collective nouns, numerical reasoning, and accurate counting skills*. Unpublished doctoral dissertation, Northwestern University, 1982.

Piaget, J. *The child's conception of number*. New York: W. W. Norton, 1965. (Translated and published in English. New York: Humanities, 1952, from original publication with A. Szemiska, *La Genese Du Nombre Chez l'Infant*, 1941.)

Polich, J. M. & Potts, G. R. Retrieval strategies for linearly ordered information. *Journal of Experimental Psychology: Human Learning and Memory*, 1977, *3*, 10–17.

Potts, G. R. Information processing strategies used in the encoding of linear orderings. *Journal of Verbal Learning and Verbal Behavior*, 1972, *11*, 727–740.

Potts, G. R. Incorporationg quantitative information into a linear ordering. *Memory and Cognition*, 1974, *2*, 533–538. (a)

Potts, G. R. Storing and retrieving information about ordered relationships. *Journal of Experimental Psychology*, 1974, *103*, 431–439. (b)

Pringle, R. *Development of the normalization strategy in comparative numerosity judgments*. Paper presented at the Annual Meeting of the Psychonomic Society, St. Louis, Mo., November, 1980.

Saxe, G. B. A developmental analysis of notational counting. *Child Development*, 1977, *48*, 1512–1520.

Saxe, G. B. Development relations between notational counting and number conservation. *Child Development*, 1979, *50*, 180–187.

Saxe, G. B. Culture and the development of numerical cognition: Studies among the Oksapmin of Papua New Guinea. In C. Brainerd (Ed.), *Progress in cognitive development*, Vol I. *Children's logical and mathematical cognition*. New York:Springer-Verlag, 1982.

Schaeffer, B., Eggleston, V. H., & Scott, J. L. Number development in young children. *Cognitive Psychology*, 1974, *6*, 357–379.

Sekuler, R., & Mierkiewicz, D. Children's judgments of numerical inquality. *Child Development*, 1977, *48*, 630–633.

Siegel, L. S. Development of the concept of seriation. *Developmental Psychology*, 1972, *6*, 135–137.

Siegel, L. S. The development of number concepts: Ordering and correspondence operations and the rule of length cues. *Developmental Psychology*, 1974, *10*, 907–912.

Siegel, L. S. The relationship of language and thought in the preoperational child: A reconsideration of nonverbal alternatives to Piagetian tasks. In L. S. Siegel & C. J. Brainerd (Eds.), *Alternatives to Piaget: Critical essays on the theory*. New York: Academic Press, 1978.

Siegel, L. S. & Brainerd, C. J. (Eds.), *Alternatives to Piaget: Critical essays on the theory*. New York: Academic Press, 1978.

Siegel, A., & Goldsmith, L. T. The development of skill in estimation problems of extent and numerosity. *Journal for Research in Mathematics Education*, in press.

Siegel, A., Goldsmith, L. T., & Madson, C. R. *How many words on this page? The development of children's estimation strategies*. Paper presented at the Annual Meeting of the Psychonomic Society, St. Louis, Mo., November 1980. Now published as Siegel, A., & Goldsmith, L. T., in press, see preceding reference.

Siegler, R. S. Developmental sequences within and between concepts. *Monographs of the Society for Research in Child Development*, 1981, *46*, 1–74.

Siegler, R. S. & Robinson M. The development of numerical understandings. In H. W. Reese and L. P. Lipsitt (Eds.), *Advances in child development and behavior* (Vol. 16). New York: Academic Press, 1982.

Silverman, I. W., & Rose, A. P. Subitizing and counting skills in 3-year-olds. *Developmental Psychology*, 1980, *16*, 539–540.

Sinclair, H. *Young children's acquisition of language and understanding of mathematics*. Paper presented at the Fourth International Congress of Mathematical Education, Berkely, Calif., August, 1980.

Starkey, P., Spelke, E., & Gelman, R. *Number competence in infants: Sensitivity to numeric invariance and numeric change*. Paper presented at the meeting of the International Conference on Infant Studies, New Haven, Conn., April 1980.

Stegge, L. P., Richard, J., & von Glasersfeld. E. *Children's counting types: Philosophy, theory, and case studies*. Monograph presented at the Conference on Interdisciplinary Research on Number, Athens, Ga., April 1981.

Stegge, L. P. Thompson P. W. & Richards, J. Children's counting in arithmetical problem solving. In T. P. Carpenter, J. M. Moser, & T. A. Romberg (Eds.), *Addition and subtraction: A developmental perspective*. Hillsdale, N.J.: Lawrence Erlbaum Associates, 1982.

Thomas, R. K., & Chase, L. Relative numerousness judgments by squirrel monkeys. *Bulletin of the Psychonomic Society*, 1980, *16*, 79–82.

Thomas, R. K. Fowlkes, D., & Vickery, J. D. Conceptual numerousness judgments by squirrel monkeys. *American Journal of Psychology*. 1980, *93*, 247–257.

Trabasso, T. Representation, memory and reasoning: How do we make transitive inferences? In A. D. Pick (Ed.), *Minnesota Symposia on Child Psychology* (Vol. 9). Minneapolis: University of Minnesota Press, 1975.

Underwood, D. J. *Developmental concerns related to place value*. Paper presented at the 55th Annual Meeting of the National Council of Teachers of Mathematics, Cincinnati, Ohio, April 1977.

von Glaserfeld, E. An attentional model for the conceptual construction of units and number. *Journal for Research in Mathematics Education*, 1981, *12*, 83–94.

Walters, J., & Wagner, S. *The earliest numbers*. Paper presented at the Biennial Meetings of the Society for Research in Child Development, Boston, April 1981.

Wang, M. C., Resnick, L. B. & Boozer, R. F. The sequence of development of some early mathematics behaviors. *Child Development*, 1971, *42*, 1767–1778.

Woocher, F. D., Glass, A. L., & Holyoak, K. J. Positional discriminability in linear orderings. *Memory and Cognition*, 1978, *6*, 165–173.

CHAPTER **3**   LAUREN B. RESNICK

A Developmental Theory of Number Understanding[1]

Research on the psychological processes involved in early school arithmetic has now cumulated sufficiently to make it possible to construct a coherent account of the changing nature of the child's understanding of number during the early school years. Earlier work, concerned largely with preschool children's informal arithmetic (e.g., Fuson & Hall, Chapter 2; Gelman & Gallistel, 1978; Ginsburg, 1977), has established the strength and the limits of the number understanding that children typically bring with them to school. My concern in this chapter will be to develop a plausible account of how number concepts are extended and elaborated as a result of formal instruction. The chapter will outline a theory of number representation for three broad periods of development: (*a*) the preschool period, during which counting and quantity comparison competencies of young children provide the main basis for inferring number representation; (*b*) the early primary period, during which children's invention of sophisticated mental computational procedures and the mastery of certain forms of story problems point to two important expansions of the number concept; and (*c*) the later

[1]The research reported here was supported by grants from the National Science Foundation and the National Institute of Education, United States Department of Education. The opinions expressed in this chapter do not necessarily reflect the position or policy of the granting agencies, and no official endorsement should be inferred.

109

primary period, during which the representation of number is modified to reflect knowledge of the decimal structure of the counting and notational system.

My account of developing number understanding is based heavily on recent work—some reported in this volume—that is providing a series of formal models of the knowledge underlying various observed arithmetic performances by children of different ages. Each of these models has been constructed to account for a particular set of performances, but there has been no systematic effort to link them into a developmental sequence. Nevertheless, an examination of the existing models strongly suggests a sequential development of mathematics competence that is characterized by (*a*) an expanding and successively elaborated set of schemata that organizes number knowledge, and (*b*) the linking of these schemata to increasingly complex procedural knowledge. In the course of the chapter I will clarify exactly what is to be understood by the terms *schematic* and *procedural* knowledge. It is important to note, however, that in stressing both procedural and schematic knowledge and their links, current theories of mathematical understanding offer promise of joining two hitherto separate and largely competing strands of research on mathematical development. These are (*a*) the behavioral, which has concentrated on number performance skills and has viewed growth in mathematical ability as the addition of successive performance skills; and (*b*) the cognitive–developmental, which has focused on changing concepts of number but has often paid little attention to the manifestation of these concepts in actual number performances.

NUMBER REPRESENTATION IN THE PRESCHOOLER: THE MENTAL NUMBER LINE

This account begins by considering what understanding of number can be assumed as the typical child enters school. Several lines of evidence point to the probability that by the time they enter school most children have already constructed a representation of number that can be appropriately characterized as a

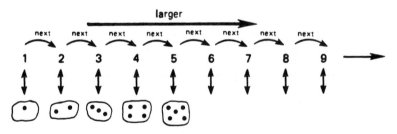

Figure 3.1 The mental number line.

mental number line. That is, numbers correspond to positions in a string, with the individual positions linked by a "successor" or "next" relationship and a directional marker on the string specifying that later positions on the string are larger (see Figure 3.1). This mental number line can be used both to establish quantities by the operations of counting and to directly compare quantities. By combining counting and comparison operations, a considerable amount of arithmetic problem solving can also be accomplished.

Counting

Several extensive studies of counting in preschool children provide the basis for inferring the number knowledge typical of children as they enter school. These include Gelman and Gallistel's (1978) study of counting and number concepts in 2- through 5-year-olds, and Fuson and Briars' (Fuson & Mierkiewicz, 1980) work on counting (see also Siegler & Robinson, 1982; Steffe, Thompson, & Richards, 1982). These investigators have shown that from a very early age, children can reliably count sets of objects and thus establish their cardinality. Greeno, Riley, and Gelman (1978) have developed a computational program that simulates the counting performances observed by Gelman and Gallistel and that is in good accord with the data reported by the other major investigators as well. This model provides the basis for my characterization of the mental number line.

At the core of the Greeno et al. model of children's counting is an ordered list of numerlogs linked by a successor (Next) relationship as shown in Figure 3.1. The program establishes the quantity of a set by a procedure that uniquely links each object in the set with one of the numerlogs and then designates the last numerlog named as the number in the set. The figure shows direct links between the smallest numerlogs and patterned set displays. These links represent the kind of knowledge that would allow children to *subitize* very small sets—that is, to quickly provide the appropriate number name without actually counting—through direct pattern recognition. This ability has been attributed to children as young as 3 or 4 by Klahr and Wallace (1976), although Greeno et al. argue that the appearance of subitizing may be a function of the rapid perceptual grouping of small sets as part of the counting process rather than as a separate means of quantifying an array. Without attempting to decide between these two accounts of rapid quantification of small sets, it seems reasonable to propose that it is through extensive practice with counting as a method of establishing quantity that the numerlog list is gradually transformed from a string of words into a *representation of quantity* in which each position (number name) in the list comes to stand for a quantity. Recent work by Comiti (1980) has shown that the counting list and its use in determining quantity is established only for relatively small numbers by the time a child enters school. For quantities in the teens and twenties, many 6-year-olds are unreliable counters and are not able to use counting to establish equivalence of sets—something they can do at a much younger age for smaller set sizes. In addition, children have difficulty for some time in starting a count at a number other than 1, indicating that individual successor links are not fully established for some parts of the string (Fuson, Richards, & Briars, 1982). It is thus clear that the number representation shown in Figure 3.1 is still developing for larger quantities once school begins.

Quantity Comparisons

A smaller but still significant body of work on magnitude comparison by children allows us to further specify the characteristics of the mental number line as the child enters school. Typically in magnitude comparison tasks, two "target" numbers are named and the subject asked to decide which is larger or "shows more." Variations of this task have been extensively used with adults (e.g., Potts, Banks, Kosslyn, Moyer, Riley, & Smith, 1979). Investigators studying children (Schaeffer, Eggleston, & Scott, 1974; Sekuler & Mierkiewicz, 1977; Siegler & Robinson, 1982) have established that children can perform this task accurately by the age of 5 or earlier—at least for small numbers.

What additions to the mental number line are necessary to account for this ability? If we were to add to the quantity representation already described a directional coding that specified that later numbers in the string represented larger quantities, a child could compare two named numbers by starting up the string from 1, noting when the first of the two target numbers was reached and then labeling the *other* number as "more" or "larger."

Although this is logically possible, it seems psychologically unlikely for at least two reasons. First, it forces the child to treat *more* as if it were the marked item in the "more–less" pair. A number of investigators, beginning with Donaldson and Balfour (1968), have demonstrated that *more* is unmarked—that is, it is more easily learned and more quickly accessed than *less*. Second and even more compelling, 5-year-old children, like adults, show a characteristic pattern of reaction times for these comparison tasks: They take *longer* to make comparison judgments the *closer* the two target numbers are. If a child were using the counting-up strategy to make comparisons, the time to make a mental magnitude comparison should be a function of the size of the smaller number and not of the size of the split between the two numbers. The existence of the split effect suggests that the child's number representation has important analog features that allow direct comparison of number positions. It is as if perceptual comparisons of positions on a measuring stick were being made; when positions are closer together, it takes longer to discriminate between them than when they are further apart.

Because of the split effect for number comparisons, we can attribute to children entering school two other features of the mental number line: (*a*) a directional marker on the line that interprets positions further along the line as "larger" (as shown in Figure 3.1), and (*b*) an ability to directly enter the positional representation for a number upon hearing its name (i.e., without counting up to it). Both of these features play a role in various kinds of informal arithmetic performances that have been observed in preschool children.

Informal Arithmetic

As just noted, the mental number line can be used both to establish quantities by the operations of counting and to directly compare quantities. By combining counting and comparison operations, the child can also accomplish a considerable amount of arithmetic problem solving. For example, Gelman (1972), in her "magic" experiments, showed that young children could recognize when the number of items in a small set had been changed while the set was hidden from view. This would involve counting the set twice, before and after the change, and then comparing the two numbers by entering them on the mental number line. Gelman and Gallistel (1978) also document some young children's ability to "fix" a set so that it has a named quantity. A child with only the number knowledge sketched thus far could build a larger set (e.g., "fix" a set of three so it has five) by counting the three objects in the presented set and then adding in more objects by "counting on" up to five. To reduce a set (e.g., "fix" a set of five so it has three), the child would have to count the objects of the set up to three and then discard the remainder. The more efficient procedure of determining in advance that two items must be added to or deleted from the set would not yet be available to the child at this stage in the development of quantity representation.

This is not to say that the child has no resources for solving addition and subtraction problems. Ginsburg (1977) has reported a variety of successful arithmetic calculation procedures employed by preschool children, all apparently invented by the children and virtually all based on counting. An example is addition by constructing sets (on fingers or with objects) to match each addend, then counting up the combined sets. A typical procedure for subtraction—one that requires no more complicated quantity representation than the one considered thus far—is to (*a*) count out a set to match the larger number (the minuend), (*b*) count out from this set the number of objects specified in the smaller number (the subtrahend), and then (*c*) count the objects remaining in the original set.

Several investigators (e.g., Carpenter & Moser, 1982; Lindvall & Gibbons-Ibarra, 1980) have shown that young children are able to solve certain classes of story problems using counting procedures. Typically in these solutions they use only forward counting, by ones, of actual countable objects. However, some

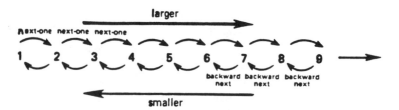

Figure 3.2 The mental number line with backward markers.

children apparently acquire the ability to use decrementing (counting backward) procedures before they enter school. This means that "backward–next" links must have been attached to adjacent numbers in their mental number line and a "smaller" (less) directional marker attached to the line as a whole (see Figure 3.2). Performances that call on backward counting include doing subtraction by counting down from the larger number. Although these performances are often used to argue that children already know important concepts of mathematics before school begins, in fact such performances require only a primitive representation of number compared to what will develop subsequently.

EARLY SCHOOL ARITHMETIC: THE PART–WHOLE SCHEMA

As long as the number line alone is used, there is no way to relate quantities to one another except as larger or smaller, further along or further back in the line. Although quantities can be compared for relative size, no precision in the relative size relationship is possible except as a specification of the number of numerlogs that must be traversed between positions in the line. Probably the major conceptual achievement of the early school years is the interpretation of numbers in terms of part and whole relationships. With the application of a Part–Whole schema to quantity, it becomes possible for children to think about numbers as compositions of other numbers. This enrichment of number understanding permits forms of mathematical problem solving and interpretation that are not available to younger children.

Figure 3.3 sketches a Part–Whole schema that plays a role in several models of children's developing number understanding (Briars & Larkin, 1981; Resnick, Greeno, & Rowland, 1980; Riley, Greeno, & Heller, Chapter 4). The schema specifies that any quantity (the whole) can be partitioned (into the parts) as long as the combined parts neither exceed nor fall short of the whole. By implication, the parts make up or are included in the whole. The Part–Whole schema thus provides an interpretation of number that is quite similar to Piaget's (1941/1965) definition of an operational number concept. To function as a tool in problem solving, the part–whole knowledge *structure* must be tied to *procedures*

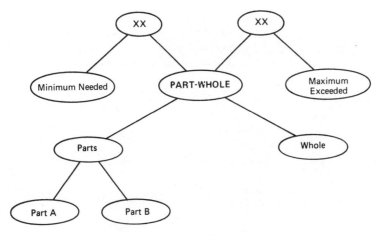

Figure 3.3 The Part–Whole schema.

for constructing or evaluating quantities. The Maximum Exceeded and Minimum Needed nodes in Figure 3.3 are connected to procedures by which deletions or additions can be made to satisfy the constraint that the sum of the parts is equivalent to the whole. For example, if the numbers in the Whole and Part A slots are known, a counting-up procedure (accessed through the Minimum Needed node) can be used to fill Part B with the number needed to keep the combined parts equal to the whole.

Story Problems

The Part–Whole schema specifies relationships among triples of numbers. In the triple 2-5-7, for example, 7 is always the whole; 5 and 2 are always the parts. Together, 5 and 2 satisfy the equivalence constraint for the whole: 7. The relationship among 2, 5, and 7 holds whether the problem is given as $5 + 2 = ?$, $7 - 5 = ?$, $7 - 2 = ?$, $2 + ? = 7$, or $? + 5 = 7$. Each of these number sentences expressing the relations among the triple 2-5-7 has one or more corresponding expressions in real-world relationships or in story problems. Figure 3.4 shows how the fundamental part–whole relationship underlies several classes of story problems as well as number sentences. In each problem the whole is coded as a dot-filled bar, whether it is a given quantity or the unknown. Similarly, each part is uniquely coded. The relationship between parts and whole for all the problems, including the number sentences, is shown in the center display. Any bar can be omitted and thus become the unknown. Although number sentences and the given words of story problems cannot be mapped directly onto one another (Nesher & Teubal, 1975), each can be mapped directly onto a more abstract part–whole representation, such as the bars shown here. The Part–Whole sche-

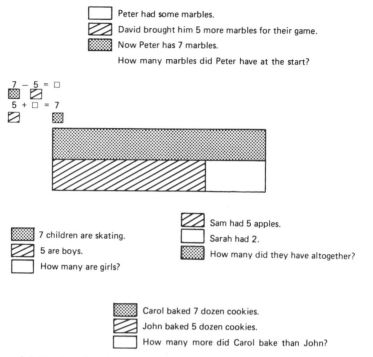

Figure 3.4 Mapping of stories and number sentences to a concrete model of Part–Whole.

ma thus provides an interpretive structure than can permit the child to either solve certain more difficult problems directly by the methods of informal arithmetic, or to convert them into number sentences that can then be solved through procedures taught in school.

Riley, Greeno, and Heller (Chapter 4) have developed a family of computational models that account for the development of competence in solving one-step addition and subtraction story problems of the kind studied by a number of investigators (e.g., Carpenter & Moser, 1982; Nesher, 1982; Vergnaud, 1982). These models suggest that it is application of the Part–Whole schema that makes it possible to solve difficult classes of story problems that children usually cannot solve until the second or third school year. These include set-change problems with the starting set unknown (e.g., *John had some marbles. Michael gave him 4 more. Now he has 7. How many did he have to start?*) and various kinds of comparison problems (e.g., *John has 4 marbles. Michael has 7. How many more does Michael have than John?*). An alternative story problem model by Briars and Larkin (1981) solves some of the more difficult problems by constructing a mental script that reflects real-world knowledge about combining and separating objects, rather than abstract part–whole relationships. The script describes the actions in the story and allows the system to keep track of the sets and subsets involved. Yet in Briars and Larkin's model, too, it proves possible to solve unknown-first problems only by instantiating a Part–Whole schema. Both theories, then, assume that story problem solution—at least for the most difficult problems—proceeds by mapping the statements in the problem into the slots of the Part–Whole schema. This allows the numbers in the problem to be assigned to either "part" or "whole" status and permits a clear identification of whether the unknown is a part or a whole. This in turn allows flexible computational strategies, including *either* direct counting solutions (for example, by counting up from Part A if Part B must be found) *or* the construction of an appropriate number sentence and then solution of the arithmetic problem specified in the number sentence.

Mental Addition and Subtraction

We have seen that preschool children using mainly forward counting procedures are capable of solving a surprising variety of arithmetic problems as long as they have actual countable objects to aid in the calculation. During the early years of school, children come to be able to solve many of the simpler arithmetic problems "in their heads"—that is, without any overt counting. It had long been assumed that when children ceased overt counting, they had switched to an adult-like performance in which the number facts (e.g., single-digit addition or subtraction problems) were simply associations, memorized and then recalled on demand. Presumably, no reasoning went on in arriving at an answer. Recent work, however, has established quite clearly that there is an intermediate period of several years during which arithmetic problems are solved by mental counting processes. These procedures appear to be children's own inventions. There is reason to believe that the Part–Whole schema plays a role in establishing these procedures, although there is no formal theory nor very direct evidence yet available to specify that role.

Research by Groen and Parkman (1972) is the point of reference for work on simple mental calculation. Working with simple addition (two addends with sums less than 10), Groen and Parkman tested a family of process models for single-digit addition. Figure 3.5 shows the general model schematically. All of the models assumed a "counter in the head" that could be set initially at any number, then incremented a given number of times and finally "read out." The specific models differed in where the counter was set initially and in the number of increments-by-one required to calculate the sum. For example, the counter can be set initially at zero, the first addend added in by increments of one, and then the second addend added by increments of one. If we assume that each increment

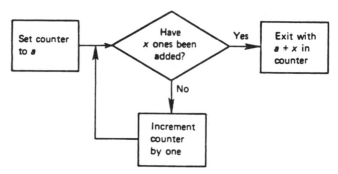

Figure 3.5 Counting model for simple addition. (From Groen & Parkman, 1972. Copyright 1972 by the American Psychological Association. Reprinted by permission.)

needs about the same amount of time to count, then someone doing mental calculation this way ought to show a pattern of reaction times in which time varies as a function of the sum of the two addends. This has become known as the *sum* model of mental addition. A somewhat more efficient procedure begins by setting the counter at the first addend and then counting in the second addend by increments of one. In this case—assuming that the time for setting the counter is the same regardless of where it is set—reaction times would be a function of the quantity of the second addend. A still more efficient procedure starts by setting the counter at the larger of the two addends, regardless of whether it is the first or the second, and then incrementing by the smaller. Obviously, this would require fewer increments. Such a procedure would produce reaction times as a function of the size of the minimum addend and has thus become known as the *min* model.

Groen and Parkman evaluated these (along with some other logically possible but psychologically implausible) models by regressing observed on predicted patterns of reaction times for each model. The finding was that children as young as first-graders used the *min* procedure. Subsequently, the *min* model has been confirmed in studies that have extended the range of problems up to sums of 18, and the ages of children from 4½ or so up to 9 or 10 (Groen & Resnick, 1977; Svenson & Broquist, 1975; Svenson & Hedenborg, 1979; Svenson, Hedenborg, & Lingman, 1976). Figure 3.6 shows a characteristic data plot. Note that problems with a minimum addend of 4 cluster together and take longer than problems with a minimum addend of 3, and so on. It is also typical that doubles (e.g., 2 + 2) do not fall on the regression line but instead are solved particularly fast. We can infer that some process other than counting is used in responding to doubles problems, a point I shall return to later.

Counting models have also been applied to other simple arithmetic tasks, especially subtraction (Svenson *et al.*, 1976; Woods, Resnick, & Groen, 1975), and addition with one of the addends unknown (Groen & Poll, 1973). In the case of subtraction, at least three mental counting procedures are mathematically correct. One procedure would involve initializing the counter in the head at the larger number (the minuend) and then decrementing by one as many times as indicated by the smaller number (the subtrahend). In this *decrementing* model, reaction times would be a function of the smaller number. A second procedure would involve initializing the counter at the smaller of the two numbers and incrementing it until the larger number is reached. The number of increments then would be read as the answer. Reaction times for this *incrementing* model would be a function of the remainder—the number representing the difference between the minuend and subtrahend. A particularly efficient procedure would involve using *either* the decrementing *or* the incrementing process for subtraction, depending upon which required fewer steps on the counter. Reaction times would be a function of the smaller of the subtrahend and the remainder. This *choice* model is what most primary school children use, although a few second-graders use the straight *decrementing* model (See Figure 3.7). Here again, note how the doubles fall below the regression line, suggesting a faster, noncounting solution method.

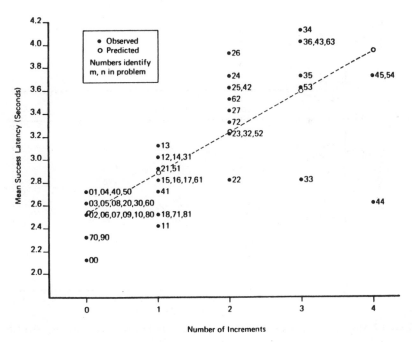

Figure 3.6 Reaction times for first graders solving addition problems. Pairs of numbers listed above or below dots stand for single-digit addition problems (e.g., 0 + 0, 0 + 1, 1 + 0). Dots indicate the average reaction times for adding each pair of numbers. (From Groen & Parkman, 1972. Copyright 1972 by the American Psychological Association. Reprinted by permission.)

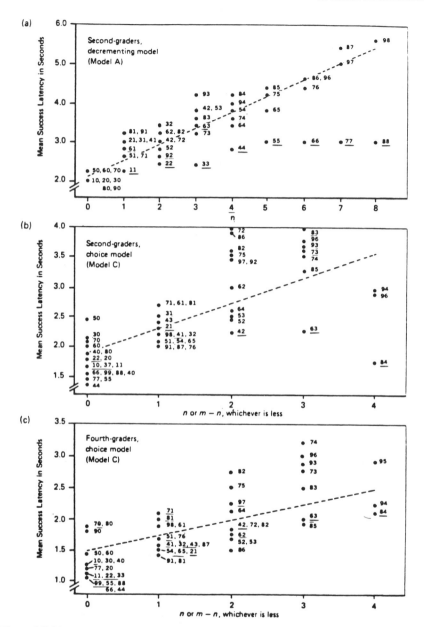

Figure 3.7 Mean reaction time patterns for three groups of children using a *decrementing* model or a *choice* model of subtraction. (Adapted from Woods *et al.*, 1975. Copyright 1975 by the American Psychological Association. Reprinted by permission.)

It is risky to attribute complex processes such as *min* and *choice* to people entirely on the basis of reaction time patterns. For this reason, it is important to ask what converging evidence exists that points to the reality of mental counting procedures. Observations of overt counting-on strategies for addition by several investigators (Carpenter, Hiebert, & Moser, 1981; Fuson, 1982; Houlihan & Ginsburg, 1981; Steffe *et al.*, 1982) suggest that the counting presumed in these models is real. Furthermore, Svenson and Broquist (1975) interviewed their subjects after each timed trial and found that on about half of the problems, children reported counting up from the larger number (by ones or in larger units). Finally, evidence comes from comparing children's reaction-time patterns for addition with those of adults, whom we can assume retrieve elementary addition and subtraction facts by some kind of direct "look-up" procedure. Adults show much faster reaction times and a far shallower slope (20 msec) when their data are fit to *min* than do children (Groen & Parkman, 1972). Their slope, which is presumably the time needed for each count, seems too fast to represent anything like a real counting procedure. Groen and Parkman suggested that this shallow slope might be an artifact of averaging over many trials in which the answers were looked up (presumably producing a flat slope) and a few trials in which they were counted. More recently, Ashcraft and Battaglia (1978) have suggested that adults do not produce a linear increase in time as the minimum addend grows, but instead produce a positively accelerating curve that is best fit by *square of the sum*. Ashcraft and Fierman (1982) tried to fit children's data to sum^2, but not until fourth grade did sum^2 provide the best fit. Younger children thus do appear to solve addition problems by counting. The converging evidence for subtraction is less rich, although some of Svenson's (Svenson & Hedenborg, 1979; Svenson *et al.*, 1976) subjects described the *choice* strategy in interviews.

It is important to note, however, that while *min* and *choice* appear to be the dominant procedures during the early school years, they are not the only ones used. Several investigators have noted the use of special shortcut mental addition strategies by children during this period. These have been documented in some detail by investigators (Carpenter & Moser, 1982; Houlihan & Ginsburg, 1981; Svenson & Hedenborg, 1979; Svenson & Sjoberg, in press) who used verbal protocols and reaction times to document strategies that made special use of addition and subtraction facts that children had committed to memory and could retrieve directly. Most common were the use of solutions with tie references (e.g., 3 + 4 is solved by saying 3 plus 3 is 6, plus 1 more makes 7; or 13 − 6 is solved by saying 12 minus 6 is 6, plus 1 is 7). Saxe and Posner (Chapter 7) found similar strategies among illiterate Africans. Less frequent, but of considerable interest because they signal a developing appreciation of the decimal number system, are solutions that depend on knowledge of tens complements. For example, 6 + 5 is converted to 6 + 4 (=10), plus 1 more. Or, for subtraction, 11 − 4 is converted to 10 − 3 + 1. These shortcut procedures provide evidence that children understand the compositional structure of numbers and are able to partition and recombine quantities with some flexibility.

The Origins of Invented Arithmetic Procedures

What must be added to the mental number line representation to account for the predominance of *min* and *choice* and for the occurrence of special tie- and complements-referenced strategies during the earliest school years? In considering this question, we should keep in mind that these strategies are not directly taught in most school programs. Extensive practice in addition and subtraction is given, some of it organized to highlight commutative pairs in addition and the complementarity of addition and subtraction. But the actual counting procedures and the conversions to make use of tie and complements facts must usually be invented by the children themselves—sometimes in the face of strictures against overt counting. Indeed, the invented character of *min* has been demonstrated directly (Groen & Resnick, 1977). We taught preschool and kindergarten children a procedure for addition that involved counting out both sets. Half of the children switched to *min* without further instruction after about 12 weeks of practice sessions.

The invented character of *min* and *choice* poses an interpretive challenge, for neither of these procedures appears to derive in a straightforward, mechanical way from the overt counting procedures observed among younger children. That is, they are not simply shortcuts, in the sense of dropping redundant steps. Indeed, in each case a new step—deciding which number to start counting from—is added. Furthermore, *min* seems to depend upon the mathematical principle of commutativity, the recognition that the sum of two numbers is the same regardless of the order in which they are added, and *choice* appears to depend upon recognition of the complementarity of addition and subtraction. Yet neither of these principles is directly taught to children in the earliest grades of school any more than the actual *min* and *choice* procedures are taught, and no study has suggested that children who use them have any verbal awareness of the general principles involved. Our interpretive task, then, is to account for the emergence of *min* and *choice* as procedures that accord with mathematical principles of commutativity and complementarity but are not systematically derived from those principles. There are several possible explanations to consider.

A "PAIR-EQUIVALENCE" ACCOUNT

The simplest account of the discovery of *min* would assert that the special relationships between certain pairs of problems (e.g., 3 + 4 and 4 + 3; 2 + 7 and

7 + 2) are noticed after extensive practice on the individual pairs, through a general learning process that looks for regularities and shortcuts after a procedure becomes at least partially automated (cf. Anderson, 1981; Klahr & Wallace, 1976). In this view, the child would notice that specific pairs of problems yielded the same answer and would infer that they could be substituted for one another. A preference for efficiency would then lead to the strategy of always starting the count at the larger number.

This seems plausible until we consider that if the child is to notice the equivalence of two problems, the result of both pairs must be present in short-term memory simultaneously so that they can be compared. This could happen in two ways. First, if commuted pairs (e.g., 7 + 3 and 3 + 7) were presented successively, the result of the first calculation might still be present when the second calculation was completed. However, in our experiment (Groen & Resnick, 1977) the children invented *min* under controlled practice conditions in which these pairings of problems did not occur. Practice on paired problems, then, cannot be a general explanation for the development of *min*, although it may play a role in some cases. A second possibility is that the result of 7 + 3 can be quickly retrieved when 3 + 7 is computed. But this would mean that 7 + 3 was already known as a retrievable addition fact. If such retrievable facts were available, however, children would not need to use counting procedures to compute the answers to simple addition and subtraction problems. It therefore appears implausible to attribute the discovery of *min* to simply noticing the common outcome of different orders of performing addition.

A modified version of the pair-equivalence account may survive, however. This version would assume that the equivalence was noticed first for very easily computable pairs (e.g., those involving an addend of 1). It seems plausible that the sum of 7 + 1 could be retrieved (or constructed) fast enough to be simultaneously present in short-term memory with the sum of 1 + 7. Having noted equivalence for a subset of the addition pairs, a child might plausibly construct a more general commutativity rule that could be applied to other pairs.

A "DEFAULT" ACCOUNT

Another possibility is that children begin by *assuming* that arithmetic operations are commutative and only gradually learn that some (for example, subtraction) are not. This would lead them to try *min* procedures in the search for less-effort processes. Since *min* "works" (i.e., the answer turns out to be correct when checked by counting the whole joint set, and adults do not comment on the result as wrong), they would retain it as the preferred procedure. In support of this possibility is the observation that children frequently attempt to commute subtraction problems. That is, when given the problem 2 − 5, they respond with

3 rather than -3, 0, or "you can't do it"—any of which would indicate recognition of the noncommutativity of subtraction. Another common attempt to commute in subtraction is shown by giving solutions such as:

$$\begin{array}{r} 348 \\ -169 \\ \hline 221 \end{array}$$

A child would arrive at this incorrect answer by "subtracting within columns" (Brown & Burton, 1978)—that is, by taking the smaller number from the larger in each column regardless of which is on top.

The Gelman and Gallistel (1978) analysis of young children's counting makes it clear that they proceed in accord with an "order-invariance" principle—that is, they recognize that *objects* can be counted in any order, although the numerlogs must be assigned in their standard sequence. A natural extension of order-invariance would allow subsets as well as individual objects to be enumerated in any order. This would allow *min* to emerge as part of a general search for low-effort solutions without requiring that the child construct any kind of commutativity rule.

Neches (1981, and personal communication) has provided a formal account of how *min* might be discovered on such a "default" basis. His computer model of addition begins by performing a *sum* solution in which both subsets are counted out and the combined set recounted. After a number of practice trials, the system notices that a *portion* of the counting process for finding the total is redundant with the original counting process for each of the subsets. In recounting for the problem $2 + 5$, for example, the first two counts are redundant with counting out the first subset, and the first five counts are redundant with the original count for the second subset. The system has some general redundancy elimination mechanisms that lead it to reuse existing computations rather than duplicate them. This means that two counting-on solutions are constructed, one for each addend. The system eventually comes to count on from the *larger* addend (thus performing the *min* procedure) because it can detect a redundancy when the smaller-addend alternative is tried.

A "PART-WHOLE" ACCOUNT

Still another possibility for the emergence of *min* is that children apply a simple Part–Whole schema to addition. For example, a child could solve addition problems by binding the given addends to the Part slots of the schema. Since the slots contain no order information, the addends can now be used in either order to discover the value of the Whole. This is an attractive explanation of *min* because it also accounts economically for the discovery of *choice*. Part–Whole puts the three terms of a complementary addition–subtraction pair into a stable relationship with one another. For the problem $9 - 7$, for example, 9 would fill the Whole slot and 7 one of the Part slots. For $9 - 2$, 9 would fill the Whole slot and 2 one of the Part slots. In finding the missing part (using the procedures attached to the Minimum Needed and Maximum Exceeded nodes of the schema), the child would become aware of the complementary relationships between $9 - 2 = 7$ and $7 + 2 = 9$. This complementary relationship could then be used to generate least-effort solution rules. Part–Whole also provides a convenient account of the basis for complement- and tie-based shortcut procedures.

Application of Part–Whole seems to be a plausible account for the emergence of *min* and *choice*, at least to the extent that it is plausible to attribute the Part–Whole schema to children at an early enough age so that it precedes *min* and *choice* as part of the knowledge structure. We have mixed evidence here. On the one hand, a fully general Part–Whole schema does not seem to be reliable until the age of 7 or 8. This is when children master Piagetian class inclusion problems (Inhelder & Piaget, 1964/1969), which are part–whole problems without a requirement of specific numerical quantification. It is also the age at which children can reliably solve those story problems that clearly depend on the part–whole structure (e.g., set-change problems with the starting set unknown). This age would be too late to account for *min*, although it is possibly an acceptable age for *choice*, which as far as we know develops later.

Still, several investigations point to an earlier understanding of certain class relationships than the Piagetian studies have suggested. For example, Markman and Siebert (1976) have shown that if the class character of the Whole set is emphasized by the wording of the problem, children can perform class inclusion problems quite early, and Smith and Kemler (1978) have shown that kindergarten children use component dimensions in certain kinds of classification tasks. Furthermore, children as early as first grade can solve comparison story problems when they are worded so as to make the part–whole relations evident (see Riley *et al.*, Chapter 4). Thus, it seems plausible that children may possess at least a simple version of the Part–Whole schema at a quite young age but may not yet have learned all of the situations where it is appropriate to apply it. Addition and subtraction of small numbers, unencumbered by story content, may be one of the easy-to-recognize situations. Indeed, application of a primitive Part–Whole schema to simple number problems may be an important step in developing a more elaborate version, including many procedural connections, that will play a role in subsequent development of number knowledge.

DEVELOPMENT OF DECIMAL NUMBER KNOWLEDGE

All of the research discussed so far has focused on small numbers—quantities up to about 20. From this work we are able to trace a probable course of development of number representation in which the fundamental relationships between numbers are units. Yet the introduction of decimal numbers, which form an important part of the primary school mathematics curriculum, demands that a new relationship among numbers be learned. This relationship is based on tens rather than units. The initial introduction of the decimal system and the positional notation system based on it is, by common agreement of educators, the most difficult and important instructional task in mathematics in the early school years. Starting in about second grade, most schools begin to teach children about the structure of two-digit numbers. Toward the end of second grade, addition (and in some schools, subtraction) with regrouping is introduced. What is known about the development of knowledge of the base ten system—its representation in written form, and the calculation algorithms that are based on it? How does the quantity representation change as skill in the posititional notation system develops? These questions are addressed below.

Numbers as Compositions of Tens and Units: Restriction and Elaboration of the Part–Whole Schema

We have already seen that an important aspect of the development of number during the early school years is the interpretation of numbers as compositions of other numbers—that is, the application of the Part–Whole schema to numbers previously defined solely in terms of position in a linear string. In story problems and simple mental arithmetic, the Part–Whole schema is applied with few restrictions and little elaboration. I will now try to show that the development of decimal number knowledge can be understood as the successive elaboration of the Part–Whole schema for numbers, so that numbers come to be interpreted by children as compositions of units and tens (and later of hundreds, thousands, etc.) and are seen as subject to special regroupings under control of the Part–Whole schema.

There is far less research to draw on in making this characterization of developing place value knowledge than there is for early number concepts, story problems, and simple arithmetic. In addition to ongoing work in our own laboratory, I will refer to empirical and theoretical work by several others in building this account of stages of development in decimal number understanding. The account must be viewed as tentative and subject to modification as further evidence on the development of understanding of the decimal number system accumulates. In particular, the later stages of this account are based on data from a small number of children who were receiving remedial instruction in our laboratory. We need to extend this data base to include more children—especially those children who acquire place value understanding without the special intervention included in our studies.

We can identify three main stages in the development of decimal knowledge. First, there is an initial stage in which a unique partitioning into units and tens (e.g., 47 is 4 tens plus 7 units) is recognized. Next, in stage two, children recognize the possibility of multiple partitionings of a quantity. This second stage occurs in two phases: Multiple partitionings are (*a*) arrived at empirically (e.g., the equivalence of 30 tens plus 17 units to 40 tens plus 7 units is established by counting), and (*b*) established directly by application of exchanges that maintain equivalence of the whole (e.g., $40 + 7 = 30 + 17$, because 1 ten can be exchanged for 10 units). Third, a formal arithmetic stage appears in which exchange principles are applied to written numbers to produce a rationale for algorithms involving carrying and borrowing.

Stage One: Unique Partitioning of Multidigit Numbers

The earliest stage of decimal knowledge can be thought of as an elaboration of the number line representation so that, rather than a single mental number line linked by the simple "next" relationship, there are now two coordinated lines, as sketched in Figure 3.8. Along the rows a "next-by-one" relationship links the numbers. This can be extended indefinitely, as shown in the top row, indicating that a units representation of number coexists with a decimal representation. Along the columns a "next-by-ten" relationship links the numbers. In a fully developed number representation this "next-by-ten" link might hold for the numbers inside the matrix as well as for those along the edges, permitting more efficient addition or subtraction of the quantity 10 than of other quantities. Earlier, and perhaps indefinitely, the "inside" links (e.g., $37 + 10 = 47$) might be constructed on each occasion of use by a procedure that decomposes the two–digit number into a tens and a units portion ($37 = 30 + 7$), then adds 10 to the tens portion ($30 + 10 = 40$), and finally adds back the units ($40 + 7 = 47$). In either case, the most important feature of this new stage of number understanding is that each of the numbers is represented as a *composition* of a tens value and a units value. This means, in effect, that two-digit numbers are interpreted in terms of the Part–Whole schema, with the special restriction that one of the parts be a multiple of 10.

There is some evidence that this compositional structure of the numbers arises first in the context of oral counting—that is, that it is not at first tightly linked to quantification of large sets of objects or to grouping of units by tens. Several investigators (Fuson *et al.*, 1982; Siegler & Robinson, 1982) found that many 4- and 5-year-olds could count orally well into the decades above 20 and that their counting showed evidence of being organized around the decade structure. For example, the most common stopping points in the children's counting were at a number ending in 9 or 0 (e.g., 29 or 40); and their omissions in the

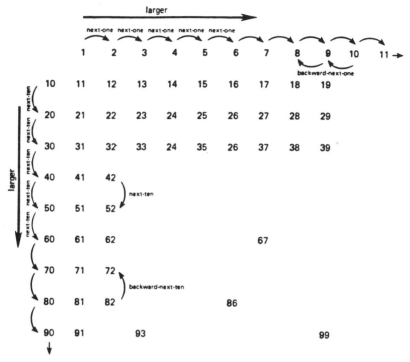

Figure 3.8 Earliest stage of decimal knowledge represented as two coordinated mental number lines.

number string tended to be omissions of entire decades (e.g., ". . . 27, 28, 29, 50 . . ."). They also sometimes repeated entire decades (e.g., " . . . 38, 39, 20, 21 . . .") and sometimes made up nonstandard number names reflecting a concatenation of the tens and the units counting strings (e.g., " . . . twenty-nine, twenty-ten, twenty-eleven . . ."). Finally, these children could usually succeed in counting on within a decade higher than their own highest stopping point when asked by the experimenter to start counting from a particular number, such as 51 or 71.

In our own work on place value, we have collected many observations of primary school children's methods of establishing the quantity shown in displays of blocks or other objects coded for decimal value (see Figure 3.9 for examples of such displays). The typical method that children use in this kind of task is to begin with the largest denomination and enumerate the blocks of that denomination using the appropriate counting string (e.g., 100, 200, 300, etc., for hundreds blocks), then add in successive denominations by counting on using the appropriate counting string. A successful quantification of the display in Figure 3.9a, for example, would produce the counting string: 100, 200, 300, 400, 410, 420, 430,

440, 450, 460, 461, 462, 463. A few children, mainly those who show the most sophisticated knowledge of other aspects of place value, count all denominations by ones and then "multiply" by the appropriate value (e.g., for Figure 3.9a: 1, 2, 3, 4, 400; 1, 2, 3, 4, 5, 6, 460; 1, 2, 3, 463). However, counting using the decimally structured number strings seems to be the earliest application of decimal knowledge to the task of quantifying sets. Furthermore, between simple oral counting competence and the successful use of the decimal-structured counting strings for quantification, there seems to be a period during which the child knows the individual strings well enough to use them separately for quantifica-

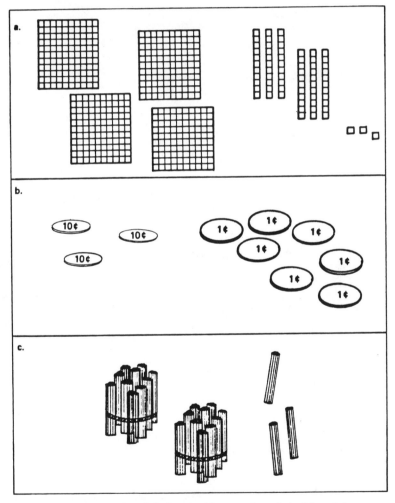

Figure 3.9 Examples of displays used in research on decimal knowledge.

tion but cannot coordinate the use of several strings within a single quantification task. In one of our studies, for example, all of the third-grade children we interviewed could count any single block denomination, but more than half of the children became confused when two or more denominations were to be quantified. Examples from the protocols of two such children appear in Figure 3.10.

Other performances characteristic of children in this early stage of decimal number knowledge suggest that children typically recognize the relative values

a) Alice
    E: Shows:

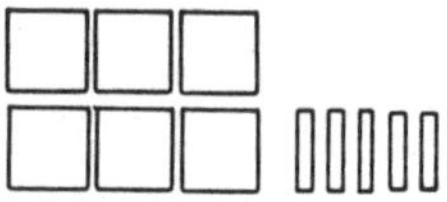

    S: (Touching the hundreds) 100, 200, 300, 400, 500, 600 . . . (touching the tens) 7, 8, 9, 10, 11 . . . 611.
    E: Let's try one more like this. How about this one?

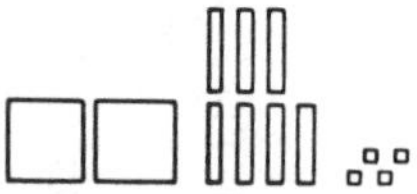

    S: (Touching the hundreds) 100, 200 . . . (touching the tens) 201, 202, 203, 204, 205, 206, 207 . . . (touching the ones) 208, 209, 210, 211.
    E: Hmm. Let's count them again. This time, why don't you count these (tens and ones).
    S: (Touching the tens) 10, 20, 30, 40, 50, 60, 70 . . . (touching the ones) 71, 72, 73, 74.
    E: How much is this (hundreds)?
    S: 200.
    E: Okay. How much is that altogether?
    S: 200 and . . .
    E: I have 200, and I add this much (a ten block) more. How much is that worth?
    S: 201.

b) Jane
    E: Good. So how much do you think this would be?

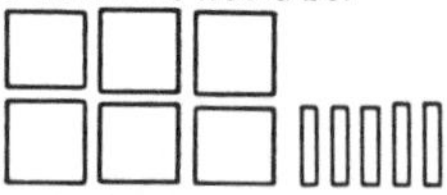

    S: (Touching the hundreds blocks) 100, 200, 300, 400, 500, 600 . . . (touching the tens blocks) 700, 800, 900, ten hundred, eleven hundred.
    E: Are these (tens) worth 100?
    S: I count them all together.
    E: But these (tens) aren't hundreds.
    S: I am counting these like tens.
    E: OK. But how much would these (tens) be worth then?
    S: Oh. 10, 20, 30, 40, 50 . . . 50 dollars.
    E: How much would this (entire display) be worth altogether?
    S: 600 . . . wait! It's 5 and 6.
    E: But how much is it altogether? This (hundred) is 6, right?
    S: Eleven hundred.

Figure 3.10 Examples of confusions in multidenominational counting.

of the different parts that make up the whole number. For example, most second-through fourth-graders we have interviewed compared numbers on the basis of the higher-value digits without reference to the lower-value positions. For example, when comparing written numerals or block displays for the numbers 472 and 427, a child would typically say 472 was larger ". . . because it has 7 tens (or 70) and the other only has 2 tens." It is interesting to note that these judgments assume that the block displays are *canonical*—that is, that they contain no more than 9 blocks of a given denomination. The assumption of canonicity disappears in the second stage of decimal knowledge, as we shall see next.

MENTAL ARITHMETIC

The most stunning displays of a compositional representation of number are in children's invented mental calculation methods. Consider the following performance by an 8-year-old, Amanda:

E: *Can you subtract 27 from 53?*
A: *34.*
E: *How did you figure it out?*
A: *Well, 50 minus 20 is 30. Then take away 3 is 27 and plus 7 is 34.*

Amanda came up with the wrong answer, but by a method that clearly displayed her understanding of the compositional structure of two-digit numbers. She first decomposed each of the numbers in the problem into tens and units, and then performed the appropriate subtraction operation on the tens components. Next she proceeded to add in and subtract out the units components. She should have subtracted 7 and added 3, but instead reversed the digits. Amanda performed on other problems without this difficulty, yielding correct answers. Other children have shown similar strategies.

We have also begun to explore decimal-based mental arithmetic using the reaction-time methods that yielded initial evidence for the *min* and *choice* procedures for smaller numbers. We now have reaction-time data from 12 second- and third-grade children on a set of problems of the form 23 + 9, 35 + 2, 48 + 5. In each problem the two-digit number was presented first and fell within the 20s, 30s, or 40s decade. Each child responded to three sets of 100 such problems; the sets consisted of all possible pairings of the units digits, with the tens digits allowed to vary randomly. The problems were presented horizontally on a videoscope, and the child responded on the digit keys of a computer terminal. Time from presentation to response was recorded.

Assuming that one is going to use a mental counting procedure for solving these problems, there are two plausible possibilities that distinguish clearly between use and non-use of the decade structure:

1. Set the counter to the two-digit number, then add in the one-digit number in increments of one. Reaction time would be a function of the single-digit number (in this case, always the second number). We call this the *min of the addends* procedure. No understanding of the decade structure of the numbers is required for this procedure. However, the child does have to know how to count over the decade barrier (e.g., " . . . 29, 30, 31 . . .") and must have a units number string that extends up through several decades.

2. Decompose the two-digit number into a tens component and a ones component, then recombine the tens component with whichever of the two units quantities is larger. Set the counter to this reconstituted number and then add in the smaller units digit in increments of one. For example, for 23 + 9, the counter would be set at 29 and then incremented three times to a sum of 32. Reaction time would be a function of the smaller of the two *units* digits, so the procedure is called *min of the units*. This procedure is a simple version of the one Amanda used. It not only uses the decade structure of the numbers but behaves in accord with principles of commutativity and associativity (e.g., 23 + 9 = [20 + 3] + 9 = 20 + [3 + 9] = 20 + [9 + 3] = [20 + 9] + 3 = 29 + 3).

We fit each of these models (along with several others that are plausible but whose use would not clearly illuminate decimal structure knowledge) to the reaction times (correct solutions only) of each of our subjects. We predicted the pattern of reaction times for a "pure" model, for a model with very fast times for doubles in the units digits, and for a model with very fast times for tens complements (i.e., pairs that add to 10, such as 3 + 7, 6 + 4, etc.). We also interviewed each child on a set of similar problems in a think-aloud format. Finally, we had reaction-time data on each child's performance on a set of single-digit addition problems. Because a purely mathematical discrimination between models is so difficult (the models themselves are highly intercorrelated), we used a combination of model fits, plausibility of the slopes (presumed counting speeds), children's think-aloud protocols, and the match between lower decade (single digits) and upper decade (two digits plus one digit) performances to tease out a story about each child's performance.

Two children, Ken and Alan, provide particularly clear illustrations of the differences between children who are in a predecimal stage of number representation and those who are clearly using a decimal representation in their mental arithmetic. Ken's reaction times on the upper decade problems were best fit by *min of the addends* ($r^2 = .761$). On the single-digit problems his data cleanly fit the *min* model, with doubles ($r^2 = .695$). The slope of the regression lines for the upper and lower decades (1.164 and .960, respectively) indicated a mental counting time of about one second per increment for both kinds of problems. This suggests that Ken was using the same basic units-counting strategy for both the single- and the two-digit problems. Ken also described the *min of the addends* counting-up procedure as his method in the think-aloud protocols.

Alan provides a contrast case. His reaction times on the upper decade problems fit best the *min of the units* model ($r^2 = .847$). He also showed a next-best fit for *min of the units with complements,* the only child to show a good fit to any complements model; and he showed a reassuringly poor fit to the *min of the addends* model. On the single-digit problems, his data best fit *min,* with doubles ($r^2 = .831$). His slopes for upper and lower decade problems were also similar (.346 for the single-digit problems; .441 for the two-digit problems), indicating a similar mental counting speed for both kinds of problems. Although this story seems very straightforward, it is also incomplete, for Alan's data also fit (although with less variance explained) other models. It seems quite likely that he was using a variety of strategies on different problems. This impression is confirmed by his interview data. He clearly described himself as using the *min of the units* strategy for some problems, but on others he described various other methods that relied on knowledge of doubles and complements. It seems reasonable to conclude that Alan was using complex representations of number relationships to generate strategies that included but were not limited to *min of the units.*

OTHER STAGE ONE TASKS

There are a number of tasks that an individual with the compositional representation of number shown in Figure 3.8 ought to be able to perform, but on which we have only impressionistic data at the present time. These include:

1. adding or subtracting 10 from any quantity more quickly than adding or subtracting other numbers (except 0 or 1, and *possibly* 2). To subtract 10 from 47, for example, an individual could enter the representation at 47 and move one step on the "tens-backward-next" link directly to 37.

2. counting up (or down) by tens from any starting number.

3. constructing mental addition and subtraction algorithms that use the ability to count by 10 from any number. For $72 - 47$, for example, enter the number representation at 72; move down the 10 string four positions to 32. Move down the ones string (crossing the tens position) seven positions to 25. This strategy is related to those (such as *min of the units* and Amanda's strategies) that partition numbers and operate separately on the tens and units, but it reflects a somewhat different use of the decimal structure.

A FORMAL THEORY OF STAGE ONE KNOWLEDGE

We are able to benefit in our analysis of the development of decimal number knowledge from a computer program that simulates the performances of a 9-year-old girl, Molly, on a number of the tasks that provide the basis for inferring place-value knowledge. The program, MOLLY, matches Molly's performance at several points before, during, and after remedial tutorial instruction aimed at establishing an understanding of the rationale for the standard, school-taught written subtraction algorithm. Prior to our instruction, Molly demonstrated the ability to perform tasks such as constructing, interpreting, and comparing block displays of two- and three-digit numbers. The knowledge structure included in the program that was used in performance of all of these tasks is shown schematically in Figure 3.11. This structure organizes conventional information about multidigit written numbers. The structure identifies columns according to their positional relationship to each other. The rightmost column is tagged as the units column, the tens column is the one that is next to the units, the hundreds is next to the tens, and so forth. Which column is being attended to can be determined by starting at the rightmost position and running through the succession of Next links. Attached to each column is a block shape (the block names are those used by Dienes, 1966, in referring to blocks such as those in Figure 3.9), a counting string, and a column value. The value specifies the amount by which a digit must be multiplied to yield the quantity represented by the digit (e.g., in the tens column, multiply by 10).

Someone who possessed this knowledge structure should be able to associate block shapes with column positions, block shapes with column values, and so on. Table 3.1 gives the number of third-grade children in one of our studies who showed reliable knowledge of each type of association at each of two interview points during the year. Since the knowledge was inferred from the method by which children solved the various problems presented, rather than by direct questioning, it was not possible to observe each child on each association in each interview. For this reason the data are given as proportions—the number of children who showed knowledge of the association over the number observed.

As can be seen, all of the children had the position-name association from the outset. That is, they could read two- and three-digit numbers aloud using the proper conventions. A position-shape association was inferred when the children constructed displays in a manner that directly matched each block shape to a digit. The children using a column-by-column match strategy typically worked on the leftmost column first and pointed to each column in succession, saying, "*n* of these." Three of our subjects worked this way successfully in their first interview, more in the second interview. All of the children we observed could apply the appropriate counting strings to block shapes as long as there was only a single block shape to be counted. When they had to switch denominations (hundreds to tens, or tens to ones), however, they had difficulty: Less than half of those observed succeeded (cf. Figure 3.10). To be counted as knowing the value of a column position, the child had to either tell us that, for example, a 9 in the tens column was "worth" 90, or select 9 tens blocks to represent that quantity. Only one child demonstrated this knowledge. Nevertheless, the children demonstrated fairly strong knowledge of the value of block shapes, as is shown in the final row of Table 3.1.

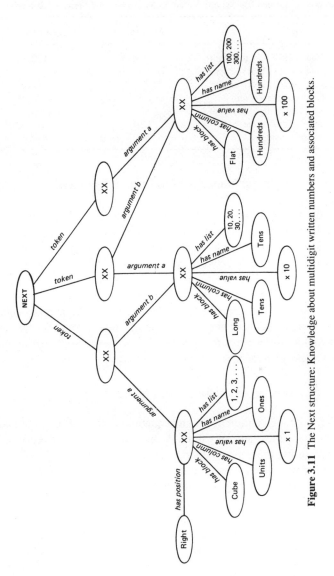

Figure 3.11 The Next structure: Knowledge about multidigit written numbers and associated blocks.

Table 3.1

Proportions of Third Grade Children Displaying Knowledge of Associations between Columns, Blocks, and Values in November and February Interviews

	November	February
Column position/Column name	10/10	10/10
Column position/Block shape	3/3	6/7
Counting strings/Block shapes		
One denomination only	6/6	7/7
Two or more denominations	2/7	3/6
Column position/value	1/10	1/7
Block shape/value	7/10	9/10

Stage Two: Multiple Partitionings of Multidigit Numbers

As long as the Next structure alone is used to interpret numbers, each written number can have only one block representation: a "canonical" representation, with no more than 9 blocks per column. In this canonical display there exists a one-to-one match between the number of blocks of a particular denomination and the digit in a column in standard written notation. Insistence on the canonical form, however, means that there is no basis for carrying and borrowing—or, in block displays, for exchanges and multiple representations of a quantity. During the next stage in development, the Part–Whole schema is applied to multidigit numbers in a manner that allows multiple partitionings and thereby a variety of noncanonical representations of quantity.

MULTIPLE PARTITIONING ARRIVED AT EMPIRICALLY

At first, although children recognize that multiple representations are possible, they can construct them only through an empirical counting process. Molly's performance during the preinstructional phase of her work with us illustrates this method. Molly was asked to use Dienes blocks to subtract 29 from 47. She began by constructing the block display that matched the larger number—that is, 4 tens and 7 units. She then tried to remove 9 units and, of course, could not. The experimenter asked if she could find any way to get more units. Molly responded by putting aside all of the units blocks and one of the tens in her display, leaving just 3 tens. She counted these by tens ("10, 20, 30") and then continued counting by ones, adding in a units block with each count, up to 47. On the next subtraction problem, $54 - 37$, Molly began with a noncanonical display of the top number. That is, she put out 4 tens and counted in units blocks until she reached 54, yielding a final display of 4 tens and 14 units. Molly thus appeared to have learned that certain problems will require noncanonical displays; she had incorporated into her plan for doing block subtraction a check for whether there were more units to be removed than the canonical display would provide. However, at this stage she was able to establish the equivalence of the canonical and noncanonical displays only by the counting process that yielded the same final number in each case.

The MOLLY program provides a formally stated theory of what Molly knew and how she used her knowledge at each of several stages. To simulate the stage of performance just described, MOLLY-1 uses several procedures that call upon the Part–Whole schema described earlier for story problems. In MOLLY-1, the schema is elaborated to include a special restriction, applied to two-digit numbers, that one of the parts be a multiple of 10. To "show 47 with more ones," MOLLY-1 first applies Part–Whole in a global fashion, concluding that if the Whole is to stay the same but more ones are to be shown, there must be fewer tens. MOLLY-1 then reduces the tens pile by a single block, the smallest possible amount to remove. Next, the schema is instantiated with 47 filling the Whole slot, and 30 in one of the Part slots. The Minimum Needed node of the schema is then used to access a procedure for finding the remaining Part by adding ones blocks and counting up until 47 is reached.

Two important concepts have been added to the number representation at this stage. First, the equivalence of several partitionings has been recognized. Second, the possibility of having more than 9 of a particular block size has been admitted. This is crucial for an eventual understanding of borrowing, where—temporarily—more than 9 of a given denomination must be understood to be present, without changing the total value of the quantity. Interviews with a number of children in addition to Molly make it clear that prior to this stage the possibility of borrowing or trading to get more blocks is rejected because it will produce an "illegal" (i.e., noncanonical) display.

PRESERVATION OF QUANTITY
BY EXCHANGES THAT MAINTAIN EQUIVALENCE

A complete understanding of the possibilities for multiple representation can be attributed to children only when they are no longer dependent upon counting to establish the equivalence of displays—that is, when they recognize a class of legal exchanges that will automatically preserve equivalence. Although Molly received no explicit instruction from us on this point, it was clear that after a certain amount of practice with the counting-up method of creating noncanonical displays, she came to recognize that 10-for-1 exchanges would retain the Whole quantity while changing the specific amounts in the Parts. At this point she stopped counting up and began simply to trade—that is, discard a tens block and count in 10 units, or discard a hundreds block and count in 10 tens. We have observed the same kind of performance in other children as well. Some children

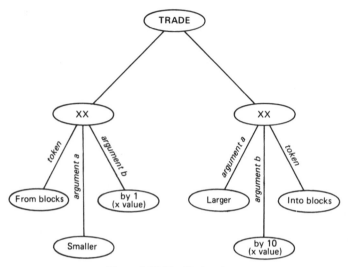

Figure 3.12 The Trade schema.

who engage in trades rather than counting up even become annoyed or amused with an experimenter who keeps asking how they know that the display still shows the same number. They indicate in various ways that they believe that if a ten-for-one trade has been made, the total quantity could not have changed.

The MOLLY-2 program provides a formal theory of Molly's knowledge at this stage. In what can be viewed as a further elaboration of the Part–Whole schema, MOLLY-2 adds to the representation for multidigit numbers an explicit 10-for-1 relationship for adjacent block sizes. This knowledge is represented by a Trade schema (Figure 3.12), which specifies a class of legal exchanges among blocks. The schema specifies that there is a "from" pile of blocks from which blocks are removed. This pile becomes smaller by one block. There is also an "into" pile of blocks that becomes larger by 10 blocks. The *value* of the blocks in the From and Into piles is established by multiplying the number of blocks removed or added by the value of the block shape (as specified in the Next structure and a separate Value schema that is also part of the program). Thus, when trades are made between adjacent block sizes, the schema specifies that the Into and the From values will be the same, even though the number of physical objects present has changed. Applied as an elaboration of the Part–Whole schema, the Trade schema allows MOLLY-2 to construct alternative partitionings of a quantity without having to count up from one of the parts.

Stage Three: Application of Part–Whole to Written Arithmetic

I turn now to children's written arithmetic—in particular, to how the elaborated Part–Whole schema is eventually applied to the interpretation of the conventions of written calculation. There is abundant evidence now available that many children learn rules for the written algorithms of subtraction and addition without linking these rules to the kind of knowledge about place value and number that I have described here. What they seem to learn is a procedure for identifying columns, operating on them, making marks (writing in little 1's, crossing out and rewriting numbers, etc.), but not a rationale that makes the procedure sensible. Brown and Burton (1978) have demonstrated that when children make errors in written arithmetic—particularly subtraction—the errors are often the result not of random mistakes, but of the systematic application of wrong ("buggy") algorithms. Figure 3.13 describes and illustrates some of the most common subtraction bugs. Elsewhere (Resnick, 1981) I have analyzed a number of the Brown and Burton bugs to show that they typically follow rules of syntax, or procedure, while ignoring or contravening the "semantics" of exchange—that is, the principles embodied in the Part–Whole, Trade, and Value schemata described here. For example, in the bug called Borrow-Across-Zero the child follows a rule specifying the need for a written-in little 1 and a crossed-out and decremented number to its left. The syntax of subtraction is largely respected. However, the semantics of exchange is violated, for the child has in fact borrowed 100 but added back only 10—thus failing to conserve the original quantity.

Brown and VanLehn (1980, 1982; VanLehn, Chapter 5) have developed a theory intended to account for the process by which buggy algorithms are invented. The theory assumes that the correct algorithm has been learned but is incomplete for certain problems, either because an incomplete algorithm was taught or because certain steps have been forgotten. When these problems— which most often contain zeros in the top number—are encountered, the attempt to apply the learned algorithm creates an impasse. The child attempts to cope with the impasse by "repairing" the learned algorithm. The repairs proceed in a "generate-and-test" mode that is shared with many other problem-solving process theories (e.g., Newell & Simon, 1972). First, a repair is generated from a very limited list of potential repairs. The list includes moving into the next column to perform an action (this would produce the Borrow-Across-Zero bug), skipping an action, copying a number, and the like. Once generated, a repair is tested against a set of "critics" that specify certain constraints that a subtraction algorithm must obey. These include rules such as acting at least once on each column, showing decrement and increment marks, and not writing more than one digit in each answer column. There is nothing in either the critic list or the repair generation list that refers to what I have been developing in this chapter as the "meaning" of decimal numbers. There is no critic that specifies that the original Whole quantity must be preserved, nor is there anything in the repair or critic lists that even identifies the value of the borrow and increment marks. The theory thus describes an almost wholly syntactic set of bug-generating processes.

Given this characterization of the origin of buggy arithmetic, it can be

1. **Smaller-From-Larger.** The student subtracts the smaller digit in a column from the larger digit regardless of which one is on top.

$$
\begin{array}{r}
3\ 2\ 6 \\
-\ 1\ 1\ 7 \\
\hline
2\ 1\ 1
\end{array}
\qquad
\begin{array}{r}
5\ 4\ 2 \\
-\ 3\ 8\ 9 \\
\hline
2\ 4\ 7
\end{array}
$$

2. **Borrow-From-Zero.** When borrowing from a column whose top digit is 0, the student writes 9 but does not continue borrowing from the column to the left of the 0.

$$
\begin{array}{r}
6\ \overset{7}{\cancel{0}}_{,}2 \\
-\ 4\ 3\ 7 \\
\hline
2\ 6\ 5
\end{array}
\qquad
\begin{array}{r}
8\ \overset{9}{\cancel{0}}_{,}2 \\
-\ 3\ 9\ 6 \\
\hline
5\ 0\ 6
\end{array}
$$

3. **Borrow-Across-Zero.** When the student needs to borrow from a column whose top digit is 0, he skips that column and borrows from the next one. (Note: this bug must be combined with either bug 5 or bug 6).

$$
\begin{array}{r}
\overset{7}{\cancel{8}}\ 0_{,}2 \\
-\ 3\ 2\ 7 \\
\hline
2\ 2\ 5
\end{array}
\qquad
\begin{array}{r}
\overset{7}{\cancel{8}}\ 0_{,}4 \\
-\ 4\ 5\ 6 \\
\hline
3\ 0\ 8
\end{array}
$$

4. **Stops-Borrow-At-Zero.** The student fails to decrement 0, although he adds 10 correctly to the top digit of the active column. (Note: this bug must be combined with either bug 5 or bug 6).

$$
\begin{array}{r}
7\ 0_{,}3 \\
-\ 6\ 7\ 8 \\
\hline
1\ 7\ 5
\end{array}
\qquad
\begin{array}{r}
6\ 0_{,}4 \\
-\ 3\ 8\ 7 \\
\hline
3\ 0\ 7
\end{array}
$$

5. **0 − N = N.** Whenever there is 0 on top, the digit on the bottom is written as the answer.

$$
\begin{array}{r}
7\ 0\ 9 \\
-\ 3\ 5\ 2 \\
\hline
4\ 5\ 7
\end{array}
\qquad
\begin{array}{r}
6\ 0\ 0\ 8 \\
-\ \ \ 3\ 2\ 7 \\
\hline
6\ 3\ 2\ 1
\end{array}
$$

6. **0 − N = 0.** Whenever there is 0 on top, 0 is written as the answer.

$$
\begin{array}{r}
8\ 0\ 4 \\
-\ 4\ 6\ 2 \\
\hline
4\ 0\ 2
\end{array}
\qquad
\begin{array}{r}
3\ 0\ 5\ 0 \\
-\ \ \ 6\ 2\ 1 \\
\hline
3\ 0\ 3\ 0
\end{array}
$$

7. **N − 0 = 0.** Whenever there is 0 on the bottom, 0 is written as the answer.

$$
\begin{array}{r}
9\ 7\ 6 \\
-\ 3\ 0\ 2 \\
\hline
6\ 0\ 4
\end{array}
\qquad
\begin{array}{r}
8\ \overset{7}{\cancel{8}}_{,}6 \\
-\ 4\ 0\ 9 \\
\hline
4\ 0\ 7
\end{array}
$$

8. **Don't-Decrement-Zero.** When borrowing from a column in which the top digit is 0, the student rewrites the 0 as 10, but does not change the 10 to 9 when incrementing the active column.

$$
\begin{array}{r}
\overset{6}{\cancel{7}}\ 0_{,}2 \\
-\ 3\ 6\ 8 \\
\hline
3\ 4\ 4
\end{array}
\qquad
\begin{array}{r}
\overset{1}{\cancel{2}}_{,}0_{,}5 \\
-\ \ \ \ 9 \\
\hline
1\ 1\ 0\ 6
\end{array}
$$

9. **Zero-Instead-Of-Borrow.** The student writes 0 as the answer in any column in which the bottom digit is larger than the top.

$$
\begin{array}{r}
3\ 2\ 6 \\
-\ 1\ 1\ 7 \\
\hline
2\ 1\ 0
\end{array}
\qquad
\begin{array}{r}
5\ 4\ 2 \\
-\ 3\ 8\ 9 \\
\hline
2\ 0\ 0
\end{array}
$$

10. **Borrow-From-Bottom-Instead-Of-Zero.** If the top digit in the column being borrowed from is 0, the student borrows from the bottom digit instead (Note: this bug must be combined with either bug 5 or bug 6).

$$
\begin{array}{r}
7_{,}0_{,}2 \\
-\ 3\ \overset{7}{\cancel{8}}\ 8 \\
\hline
4\ 5\ 4
\end{array}
\qquad
\begin{array}{r}
5\ 0_{,}8 \\
-\ 4\ \overset{7}{\cancel{8}}\ 9 \\
\hline
1\ 0\ 9
\end{array}
$$

Figure 3.13 Descriptions and examples of Brown and Burton's (1978) common subtraction bugs. (Adapted from Resnick, 1982. Copyright 1982 by Lawrence Erlbaum Associates. Reprinted by permission.)

argued that one of the important tasks of primary school arithmetic learning is the development of knowledge structures that provide a ''semantic justification'' for procedures of written borrowing and carrying. As we have seen earlier in this discussion, there is evidence that children have or can relatively easily acquire substantial semantic knowledge—in the form of Part–Whole and Trade schemata and associated procedures—applied to concrete representations of number. It therefore seems likely that a useful method for assisting children in the development of a semantic interpretation of written arithmetic would be to call their attention to correspondences between the steps in written arithmetic and the performance of addition and subtraction with concrete materials (cf. Dienes, 1966). In an earlier work (Resnick, 1981) I described one method for doing this, via what was termed *mapping instruction*. In this instruction the child is required to perform the same problem in blocks and in writing, alternating steps between the two. Under these conditions the written notations can be construed as a

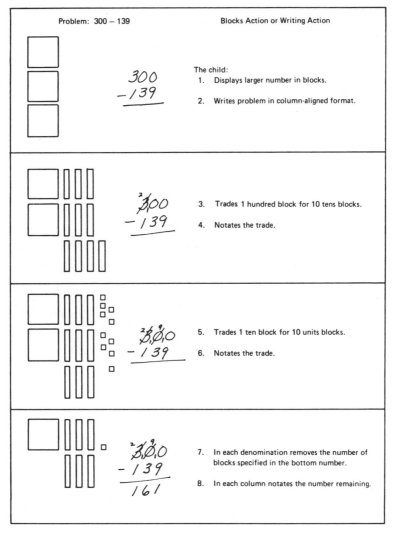

Figure 3.14 Outline of mapping instruction for subtraction. (From Resnick, 1982. Copyright 1982 by Lawrence Erlbaum Associates. Reprinted by permission.)

"record" of actions on the blocks. Figure 3.14 summarizes the process for a subtraction problem.

Mapping instruction has been successfully used with several children who had buggy subtraction algorithms. Not only did their bugs disappear, but the children demonstrated that they had acquired an understanding of the semantics of the written algorithm. Once again, Molly's performance and our simulation of it provide both a clear example of typical behavior and a theoretical account of the mental processes involved.

VALUE OF CARRY AND BORROW MARKS

We have seen that rather early in their development children can recognize the values of digits in various columns of standard notation, using the Next structure only. There is evidence in our data, however, that this ability to assign value does not extend to the notations made in the course of carrying and borrowing. In one of our studies, third-grade children were asked to tell us the value of the carry and borrow digits in written addition and subtraction. In virtually every case they simply named the digit rather than its actual value. For example, when they were shown the solved problem in Figure 3.15, the little 1 at *a* was assigned a value of 1 instead of 10, and the little 1 at *b* was assigned a value of 1 instead of 100. When asked to select the block(s) that would represent these 1 marks, the children typically selected a single units block. By contrast, after instruction Molly and others who had been taught via mapping assigned a value of 10 to the 1 at *a* and 100 to the 1 at *b,* and selected blocks accordingly.

EXPLAINING THE WRITTEN BORROWING ALGORITHM

Molly's most stunning display of understanding written borrowing came in a follow-up interview about four weeks after instruction. During this time she had had no direct instruction on subtraction. When asked to do problems in writing in this follow-up interview, Molly did not use exactly the procedure she had learned from us. That is, on problems with 0 in the top number, she did not begin by decrementing in the hundreds column and changing the 0 in the tens

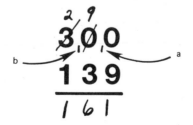

Figure 3.15 Solved problem showing carry and borrow marks.

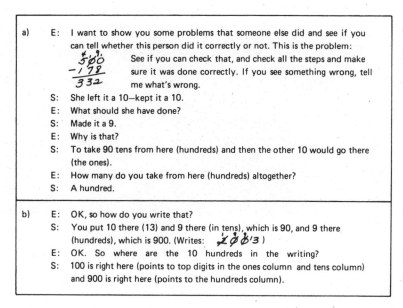

Figure 3.16 Two extracts of Molly's explanations.

column to 10, then decrementing this 10 to produce 9 as part of the exchange into the units column. Instead, she used the "school algorithm," going right to left and changing each 0 directly to 9.

This algorithm cannot be directly mapped onto blocks, and thus one cannot explain why it works by simply describing exchanges as if they had been done with blocks. Thus, any justification Molly was able to offer for her written work would have to depend on her schematic knowledge. Figure 3.16 gives two extracts of Molly's explanations. In the first case Molly was asked to check another child's work. She knew the 10 in the tens column should be changed to 9, but she did not justify this as the outcome of a trade. Instead, she gave an explanation in terms of the values of the decrement and increment marks (9 tens in the tens column plus 1 ten in the units column), with the clear implication that a whole-preserving exchange had been made (otherwise she would not have sought the "other ten"). In the second extract, Molly shows even more clearly that she was searching for parts to make up the 1000 that she recognized had been borrowed in the course of decrementing the thousands column.

MOLLY-3 provides a theory of how these explanations were constructed. To construct analogous explanations, MOLLY-3 uses an Exchange schema (Figure 3.17) that develops by interpreting borrowing as an analog of trading. The Trade and the Borrow portions of the Exchange schema have analogous elements. As a result, for written borrowing there is a From column that gets smaller by 1 and an Into column that gets larger by 10. The *values* of these

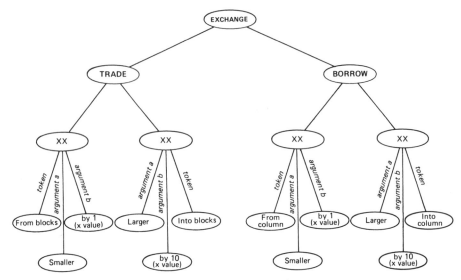

Figure 3.17 The Exchange schema.

decrements and increments are, as in the case of trading, determined by multiplying by the column value. In the units column, the increase of 10 in the Into column is multiplied by 1; but in the tens column it is multiplied by 10 to yield a value of 100. As a result, when interpreted under control of the Exchange schema, the increment marks would be represented by tens or hundreds blocks, never by unit cubes. The effect of having the Exchange schema is to allow MOLLY-3 to interpret borrowing as it had trading: as an exchange among parts that maintains the value of the whole.

MOLLY-3 uses its newly constructed Exchange schema to construct explanations for the standard school borrowing algorithm that parallel those of Molly. For example, for the problem 403 − 275, MOLLY-3 handles several questions about increments and decrements as follows: It keeps track of its actions by building a temporary Changes structure that specifies old and new values in particular columns. The Changes structure also records whether the new value is larger or smaller than the original. Faced with the question, *Where did the 13 in the units column come from?*, the program examines its current Changes structure, searching for a 13 as a *new value* in the units column. Finding this, it can determine that the 13 is *larger* than the original value for that column. Now it looks for a place in its knowledge where *larger* is linked with a column *into* which something is added. It finds the Borrow schema. It instantiates this schema, with the units column as the Into column. It can then "read out" the answer from the instantiated schema as: *It comes from borrowing 1 ten from the tens column for the units column.*

Now given the question, *Where is the 100 you borrowed from the hundreds column?*, MOLLY-3 uses its Changes structure to determine that the hundreds column is *smaller*. As a result, it searches for a structure in its own knowledge base in which a column is made *smaller* by taking something *from* it. This leads it to the Borrow schema, which it activates and tries to instantiate. It fills the From column with the hundreds column, and it knows this column has gotten smaller by one times the column value (of 1000). It must now fill the slots on the Into side. To do this it tries at first to find a column made *larger* by 10 times a column value of 10, but it cannot find such a column in the written notation. Instead, it finds a value of 90 shown in the tens column. At this point it calls on the Part–Whole schema, sets the Whole equal to 100 and Part A equal to 90. From this it can determine that Part B must equal 10. Now it inspects the written notation again, looking for a column that shows an increment with a value of 10. It is able to find this in the units column of the written notation. As a result it can conclude that: *The 100 from the hundreds column has been made into the 90 in the tens column plus the 10 in the units column.* MOLLY-3 can answer the analogous question for borrowing across two zeros (for example, when 2003 is the top number in a problem) by iterating through the Part–Whole schema twice, first setting the Whole slot equal to 1000 and Part A to 900, then setting the Whole to 100 and Part A to 90. It then answers: *The 1000 from the thousands column has been made into the 900 in the hundreds column, plus the 90 in the tens column, plus the 10 in the units column.*

CONCLUSION

Other topics in mathematics (multiplication, division, fractions), of course, will have been introduced by the end of the early school years and will have induced changes in representation not considered here. Nor can it be expected that all children by the end of primary school will have achieved the level of understanding represented by Molly. Yet such understanding is certainly an important goal of early instruction in place value. Thus, it seems a suitable point at which to conclude this account of the cognitive development that accompanies early school arithmetic learning. What general conclusions about the nature of number understanding and its development can be drawn from this account?

The Centrality of the Part–Whole Schema in Number Understanding

First, it seems clear that a reasonable account of the knowledge underlying changing mathematics competence can be given in terms of a few schemata and their successive elaborations. As we have seen, the Part–Whole schema plays a central role. Although I have not attempted here to explain the origin of the Part–Whole schema, it seems likely that it arises in connection with various real-life situations in which partitions must be made but no exact quantification is required. Such situations are easy to imagine in the life of the young child. For example, a hole in an otherwise complete puzzle means that a *part* is missing; food is shared with the recognition that the individual portions together represent *all* (the *whole*) that is available; or a child gives *some* (but not *all*) of her candy to her brother.

I have pointed to evidence that Part–Whole in this primitive form is available to children before school begins. I have also suggested that its systematic application to *quantity* characterizes the early years of school. A first elaboration of the basic Part–Whole schema, in this view, is its attachment to procedures for counting up (the procedures attached to the Minimum Needed node) and taking away (the procedures attached to the Maximum Exceeded node). These procedures, which are based on the units number string, produce a quantitative interpretation of Part–Whole. The schema in turn allows numbers to be interpreted *both* as positions on the mental number line *and, simultaneously,* as compositions of other numbers. This interpretation of number appears to underlie both story–problem solution and the invented mental arithmetic procedures for small numbers that characterize the earliest school years.

Further elaboration of the Part–Whole schema appears to characterize subsequent development of an understanding of the place-value system of notation and the calculation procedures based on it. Children apparently find it easy to place a special restriction on Part–Whole such that one of the parts must always be a multiple of 10. This initial elaboration generates an interpretation of multidigit numbers as compositions of units, tens, hundreds, and so on. This in turn permits invention of several quite elegant mental calculation shortcuts. However, further elaborations—those specified in the Trade and later the more abstract Exchange schemata—are required before multiple partitionings of quantity can be recognized and the rules of written arithmetic interpreted. Since Trade and Exchange are always called upon by Part–Whole, it seems reasonable to view them as elaborations of the more general schema for partitioning quantity.

Microstages in Development

Many readers will have noted parallels between the analysis offered here and interpretations of the number concept proposed by Piaget and others working in the Genevan tradition. Indeed, this analysis shares two central emphases with the Piagetian view: (*a*) an emphasis on part–whole (class inclusion, for Piaget) relationships as a defining characteristic of number understanding, and (*b*) the proposal that ordinal (counting) and cardinal (class inclusion or part–whole) relationships must be combined in the course of constructing the concept of number.

It is especially pleasing to have arrived at this convergence because the present analysis was conducted quite independently of Piaget's work. I did not set out to either support or disconfirm Piaget's theory of number understanding but rather to build a plausible account, from a current cognitive science point of view, of what number knowledge must underlie the various arithmetic performances observed in young school children. In doing this, I drew on formal theoretical analyses that worked *from* task performances *to* the kind of knowledge children "must have" in order to engage in the performances observed. This effort to build a theory of understanding on the basis of detailed analyses of procedures used in performing tasks is quite different from the Piagetian method of hypothesizing a mental structure and then seeking tasks that might reveal its presence or absence. One might well characterize the methods used here as more bottom-up than those of Piaget.

One result of these more bottom-up task- and performance-driven methods is that we are able to detect—indeed, are forced to recognize—relatively small changes in cognitive structures. In a sense, we have been able to produce a *microstage* theory for number understanding, a theory that specifies many small changes in number representation and schematic interpretation of number in a period of development for which the Piagetian analysis recognized only the *macrostages* of preoperativity and concrete operativity. This enriched theory of changes in number knowledge is of clear importance to those concerned with instruction, for it specifies "what to teach" at successive stages of learning or development. The microstages of understanding developed here also permit us to give a more precise psychological interpretation to certain key mathematical concepts than has heretofore been possible.

An Interpretation of Cardinality

One example of such interpretation is the one that is now possible for the development of an understanding of cardinality. Gelman and Gallistel (1978) included in their principles of counting a cardinality principle, which specifies that the final count word reached when a set of objects is being enumerated is the total number in the set—that is, the set's cardinality. For the preschool child, who has not yet come to interpret quantity in terms of a fully developed Part–Whole schema, this is the only meaning of cardinality available. This criterion of understanding cardinality has been criticized (e.g., Bessot & Comiti, 1981) as too weak and in particular as not reflecting the Piagetian definition of cardinality. We can now see that a higher stage of cardinality understanding can be recognized in the child's subsequent application of the Part–Whole schema to number. Although a primitive form of partitioning is clearly present in early counting behavior (this is what is required to keep counted and not-yet-counted objects separate), the Part–Whole schema used later in solving story problems yields the understanding that a total (whole)quantity remains the same even under variant partitionings.

The meaning of cardinality is further elaborated when the place-value schemata outlined here are acquired. When the Part–Whole schema with the multiple-of-10 restriction is applied to two-digit numbers, the amount represented by the number becomes subject to multiple partitioning without a change in quantity. This is exactly parallel to the new understanding of cardinality for smaller numbers that was achieved when the Part–Whole schema was applied to them. Without application of the Part–Whole schema, the cardinality of a number resides in the specific display set and the number attached to it through legal counting procedures. With Part–Whole, cardinality resides in the total quantity, no matter how it is displayed or partitioned.

The Trade and Exchange stages of multidigit number representation show yet a higher level of understanding of cardinality. At these stages it is recognized that cardinality is not altered by a specified set of legal exchanges. An analogy can be drawn with the earlier recognition of quantity as unchanged under various physical transformations (such as spreading out a display of objects—the classic Piagetian test of conservation). However, the transformations produced under control of the Trade schema do in fact involve a change in the actual number of *objects* present. Thus, recognition that the value of the total quantity remains unchanged requires a level of abstraction concerning the nature of cardinality that was not required for earlier stages of understanding.

Procedural Knowledge and Understanding

An important characteristic of the account of number development offered here is the close link between procedural skill and understanding. It has been characteristic of many past efforts to promote understanding of mathematics to speak as if understanding and procedural skill were somehow incompatible. Wertheimer (1945/1959), for example, in pressing for structural understanding as the goal of education, attacked the teaching of algorithms and other aspects of "mindless drill." Piaget, too, was largely disinterested in procedural skills, despite the role that "reflective abstraction"—the process of reflecting on one's own procedures to draw out principles—plays in his theory of development (Piaget, 1967/1971). Many educators inspired by Piaget's emphasis on understanding have actively argued against any kind of procedural emphasis in mathematics instruction.

The present analyses, by contrast, suggest that procedural skill often underlies understanding. For example, the account proposed here for the invention of the *min* and *choice* calculation procedures suggests that inventions reflecting an understanding of number can come about only when procedures become well enough established that their results can be inspected and compared. Similarly, children apparently learn about the decade structure of the number system through what must be, at first, rather "mindless" repetition of conventional counting strings.

We do not yet have a full theory to propose about exactly *how* practice in counting and other arithmetic procedures interacts with existing schematic knowledge to produce new levels of understanding. Nevertheless, it already seems clear that a detailed theory of how new levels of number understanding are achieved will reveal active interplay between schematic and procedural knowledge.

REFERENCES

Anderson, J. R. (Ed.) *Cognitive skills and their acquisition*. Hillsdale, N. J.: Lawrence Erlbaum Associates, 1981.

Ashcraft, M. H., & Battaglia, J. Cognitive arithmetic: Evidence for retrieval and decision processes in mental addition. *Journal of Experimental Psychology: Human Learning and Memory*, 1978, *4*(5):527–538.

Ashcraft, M. H., & Fierman, B. A. Mental addition in third, fourth, and sixth graders. *Journal of Experimental Child Psychology*, 1982, *33*(2); 216–234.

Bessot, A., & Comiti, C. Etude du fonctionnement de certaines propriétés de la suite des nombres dan le domaine numérique [1,30] chez des élèves de fin de première année de l' école obligatoire en France. *Proceedings of the Fifth Conference of the Internationl Group for the Psychology of Mathematics Education*. Grenoble, France: July, 1981. Available from Dean Robert Karplus, College of Education, University of California, Berkeley, CA 94720.

Briars D. J., & Larkin, J. H. *An integrated model of skill in solving elementary word problems*. Paper presented at the annual meeting of the Society for Research in Child Development, Boston, April 1981.

Brown, J. S. & Burton, R. R. Diagnostic models for procedural bugs in basic mathematical skills. *Cognitive Science*, 1978, *2*,(2), 155–192.

Brown, J. S., & VanLehn, K. Repair theory: A generative theory of bugs in procedural skills. *Cognitive Science*, 1980 *4*(4), 379–426.

Brown, J. S. & VanLehn, K. Toward a generative theory of bugs in procedural skills. In T. Carpenter, J. Moser, & T. Romberg (Eds.), *Addition and subtraction: A cognitive perspective*. Hillsdale, N.J.: Lawrence Erlbaum Associates, 1982.

Carpenter, T., & Moser, J. The development of addition and subtraction problem-solving skills. In T. Carpenter, J. Moser, & T. Romberg (Eds.), *Addition and subtraction: A cognitive perspective*. Hillsdale, N.J.: Lawrence Erlbaum Associates, 1982.

Carpenter, T. P., Hiebert, J., & Moser, J. M. The effect of problem structure on first graders' initial solution processes for simple addition and subtraction problems. *Journal for Research in Mathematics Education*, 1981, *12*(1), 27–39.

Comiti, C. Les premières acquisitions de la notion de nombre par l'enfant. *Educational Studies in Mathematics*, 1981, *11*, 301–318.

Dienes, Z. P. *Mathematics in the primary school.* London: MacMillan, 1966.

Donaldson, M., & Balfour, G. Less is more: A study of language comprehension in children. *British Journal of Psychology*, 1968, *59*, 461–471.

Fuson, K. C. The counting-on solution procedure: Analysis and empirical results. In T. Carpenter, J. Moser, & T. Romberg (Eds.), *Addition and subtraction: A cognitive perspective.* Hillsdale, N.J.: Lawrence Erlbaum, Associates, 1982.

Fuson, K. C. & Mierkiewicz, D. B. *A detailed analysis of the act of counting.* Paper presented at the annual meeting of the American Education Research Association, Boston, April 1980.

Fuson, K. C., Richards, J., & Briars, D. J. The acquisition and elaboration of the number word sequence, In C. Brainerd (Ed.), *Progress in logical development: Children's logical and mathematical cognition* (Vol. 1). New York: Springer-Verlag, 1982.

Gelman, R. Logical capacity of very young children: Number invariance rules. *Child Development,* 1972, *43,* 75–90.

Gelman, R. & Gallistel, C. R. *The child's understanding of number.* Cambridge Mass: Harvard University Press, 1978.

Ginsburg, H. *Children's arithmetic: The learning process.* New York: D. Van Nostrand, 1977.

Greeno, J. G. Riley, M. S., & Gelman R. *Young children's counting and understanding* Paper presented at the annual meeting of the Psychonomic Society, San Antonio November 1978.

Groen, G. J., & Parkman, J. M. A chronometric analysis of simple addition. *Psychological Review,* 1972, *79*(4), 329 –343.

Groen, G. J., & Poll, M. Subtraction and the solution of open sentence problems. *Journal of Experimental Child Psychology,* 1973, *16,* 292–302.

Groen, G. J., & Resnick, L. B. Can preschool children invent addition algorithms? *Journal of Educational Psychology, 1977, 69,* 645–652.

Houlihan, D. M., & Ginsburg, H. P. The addition methods of first- and second-grade children. *Journal for Research in Mathematics Education,* 1981, *12*(2), 95–106.

Inhelder, B., & Piaget, J. *The early growth of logic in the child.* New York: Norton, 1969. (Original English translation 1964).

Klahr, D., & Wallace, J. G. *Cognitive development: An information-processing view.* Hillsdale, N.J.: Lawrence Erlbaum Associates, 1976.

Lindvall, C. M., & Gibbons-Ibarra, C. G. *A clinical investigation of the difficulties evidenced by kindergarten children in developing "models" for the solution of arithmetic story problems.* Paper presented at the annual meeting of the American Educational Research Association, Boston, April 1980.

Markman, E. M., & Siebert, J. Classes and collections: Internal organization and resulting holistic properties. *Cognitive Psychology,* 1976, *8,* 561–577.

Neches, R. *Models of heuristic procedure modification.* Unpublished doctoral dissertation, Carnegie-Mellon University, Department of Psychology, Pittsburgh, Pa., 1981.

Nesher, P. Levels of description in the analysis of addition and subtraction word problems. In T. Carpenter, J. Moser, & T. Romberg (Eds.), *Addition and subtraction: A cognitive perspective.* Hillsdale, N.J.: Lawrence Erlbaum Associates, 1982.

Nesher, P., & Teubal, E. Verbal cues as an interfering factor in verbal problem solving. *Educational Studies in Mathematics,* 1975, *6,* 41–51.

Newell, A., & Simon, H. A. *Human problem solving.* Englewood Cliffs, NJ: Prentice-Hall, 1972.

Piaget, J. *The child's conception of number.* New York: Norton, 1965. (Originally published 1941).

Piaget, J. *Biology and knowledge.* Chicago: University of Chicago Press, 1971. (Originally published 1967).

Potts, G. P., Banks, W. P., Kosslyn, S. M., Moyer, R. S., Riley, C. A., & Smith, K. H. Encoding and retrieval in comparative judgments. In J. N. Castellan (Ed.), *Cognitive theory* (Vol. 3). Hillsdale, NJ: Lawrence Erlbaum Associates, 1979.

Resnick, L. B. Syntax and semantics in learning to subtract. In T. Carpenter, J. Moser, & T. Romberg (Eds.), *Addition and subtraction: A cognitive perspective.* Hillsdale, NJ: Erlbaum, 1982.

Resnick, L. B., Greeno, J. G., & Rowland, J. *MOLLY: A model of learning from mapping instruction.* Unpublished manuscript, University of Pittsburgh, Learning Research and Development Center, Pittsburgh, PA, 1980.

Schaeffer, B., Eggleston, V. H., & Scott, J. L. Number development in young children. *Cognitive Psychology,* 1974, *6,* 357–379.

Sekuler, R., & Mierkiewicz, D. Children's judgment of numerical inequality. *Child Development,* 1977, *48,* 630–633.

Siegler, R. S., & Robinson, M. The development of numerical understanding. In H. W. Reese & L. P. Lipsitt (Eds.), *Advances in child development and behavior* (Vol. 16). New York: Academic Press, 1982.

Smith, L. B., & Kemler, D. G. Levels of experienced dimensionality in children and adults. *Cognitive Psychology,* 1978, *10,* 502–532.

Steffe, L., Thompson, P., & Richards, J. Children's counting and arithmetical problem solving. In T. Carpenter, J. Moser, & T. Romberg (Eds.), *Addition and subtraction: A cognitive perspective.* Hillsdale, NJ: Erlbaum, 1982.

Svenson, O., & Broquist, S. Strategies for solving simple addition problems: A comparison of normal and subnormal children. *Scandinavian Journal of Psychology,* 1975, *16,* 143–151.

Svenson, O., & Hedenborg, M. L. Strategies used by children when solving simple subtractions. *Acta Psychologica,* 1979, *43,* 1–13.

Svenson, O., Hedenborg, M. L., & Lingman, L. On children's heuristics for solving simple additions. *Scandinavian Journal of Educational Research,* 1976, *20,* 161–173.

Svenson, O., & Sjoberg, K. Solving simple subtractions during the first three school years. *Journal of Experimental Education,* in press.

Vergnaud, G. A classification of cognitive tasks and operations of thought involved in addition and subtraction problems. In T. Carpenter, J. Moser, & T. Romberg (Eds.), *Addition and subtraction: A cognitive perspective.* Hillsdale, N.J.: Lawrence Erlbaum Associates, 1982.

Wertheimer, M. *Productive thinking* (Enlarged ed.). New York: Harper & Row, 1959. (Originally published in 1945.)

Woods, S. S., Resnick, L. B., & Groen, G. J. An experimental test of five process models for subtraction. *Journal of Educational Psychology,* 1975, *67*(1), 17–21.

MARY S. RILEY
JAMES G. GREENO
JOAN I. HELLER

CHAPTER **4**

Development of Children's Problem-Solving Ability in Arithmetic[1]

This chapter is concerned with the development of an important aspect of children's problem-solving skill in arithmetic—the ability to solve arithmetic word problems. There are several factors that might enable older children to perform better in problem-solving tasks than younger children, including the complexity of conceptual knowledge about the problem domain and the sophistication of problem-solving procedures. The studies reviewed here suggest that, with age, children's improved ability to solve word problems primarily involves an increase in the complexity of conceptual knowledge required to understand the situations described in those problems. We will describe these findings in this chapter and consider some general issues about the development of problem-solving skill.

CONCEPTUAL AND PROCEDURAL KNOWLEDGE IN PROBLEM SOLVING

One major issue we will address concerns the relationship between conceptual and procedural knowledge in performance and development. In the past few years much has been learned about how improvements in either conceptual

[1]The authors' research reported herein was supported by the Learning Research and Development Center and, in part, by funds from the National Institute of Education (NIE), the United States Department of Health, Education and Welfare. The opinions expressed do not necessarily reflect the position or policy of NIE, and no official endorsement should be inferred.

153

knowledge (e.g., Chi, 1978; Gentner, 1975; Stein & Trabasso, 1981) or problem-solving procedures and strategies (e.g., Baylor & Gascon, 1974; Brown, 1978; Groen & Resnick, 1977; Klahr & Robinson, 1981; Young & O'Shea, 1981) separately contribute to improvements in performance as children get older. However, most tasks that children perform involve an interaction between both knowledge and procedures, and the nature of this interaction is a significant theoretical question for models of cognitive development. Some progress is being made. Recent studies by Siegler (1976) and Siegler and Klahr (1981) have identified the importance of understanding the relevant features and relations of a particular task in the acquisition of more advanced problem-solving procedures. Chi (1982) discusses the interactive role of domain-specific knowledge and various kinds of strategies in affecting children's memory performance. Also, Resnick (1981; Chapter 3, this volume) has related increased knowledge of relevant goals and constraints in subtraction to improvements in children's computational procedures using blocks. This knowledge can then be used to learn similar procedures in another context where these goals and constraints are less salient, as in the formal syntax of arithmetic. The theoretical and empirical studies of word problem solving that we will describe provide further analyses of interactions between conceptual and procedural knowledge.

A second and related issue concerns the knowledge (either conceptual or procedural) that we attribute to children on the basis of their problem-solving performance. Frequently children are said to understand a concept if their performance on some task is consistent with that concept. Children whose performance is inconsistent are said to lack understanding. We will argue that such an all-or-none view of children's understanding is too limiting. Our argument will be based on two lines of evidence. First, children who appear to lack understanding of a concept on one task often show performance that is consistent with that concept on other tasks (e.g., Gelman & Gallistel, 1978; Trabasso, Isen, Dolecki, McLanahan, Riley, & Tucker, 1978), thus implying some understanding of the concept. Second, even when several children perform successfully on the same problem-solving task, this does not necessarily imply that they share the same underlying knowledge (e.g., Dean, Chabaud, & Bridges, 1981). For example, children may differ in their representations of the problem, and this can affect the kinds of procedures required for solution, as well as the ability to solve related problems. We will discuss some recent theoretical analyses of children's understanding that provide explicit descriptions of the knowledge underlying different stages of problem-solving skill.

Our discussion of these issues will be based on the development of a specific hypothesis about the nature of children's skill in solving arithmetic word problems. According to this hypothesis, improvement in performance results mainly from improved understanding of certain conceptual relationships. This is not to say that knowledge of formal arithmetic lacks importance for children.

Indeed, an important possibility is that acquisition of certain conceptual structures depends upon the knowledge of formal arithmetic that children acquire through school instruction. However, we are unable to conceptualize knowledge of formal arithmetic in a way that makes it sufficient for solving word problems. Furthermore, problems with the same arithmetic structure but different conceptual structures differ substantially in their difficulty for children. We take this as evidence that conceptual understanding is required if the texts of word problems are to be mapped onto arithmetic relationships and operations. Our goal is to emphasize the importance of informal concepts in problem solving and to provide a detailed analysis of informal concepts that are needed for a class of arithmetic word problems.

APPROACHES TO ANALYZING KNOWLEDGE IN PROBLEM SOLVING

The analyses we will describe have been influenced to a large extent by recent cognitive theories of problem solving and language understanding. These theories have provided increasingly rigorous concepts and methods for understanding the knowledge underlying problem-solving performance. Early analyses of problem solving around 1950 focused on fairly general connections between actions performed during problem solving. Behaviorists (e.g., Maltzman, 1955) and associationists (e.g., Underwood & Richardson, 1956) analyzed solutions of problems using concepts such as strength of associations and competition between responses. More recent information-processing analyses (e.g., Newell & Simon, 1972) have provided more specific concepts and more rigorous methods for analyzing performance in problem-solving situations. We now can analyze the cognitive processes required for solving problems in considerable detail, providing hypotheses about specific cognitive procedures as well as the general strategies that are involved in successful performance.

Recent developments in cognitive theory have also made possible a more detailed and rigorous analysis of the role of conceptual knowledge in problem solving. The importance of conceptual knowledge for understanding and representing problems has long been recognized. Gestalt theorists such as Duncker (1945), Kohler (1927), and Wertheimer (1945/1959) conceptualized solution of a problem as achievement of understanding the problem as a whole and as the relations of problem elements and solution procedures to the whole. A primary contribution of recent cognitive theories has been the development of concepts and methods that allow more specific hypotheses about the conceptual knowledge required to solve complex problems in domains such as physics (McDermott & Larkin, 1978; Novak, 1976) and high school geometry (Anderson, Greeno, Kline, & Neves, 1981; Greeno, 1977, 1978). An important theoretical resource in this development has come from cognitive theories of language understanding (e.g., Anderson, 1976; Norman & Rumelhart, 1975; Schank & Abelson, 1977). In these theories, understanding a sentence or a story is viewed as the construction of a coherent representation of the various elements in the message, with individual elements interconnected in a network of relationships. Understanding in problem solving is characterized similarly as a process of representing problem information or solution components in coherent relational networks constructed on the basis of general conceptual knowledge.

We now sketch the contents of the remaining sections of this chapter. The next section characterizes a class of addition and subtraction word problems. We review the major factors that have been used to characterize aspects of word problems, including their relative difficulty. The next three sections concern the knowledge underlying children's performance on these problems. The section entitled ''A Theory of the Knowledge Required to Solve Word Problems'' presents a theoretical analysis of the knowledge and strategies that we hypothesize to be involved in successful performance. The section on the locus of improvement in problem-solving skill presents evidence that even very young children have available a range of strategies for solving word problems, but they differ from older children in the conceptual knowledge required to apply those strategies. The section on the stages of conceptual knowledge specifies some of these differences in conceptual knowledge as they relate to differences in the success, efficiency, and generality of problem-solving performance. In the section entitled ''Related Analyses of Conceptual Understanding in Problem-Solving we present a brief summary of analyses of word problem solving in domains other than elementary arithmetic. Finally, the discussion section relates children's performance on word problems to more general issues in developmental theory and methodology.

REVIEW OF RESEARCH ON CHILDREN'S WORD PROBLEM SOLVING

Our review of research on addition and subtraction word problems and factors that influence their difficulty has two main sections. First, we summarize findings of studies concerned with global factors. These include studies of general structural features, such as the grammatical complexity of the problem statement, and studies concerned with the effect of having materials such as blocks available as aids for problem representation. Second, we review literature on more detailed semantic factors. We present a survey of analyses that have categorized problems based on semantic relationships among quantities in the problem situation. Then we summarize empirical studies that have compared the difficulty of problems differing in their semantic characteristics.

Global Factors

STRUCTURAL FEATURES OF PROBLEM STATEMENTS

Several studies have considered general surface characteristics of problem statements as factors influencing problem difficulty. Variables such as problem length, grammatical complexity, and order of problem statements have been shown to have significant effects on ease of solution (Jerman, 1971, 1973–1974; Jerman & Rees, 1972; Loftus, 1970). Regression analyses indicate that a large proportion of variance in problem difficulty can be accounted for by these factors (Loftus & Suppes, 1972).

The type of number sentence represented by the relations among quantities in the problem has also been related to problem difficulty. (Grouws, 1972; Lindvall & Ibarra, 1980a; Rosenthal & Resnick, 1974). Problems represented by sentences where the unknown is either the first (? + a = b) or second (a + ? = b) number are more difficult than problems represented by equations where the result is the unknown (a + b = ?). Rosenthal and Resnick provided an explicit model to account for these differences in difficulty. Their model focused on the process of translating the problem text into an equation, and difficulty was predicted as a function of the number and kinds of transformations required to translate the equation into its canonical form for solution (e.g., either a + b = ? or a − b = ?).

CONCRETE AIDS

Another factor that has been examined is the availability of concrete materials such as blocks as aids in solving word problems. Several studies have found that the availability of blocks (e.g., Bolduc, 1970; Hebbeler, 1977; LeBlanc, 1968; Steffe, 1968, 1970; Steffe & Johnson, 1971) and/or reference dolls or pictures (e.g., Harvey, 1976; Ibarra & Lindvall, 1979; Marshall, 1976; Shores & Underhill, 1976) facilitate solution of problems, particularly for young children. As an illustration, Tables 4.1 and 4.2 show data from children's performance with and without concrete objects. The exact nature of the "change" and "combine" problem types in these tables will be discussed later in this section. The data in Table 4.1 are from a study by Riley (Greeno & Riley, 1981; Riley, 1981) in which kindergarten children solved problems that described a change in some quantity. Table 4.2 shows data from a study in which Steffe and Johnson (1971) asked first-graders to solve problems like those in Table 4.1, as well as problems involving combinations of quantities. The point to be made here is that in both studies there was a general improvement in children's performance when they used objects to solve problems. The fact that concrete aids did not facilitate

Table 4.1
Proportions of Kindergartners Who Performed Correctly on Arithmetic Tasks[a]

Problem type	Without objects	With objects
Change (1)	.70	.87
Change (2)	.61	1.00
Change (3)	.22	.61
Change (4)	.30	.91
Change (5)	.09	.09
Change (6)	.17	.22

[a]From Riley, 1981.

kindergartners' performance on change (5) and change (6) problems relates to some interesting theoretical issues that we will discuss later.

Additional support for the facilitation effect of objects was obtained by Carpenter, Hiebert, and Moser (1981) who showed that, given a choice between solving word problems with or without blocks, first-graders preferred to use blocks. Some data indicate that merely observing (not manipulating) concrete representations of problem solutions improves word problem performance (e.g., Buckingham & MacLatchy, 1930; Gibb, 1956; Ibarra & Lindvall, 1979).

OTHER GLOBAL FACTORS

In addition to the factors we have discussed here, the ability of individual children to solve word problems has been studied in relation to their general reading ability and various instructional methods. Extensive reviews of this research have been provided by Aiken (1971, 1972) and Barnett, Vos, and Sowder (1979).

Table 4.2
Proportions of First-Graders Who Performed Correctly on Arithmetic Tasks[a]

Problem type	Without objects	With objects
Change (1)	.67	.85
Change (2)	.43	.61
Change (3)	.41	.67
Change (5)	.41	.58
Combine (1)	.67	.80
Combine (2)	.35	.55

[a]From Steffe & Johnson, 1971.

Specific Factors

Although the analyses of global characteristics of word problems has provided a basis for reasonably accurate predictions of problem difficulty, significant differences have been found between problems for which these factors are held constant (e.g., Gibb, 1956; LeBlanc, 1968; Schell & Burns, 1962). Furthermore, the effect of certain words, such as *altogether* and *less*, has been shown to depend upon whether the operation suggested by the word matches the operation required for problem solution (e.g., Dahmus, 1970; Jerman, 1971; Linville, 1976; Nesher & Teubal, 1974). For these reasons and others, many recent analyses of word problems have focused on specific problem characteristics involving the relationships among quantities described in the problem. The main finding from these analyses is that the understanding of quantitative relationships in problems involves factors other than the arithmetic equations that express the relationships; the conceptual structure of the problem must also be taken into account. In the remainder of this section, we review the kinds of conceptual relations that describe simple addition and subtraction problems; we also review empirical studies of differences in problem difficulty that are associated with these semantic variables.

PROBLEM TYPES

A word problem identifies some quantities and describes a relationship among them. Table 4.3 shows examples of several kinds of word problems that have been included in various research studies. Each of these problems describes a simple situation involving either addition or subtraction. The categories in Table 4.3 include the change, combine, and compare categories used in an analysis by Heller and Greeno (1978). These categories are representative of categorical schemes that have been used by several investigators (e.g., Carpenter & Moser, 1981; Fuson, 1979; Nesher, 1981; Vergnaud, 1981) in analyses of simple addition and subtraction problems, although the names used to refer to the categories have varied. The equalizing category in Table 4.3 is from the work of Carpenter and Moser (1981).

Semantic Structure.   One of the ways the problems in Table 4.3 differ is in the semantic relations used to describe the problem situation. By *semantic relations* we refer to conceptual knowledge about increases, decreases, combinations, and comparisons involving sets of objects.

The first two problem categories shown in Table 4.3—change and equalizing—describe addition and subtraction as actions that cause increases or decreases in some quantity. For example, in change (1) the initial quantity or *start set* of Joe's three marbles is increased by the action of Tom giving Joe five more marbles (the *change set*). The resulting quantity or *result set* is eight. Equalizing

Table 4.3
Types of Word Problems[a]

Action	Static
CHANGE	COMBINE

CHANGE

Result unknown
1. Joe had 3 marbles.
   Then Tom gave him 5 more marbles.
   How many marbles does Joe have now?
2. Joe had 8 marbles.
   Then he gave 5 marbles to Tom.
   How many marbles does Joe have now?

Change unknown
3. Joe had 3 marbles.
   Then Tom gave him some more marbles.
   Now Joe has 8 marbles.
   How many marbles did Tom give him?
4. Joe had 8 marbles.
   Then he gave some marbles to Tom.
   Now Joe has 3 marbles.
   How many marbles did he give to Tom?

Start unknown
5. Joe had some marbles.
   Then Tom gave him 5 more marbles.
   Now Joe has 8 marbles.
   How many marbles did Joe have in the
   beginning?
6. Joe had some marbles.
   Then he gave 5 marbles to Tom.
   Now Joe has 3 marbles.
   How many marbles did Joe have in the
   beginning?

EQUALIZING
1. Joe has 3 marbles.
   Tom has 8 marbles.
   What could Joe do to have as many
   marbles as Tom?
   (How many marbles does Joe need to
   have as many as Tom?)
2. Joe has 8 marbles.
   Tom has 3 marbles.
   What could Joe do to have as many
   marbles as Tom?

COMBINE

Combine value unknown
1. Joe has 3 marbles.
   Tom has 5 marbles.
   How many marbles do they have
   altogether?

Subset unknown
2. Joe and Tom have 8 marbles altogether.
   Joe has 3 marbles.
   How many marbles does Tom have?

COMPARE

Difference unknown
1. Joe has 8 marbles.
   Tom has 5 marbles.
   How many marbles does Joe have more
   than Tom?
2. Joe has 8 marbles.
   Tom has 5 marbles.
   How many marbles does Tom have less
   than Joe?

Compared quality unknown
3. Joe has 3 marbles.
   Tom has 5 more marbles than Joe.
   How many marbles does Tom have?
4. Joe has 8 marbles.
   Tom has 5 marbles less than Joe.
   How many marbles does Tom have?

Referent unknown
5. Joe has 8 marbles.
   He has 5 more marbles than Tom.
   How many marbles does Tom have?
6. Joe has 3 marbles.
   He has 5 marbles less than Tom.
   How many marbles does Tom have?

[a]From Riley, 1981.

problems involve two separate quantities, one of which is changed to be the same as the other quantity. In equalizing (1) the problem solver is asked to change the amount of Joe's set to be the same as the amount of Tom's set.

The remaining categories—combine and compare—involve static relations between quantities. In combine (1) there are two distinct quantities that do not change—Joe's three marbles and Tom's five marbles—and the problem solver is asked to consider them in combination: *How many marbles do Joe and Tom have altogether?* Compare (1) also describes two quantities that do not change, but this time the problem solver is asked to determine the difference between them: *How many marbles does Joe have more than Tom?* Since in this case Joe's marbles are being compared to Tom's, Joe's marbles are called the *compared set* and Tom's marbles are called the *referent set*. If the question had been *How many marbles does Tom have less than Joe?* then Tom's marbles would have been the compared set and Joe's would have been the referent set.

Identity of the Unknown Quantity. In addition to the various semantic relations, there are other ways in which the problems in Table 4.3 differ. In each kind of problem—change, equalizing, combine, and compare—there are three items of information. Different problems can be formed by varying the items of information given and those to be found by the problem solver. In change problems, the three items of information are the start, change, and result sets. Any of these can be found if the other two are given, yielding three different cases: The unknown may be the start, the change, or the result. Furthermore, the direction of change can either be an increase or a decrease, so there are a total of six kinds of change problems. Change problems involving increases are referred to collectively as change/join problems; change problems involving subtraction are referred to as change/separate problems.

A similar set of variations exists for compare problems, where the direction of difference may be more or less and the unknown quantity may be the amount of difference between the referent set and the compared set, or either of the two sets themselves. Equalizing problems usually restrict the unknown to the difference between the given quantity and the desired quantity, although a total of six variations are possible. In combine problems there are fewer possible variations: The unknown is either the combined set or one of the subsets.

RELATIVE DIFFICULTY

There have been many empirical studies concerned with the relative difficulty of problems similar to those in Table 4.3. The basic procedure usually involves having children individually solve selected problems that are read to them by the experimenter. Memorial and computational difficulties are kept at a minimum by reading the problems slowly, repeating them if necessary, and by restricting the size of the numbers in the problems such that sums are less than

10. In addition, concrete objects are often provided for children to use in solving the problems.

The main findings from these studies are summarized next. In general, older children perform better than younger children, which is not surprising. Both semantic structure and identity of the unknown consistently influence relative problem difficulty.

Semantic Structure.   Evidence for the influence of semantic structure in problem solution comes from studies showing that problems described by different semantic structures are not equally difficult, even when they require the same operation for solution. This suggests that solving a word problem requires more than just knowing the operations and having some general skill in applying them.

Tables 4.4, 4.5 and 4.6 show the results from three separate studies in which children solved sets of word problems using blocks. All studies followed a procedure like the one just described, with the exception that in the Carpenter *et al.* (1981) study, sums of the given numbers were between 11 and 15. The following is a summary of the main findings from these three studies.

Compare problems (3) and (6) are more difficult than either change (1) or combine (1) problems, although all four problem solutions involve a simple addition. Similarly, problems involving subtraction can also vary in difficulty across semantic structures. Combine (2) problems and virtually all compare problems involving subtraction are, in general, more difficult than change problems (2) and (4). These findings agree with those from other studies. Compare (1) problems have been consistently shown to be more difficult than change (2) problems for first-graders (e.g., Gibb, 1956; Schell & Burns, 1962; Shores & Underhill, 1976). Combine (2) problems are, in general, more difficult than change (2) for kindergartners and first-graders (e.g., Gibb, 1956; Ibarra &

Table 4.4
Proportions of First-Graders Who Performed Correctly[a]

Problem type	With objects
Change (1)	.95
Change (2)	.91
Change (3)	.72
Equalizing (1)	.91
Equalizing (2)	.91
Combine (1)	.88
Combine (2)	.77
Compare (1)	.81
Compare (3)	.28

[a]From Carpenter, Hiebert, & Moser, 1981.

Table 4.5
Proportions of Children Who Performed Correctly Using Objects[a]

Problem type	Grade			
	K	1	2	3
Change (1)	.87	1.00	1.00	1.00
Change (2)	1.00	1.00	1.00	1.00
Change (3)	.61	.56	1.00	1.00
Change (4)	.91	.78	1.00	1.00
Change (5)	.09	.28	.80	.95
Change (6)	.22	.39	.70	.80
Combine (1)	1.00	1.00	1.00	1.00
Combine (2)	.22	.39	.70	1.00
Compare (1)	.17	.28	.85	1.00
Compare (2)	.04	.22	.75	1.00
Compare (3)	.13	.17	.80	1.00
Compare (4)	.17	.28	.90	.95
Compare (5)	.17	.11	.65	.75
Compare (6)	.00	.06	.35	.75

[a]From Riley, 1981.

Lindvall, 1979; LeBlanc, 1968; Nesher & Katriel, 1978; Vergnaud, 1981), but are slightly easier than compare (1) problems (Schell & Burns, 1962). Interestingly, children in the Carpenter *et al.* study performed relatively well on combine (2) and compare (1) problems for reasons we will discuss later.

Another source of evidence for the influence of semantic structure on problem difficulty comes from children's solution procedures with blocks. Carpenter *et al.* (1981) report that the dominant factor in determining the children's solu-

Table 4.6
Proportions of Kindergartners Who Performed Correctly[a]

Problem type	With objects
Change (1)	.89
Change (2)	.91
Change (3)	.08
Change (4)	.64
Change (5)	.32
Change (6)	.12
Combine (1)	.83
Combine (2)	.18

[a]From Tamburino, 1980.

tion strategy was the structure of the problem. For example, change (2), change (3), and compare (1) all require the child to find the difference between the two numbers given in the problem; however, the strategies children used to solve each of these problems were quite different. Almost all children used a subtraction strategy (separating or counting down) to solve change (2). For change (3) almost all children used an addition strategy (adding on or counting up). For compare (1) a matching strategy was frequently used.

The final source of evidence we present for the influence of semantic structure comes from studies showing children's lack of reliance on formal arithmetic in solving word problems. There are data showing that very young children can solve some word problems before they have received any formal introduction to the syntax of arithmetic (e.g., Buckingham & MacLatchy, 1930; Carpenter et al., 1981; Carpenter & Moser, 1981; Ibarra & Lindvall, 1979). There also is evidence that translating simple word problems into equations is not a necessary, or even usual, step in the solution processes of most children who have studied the formal notation of arithmetic. Second-grade children sometimes find it difficult or impossible to write equations for problems they have already solved (Lindvall & Ibarra, 1980a; Riley, 1981). Carpenter (1980) found that about one-fourth of first-graders solved the problem before they wrote a number sentence, in spite of instructions to the contrary. Together these studies suggest that children base solutions on an understanding of the semantic relations in the problem situation and do not need to employ standard, written procedures.

We have presented three main kinds of evidence for the influence of a problem's semantic structure on children's solutions to word problems. As mentioned earlier, we believe the various semantic structures correspond to specific concepts—concepts of quantitative change, equalization, combination, and comparison. On the basis of the aforementioned findings, it might be tempting to speculate that these four concepts emerge at different times in cognitive development. For example, at a certain age, a specific child might have the concepts of change and combination, but not the concept of comparison. The findings we present next suggest that is too simplistic.

Identity of the Unknown Quantity.   The main source of evidence that these concepts are not acquired in a sequential, all-or-none fashion comes from studies showing that problems having the same semantic structure also vary in difficulty.

Referring again to the change problems in Tables 4.4, 4.5, and 4.6, children had no difficulty solving change problems when the start and change amounts were given and they were asked for the result. Even preschool children can solve these problems (e.g., Buckingham & MacLatchy, 1930; Hebbeler, 1977). However, many kindergartners and first-graders had difficulty if the start and the result were given and they were asked to find the amount of change. Problems like (5) and (6), where the result and change were given with the start set unknown, were difficult at all grade levels (see also, Hiebert, 1981; Lindvall &

Ibarra, 1980a; Vergnaud, 1981)—even more difficult than the combine (2) and compare (1) problems previously discussed.

As with change problems, the difficulty of combine and compare problems also varies depending on which value in the problem is unknown. Combine (2) problems in which one of the subsets is unknown are significantly more difficult than combine (1) problems in which the two subsets are known and the problem solver is asked to determine their combined value. Compare problems (5) and (6) in which the referent is unknown are more difficult than any of the other compare problems.

Clearly we must consider more than just a problem's semantic structure in our effort to understand the problem-solving skills of children at different ages. Specific features within each semantic structure, like the identity of the unknown quantity, must also be taken into account.

In summary, word problems differ both in the semantic relations used to describe a particular problem situation and in the identity of the quantity that is left unknown. The resulting problem types have been related to fairly systematic differences in children's performance at various grade levels. Some problems are relatively easy for preschoolers, whereas other problems remain difficult for many third-graders, even when concrete aids are made available. However, simply identifying which problems are more difficult than others tells us little about why they are difficult. In the following sections we present a theoretical analysis that has attempted to relate differences in performance on word problems to the knowledge children have available at different ages.

A THEORY OF THE KNOWLEDGE REQUIRED TO
SOLVE WORD PROBLEMS

In this section we describe the current version of a theoretical analysis of word problem solving. The analysis is in the form of computer simulation models that solve word problems like the ones in Table 4.3. The conceptual knowledge and procedures represented in these models represent specific hypotheses about the knowledge required to solve word problems. The categories of knowledge we propose are similar to those found in other analyses (e.g., Fuson, 1979; Nesher, 1981; Vergnaud, 1981) that distinguish between semantics of problems and the semantics of addition and subtraction operations. In our analysis, we distinguish three main kinds of knowledge during problem solving: (*a*) problem schemata for understanding the various semantic relations discussed earlier, (*b*) action schemata for representing the model's knowledge about actions involved in problem solutions; and (*c*) strategic knowledge for planning solutions to problems. When a model is given a word problem to solve, it uses its knowledge of problem schemata to represent the particular problem situation being described.

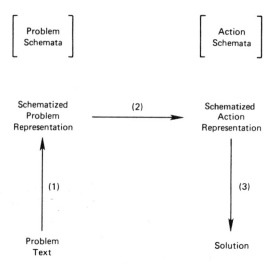

Figure 4.1. Framework of a model of problem understanding and solution. Arrows represent processes of (1) comprehension, (2) mapping from conceptual relations to quantitative procedures, and (3) execution of procedures.

The model's planning procedures then use action schemata to generate a solution to the problem. The general framework for this solution process is shown in Figure 4.1.

Problem Schemata

Our use of schemata is similar to the uses of that term in recent theories of language understanding (e.g., Anderson, 1976; Norman & Rumelhart, 1975; Schank & Abelson, 1977). In these theories, schemata have been used to organize the information in a sentence or story and to expand the representation of the message to include components that were not explicitly mentioned, but are nevertheless required to make the representation coherent and complete. Similarly, we view the process of understanding a word problem as fitting the components of the problem into a coherent structure.

The analysis that we have developed proposes three main types of problem schemata for understanding simple change, combine, and compare word problems. The representations have the form of semantic network structures consisting of elements and relations between those elements. For example, Figure 4.2 shows a representation of change (2). The representation has three main components. First, there is an initial quantity that represents the start set (1) of Joe's eight marbles. Second, there is some event that causes a change, in this case a

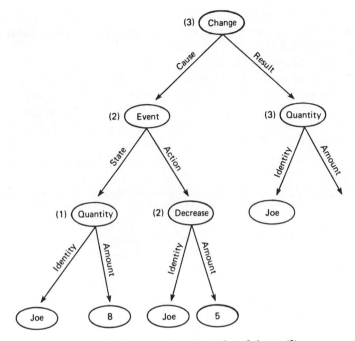

Figure 4.2. Schematized representation of change (2).

decrease, in the start set; the amount of this change is called the change set (2). The result of this change is represented as an unknown final quantity of marbles, or the result set (3).

The model that correctly solves all the change problems builds the first component (1) when it receives *Joe had 8 marbles*. When the sentence *Then he gave 5 marbles to Tom* is received, the model infers that the problem is about a change and constructs the rest of the structure in Figure 4.2, explicitly indicating that it expects to hear about a result. Finally, when the question *How many marbles does Joe have now?* is received, the model understands this as a request to determine the amount of the result set. At this point the model refers to its knowledge of action schemata and planning procedures.

Action Schemata

Once a model has represented a problem situation, it must have some way of relating this representation to its problem-solving procedures. We believe this requires a second type of schema that represents knowledge about actions used in planning solutions to problems. These action schemata are associated with the problem representation during problem solving and mediate the choice of an action or operation to solve the problem. Examples of these action schemata are shown in Figure 4.3.

The organization of action schemata into prerequisites and consequences is patterned after Sacerdoti's (1977) model of planning in problem solving called NOAH. Prerequisites are conditions in the problem situation that must be present for an action to be performed. Consequences are those conditions that will exist in the problem situation once the action is carried out. We should point out that the format we have used to illustrate these action schemata is simply a matter of convenience; they just as easily could have been drawn as network structures similar to the change schema in Figure 4.2.

Referring to Figure 4.3, make-set is the schema associated with the model's action for building a set X with an amount N. The prerequisite of this action is that the model begin with an empty set: The availability of objects and locations on which to put them is assumed. The consequence is a set X containing N objects. Thus, the action make-set (Joe, 8) would result in a set of eight objects belonging to Joe. Other action schemata include put-in, take-out, and count-all. Put-in adds N objects to an existing set X. Its prerequisite is that X exists with an amount M, and its consequence is that X now contains N more objects. Take-out removes N objects from set X. Count-all is an action that determines the number of members of set X by counting all the members of X (as opposed to counting on from a subset of X). In the diagram, all the action schemata have labels— "make-set," "take-out," etc.—although a child could have schemata without labels.

MAKE-SET: X, N
    Prerequisite: X,
    Consequence: X, N

PUT-IN: X, N
    Prerequisite: X
    Consequence: X, N More

TAKE-OUT: X, N
    Prerequisite: X
    Consequence: X, N Less

COUNT-ALL: X, ?
    Prerequisite: X
    Consequence: X, N

Figure 4.3. Action schemata.

COMPARE X, More, Y
(X has N1, Y has N2)

1. MAKE-SET: X, N1

2. MAKE-SET: Y, N2

3. MATCH: Z in X match Y
   a. Form subset of X equal to Y
   b. Z = matching subset

4. GET-REMAIN: W in X from Z
   a. SEPARATE X: Z and remainder
   b. W = remainder

5. COUNT-ALL: W

Figure 4.4. ''Compare'' action schema.

Action schemata are organized into different levels to enable efficient planning. This is because some schemata are actually composites of several other schemata and are therefore more global. For example, Figure 4.4 shows a global schema called *compare* that determines the amount of difference between the numbers of members of two sets. This particular procedure depends on counting, rather than subtraction of numbers. However, an alternative assumption might be that compare's subschemata involve subtracting numbers. This change would not alter compare's role in the knowledge structure; it would merely modify the way in which compare would be executed. More will be said about how compare is executed in the next section; the focus here is on its general structure.

Compare is composed of four specific actions—make-set, match, get-remain, and count-all—and is therefore more global than they are; get-remain is in turn more global than its component action, separate. Although not shown explicitly, each of these action schemata also contains information about its corresponding prerequisites and consequences. As will be described, the organization of action schemata into these levels of generality allows the model to consider global solution methods before taking into account all the details of implementing any particular method.

Strategic Knowledge

In addition to problem schemata and action schemata, the models also have strategic knowledge for planning solutions to problems. Strategic knowledge is

represented by production rules organized in a way that permits top-down planning of the kind studied by Sacerdoti (1977) and implemented in many current
models of human problem solving (e.g., Chi, Feltovich, & Glaser, 1981;
Greeno, Magone, & Chaiklin, 1979; Polson, Atwood, Jeffries, & Turner, 1981).
Knowledge for planning involves the action schemata just discussed as well as
simple associations between goals and procedures relevant to attaining those
goals. When a model is given a problem to solve, it sets a goal either to make the
external situation correspond to some given information or to obtain some requested information. The model then uses its knowledge about actions to plan
how to achieve that goal in the current problem situation.

Planning involves working out a solution from the top down, that is, choosing a general approach (e.g., match) to a problem, then deciding about actions
that are somewhat more specific, and only then working out the details. After a
plan is selected, the model tries to carry out the actions associated with that plan
in an attempt to achieve the current goal. If the action prerequisites are satisfied
in the current problem situation, then the plan can be carried out immediately. If
not, some further work must be done, and this requires setting one or more
subgoals. The model's planning knowledge includes knowledge of subgoals that
are useful in achieving a plan. Once generated, the new subgoal replaces the
earlier goal, but the previous goal is stored in memory to be retrieved when the
new subgoal is either achieved or is determined to be impossible. This process of
setting goals and subgoals and planning how to achieve them continues until the
problem is solved.

THE LOCUS OF IMPROVEMENT IN
PROBLEM-SOLVING SKILL

We have identified three main components of knowledge needed for successful performance in the domain of word problems. Children's difficulties in
solving some problems may be caused by the absence of one or more of these
components of knowledge. In the next section we will present an analysis of
different levels of children's problem-solving skill in which the major factor is
assumed to be acquisition of an improved ability to represent problem information. In this section, we present some evidence for this hypothesis, comparing it
with an alternative hypothesis that a main source of children's difficulty is their
lack of knowledge about the actions required to solve certain word problems.

We discuss evidence from studies of compare problems and combine problems. The gist of the findings is that problems that are difficult in their usual
wording are made much easier by changing the wording in appropriate ways.
These findings are similar to those obtained in recent studies of class inclusion
(Dean *et al.*, 1981; Markman, 1973; Trabasso *et al.*, 1978) and, as in those

studies, argue against the hypothesis that children lack the action schemata required to solve the problem. In our analysis the main locus of children's improvement in problem-solving skill is in the acquisition of schemata for understanding the problem in a way that relates it to already available action schemata.

Compare Problems

We present a brief summary of results obtained by Hudson (1980) in a study of young children's performance on compare problems of the kind called compare (1) in Table 4.3. Recall that problems involving comparison are difficult, at least for young children. One possibility is that children lack the action schemata required to plan a solution to the problem. Indeed, the compare procedure is fairly complex, as was shown in Figure 4.4. It involves first using make-set to create two sets to be compared. Then match finds a subset Z of the larger set X whose elements are in one-to-one correspondence with the elements of the smaller set Y. The procedure get-remain then uses separate to identify the difference between the two sets. Separate removes all the elements in X that are not part of Z, and identifies these elements as the remainder W. Finally, count-all determines the number of members of W. It is tempting to conclude that younger children have not yet acquired this relatively complex procedure and that this lack of a problem-solving method is responsible for their poor performance on problems involving comparisons.

This interpretation is contradicted by data collected by Hudson (1980), who presented problems of the kind shown in Figure 4.5 to 12 nursery-school, 24 kindergarten, and 28 first-grade children. Two different questions were asked. One was the usual comparative question, in this case, *How many more birds than worms are there?* The other question was an alternative that Hudson devised: *Suppose the birds all race over and each one tries to get a worm! Will every bird get a worm? . . . How many birds won't get a worm?*

The results were striking, as shown in Table 4.7. Hudson gave eight questions of each type, and the proportions here are for children who gave six or more correct responses. A correct response was the difference between the sets—for example, *One more bird than worms,* or *One bird won't get a worm.* The most

Table 4.7
Proportions of Children with Consistent Correct Responses

Grade	How many more?	How many won't get?
Nursery school	.17	.83
Kindergarten	.25	.96
First	.64	1.00

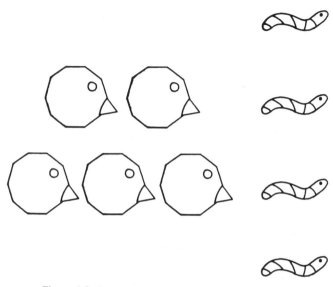

Figure 4.5. Example of problems used in Hudson's study.

frequent incorrect response was the number in the larger set—for example, *Five,* or *Five birds.* Another frequent error was to give both set sizes—for example, *Five birds and four worms.* Very few of the nursery-school or kindergarten children answered the *How many more?* questions by giving the difference between the sets. However, nearly all the children of all three ages answered the *How many won't get?* questions correctly.

Thus, Hudson's data do not support the hypothesis that children lack a procedure for finding a set difference; to answer the *Won't get* questions, children used a match procedure to form a correspondence between the two sets and to count the remaining subset of the larger set.

Combine Problems

Similar effects of rewordings have been obtained for combine (2) problems like the one in Table 4.1:

Joe and Tom have 8 marbles altogether.
Joe has 5 marbles.
How many marbles does Tom have?

As noted earlier, these problems are quite difficult for young children even though the solution procedure involves three relatively simple actions: make-set, take-out, and count-all. Here make-set counts out a set X of eight blocks to represent the marbles that Joe and Tom have altogether; take-out removes five blocks from X to represent Joe's marbles; and count-all counts the number of blocks remaining in X to determine how many marbles Tom has. We can assume that most children have this solution procedure readily available since they use it to solve change (2) problems, which also require subtraction. Furthermore, as in the Hudson study, it appears that slight rewordings of the combine (2) problem enable many children to solve it correctly using this procedure. Carpenter *et al.* (1981) report that 33/43 first-graders did in fact solve combine (2) problems correctly when asked,

> *There are 6 children on the playground.*
> *4 are boys and **the rest** are girls.*
> *How many girls are on the playground?*

Lindvall and Ibarra (1980b) found that combine (2) problems like,

> *Together, Tom and Joe have 8 apples.*
> *Three **of these** apples belong to Tom.*
> *How many **of them** belong to Joe?*

are significantly easier for kindergarten children than the combine (2) problems like the one in Table 4.1. We will return to the specific nature of the facilitation effect of such rewordings later. The point to be made here is that once again we cannot attribute children's problem-solving difficulties to a deficiency involving problem-solving actions. Instead, we hypothesize that acquisition of skill is primarily an improvement in children's ability to understand problems—that is, in their ability to represent the relationships among quantities described in problem situations in a way that relates to available solution procedures.

STAGES OF CONCEPTUAL KNOWLEDGE

In this section we present specific hypotheses about the nature of conceptual development that results in improved skill in solving arithmetic word problems. The hypotheses are based on data obtained by Riley (1981) in a developmental study of performance on word problems like the change, combine, and compare types in Table 4.3. Riley designed computational models of word problem solving that include processes of representing the problem information. The models were intended to simulate children's performance at different levels of skill. Levels of skill corresponded to different patterns of performance typical of children at different ages. Within each of the three semantic

Table 4.8
Patterns of Performance on Change Problems

	Levels of performance		
Example of problems	1	2	3
Result unknown			
1. Joe had 3 marbles.			
Then Tom gave him 5 more marbles.			
How many marbles does Joe have now?	+	+	+
2. Joe had 8 marbles.			
Then he gave 5 marbles to Tom.			
How many marbles does Joe have now?	+	+	+
Change unknown			
3. Joe had 3 marbles.			
Then Tom gave him some more marbles.			
Now Joe has 8 marbles.			
How many marbles did Tom give him?	"8"	+	+
4. Joe had 8 marbles.			
Then he gave some marbles to Tom.			
Now Joe has 3 marbles.			
How many marbles did he give to Tom?	+	+	+
Start unknown			
5. Joe had some marbles.			
Then Tom gave him 5 more marbles.			
Now Joe has 8 marbles.			
How many marbles did Joe have in the beginning?	"5"	"5"	+
6. Joe had some marbles.			
Then he gave 5 marbles to Tom.			
Now Joe has 3 marbles.			
How many marbles did Joe have in the beginning?	NA	NA	+

Table 4.9
Proportions of Patterns Consistent with Models

		Grade			
	Level	K	1	2	3
Change problems	1a	.04	.22	0	0
	1b	.30	.17	0	0
	2	.39	.17	.10	.05
	3a	.09	.17	.30	.15
	3b	.09	.22	.60	.80
	Residual	.09	.05	0	0

categories included in the study, three levels of skill were identified, each associated with a distinctive pattern of correct responses and errors on the six problems of the type.

The patterns of performance for the change problems are shown in Table 4.8. A "+" means the child answered correctly, "NA" indicates no answer, and numbers indicate the characteristic error for that problem; for example, on change (3) the specific number that was given was the result set, whatever that number was. (Different children solved these problems with different numbers involved.) Thus, reading vertically, a child at Level 1 would answer change problems 1, 2, and 4 correctly, respond with the result set for change (3) problems, give the change set for change (5), and no reply for change (6).

Table 4.9 shows the proportions of children in each of four grades whose performance was consistent with the identified patterns. These data are from performance when blocks were available. Level 1a children responded correctly on change problems 1 and 2 only. Level 1b children responded correctly on change problems 1, 2, and 4, as specified by the pattern for Level 1 in Table 4.8. Children who responded correctly on all problems except change (5) *or* change (6) were classified as being in Level 3a in Table 4.9. A child classified at Level 3b was correct on all problems. The proportions of children in the residual columns of Table 4.9 are those whose performance was not consistent with any of the patterns.

We should point out that some of the children identified in the residual column actually did respond consistently, but in ways not accounted for by our models. For example, there were a few children who consistently put out an arbitrary number of blocks for problem statements involving the word *some*. This can lead to predictable confusions when the arbitrary set does not correspond to the actual answer, but at the same time occasionally allows for fortuitous correct responses to difficult problems like change (5). A more detailed discussion of this behavior is provided by Tamburino (1980) and Lindvall and Tamburino (1981) in their account of why change (5) problems were easier than change (3) problems for some children in Tamburino's study (see Table 4.6), although the reverse is usually true in the literature.

Riley designed models to simulate each of the performance patterns in Table 4.8. The knowledge structures and procedures represented in these models represent hypotheses about the kinds of information-processing components needed to explain the different patterns of performance on the various problems. The models that simulate the three levels of change problem performance have been implemented, and we will describe their characteristics in some detail. The models for combine and compare performance have been designed, but not implemented. We will summarize their main features.

Development of Schemata for Change Problems

Riley's analysis of processes for change problems consists of three models that simulate the different levels of children's performance on these problems. That is, each model solves the six change problems in a way that leads to one of the different patterns of performance that Riley identified. All of the models employ the same general approach to problem solving, as was shown in Figure 4.1. That is, problem schemata are used to represent the current problem situation, and knowledge of action schemata and planning procedures are used to determine a solution. The main differences between the models relate to the ways in which information is represented and the ways in which quantitative information is manipulated. Models with more detailed representational schemata and more sophisticated action schemata represent the more advanced levels of problem-solving skill. Model (1) understands quantitative relations by means of a simple schema that limits its representations of change problems to the external displays of blocks. Model (2) has a change schema for maintaining an internal representation of increases and decreases in the sets of blocks it manipulates; the process of building this representation is still relatively "bottom-up" in the sense it depends upon the external display of objects. Model (3) also has a change schema for representing features internally, but can use its change schema in a more "top-down" way than model (2) to direct understanding independent of the external display of blocks. Models (2) and (3) also have a richer set of action schemata for producing and manipulating quantitative information and a richer understanding of certain relations between numbers; for example, model (3) has an understanding of part–whole relations. We will discuss the relationship between these different kinds of knowledge during performance and development in the next section.

MODEL (1)

The lowest level of performance on the change problems is represented by model (1). The knowledge that model (1) has available for problem solving includes the action schemata in Figures 4.3 and 4.4, procedures for planning in the way described on pp. 169–170, and the simple schema for representing quantitative information shown in Figure 4.6. This knowledge is sufficient to solve change problems (1), (2), and (4), but leads to predictable errors on change problems (3), (5), and (6). The first three problems share two main characteristics: The actions required to solve the problem can be selected on the basis of local problem features, and the solution set is available for direct inspection at the time the question is asked. For example, solving change (4) involves reducing Joe's initial set of eight blocks to three blocks in response to *Now Joe has 3 marbles,* with the effect that the change and result sets are now physically

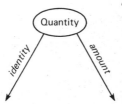

Figure 4.6. Model (1)'s schema for representing quantitative information.

separate. Thus, the model can easily identify the change set when asked *How many marbles did Joe give Tom?* and responds correctly even though it did not keep a memory record of the structural relationships in the problem.

Now consider how model (1) solves change (3) in which the solution set is not available for direct inspection. The model has no difficulty carrying out the correct procedures to solve the problem; failure is due to problem representation. The model counts out three blocks in response to *Joe has 3 marbles,* and uses the simple schema in Figure 4.6 to represent these blocks as a quantity whose identity is Joe and amount is three.

Next the model attempts to put in more blocks in response to *Then Tom gave him some more marbles.* But, since it does not yet know exactly how many to put in, it does nothing and therefore does not change its representation of the problem situation. The next sentence, *Now Joe has 8 marbles,* results in a goal to create a set of eight blocks. The model counts the three blocks, then continues to add in additional blocks until there are eight blocks total. The resulting set is represented as a quantity whose identity is Joe and amount is eight.

The difficulty arises when model (1) is asked to determine the number of marbles that were added in to change the initial set. Since the start set and change set are not distinguished in model (1)'s final representation, the question is simply interpreted as a request to determine the total number of marbles in the set. It therefore counts all the marbles and incorrectly answers *Eight.* To solve change (3) problems correctly, the child would have to represent, internally, additional information about the sets in the problem. This probably accounts for why change (3) problems are generally more difficult for young children than change (4) problems, even though both problems involve an unkonwn change set (see also Hiebert, 1981; Tamburino, 1980).

The idea that children's failure on change (3) is due to a failure to represent the separate start and change sets is consistent with several findings. Many studies have shown that even when children have little difficulty selecting and carrying out the appropriate actions to solve change (3) problems using blocks, many of them give the value of the result set as their answer (e.g., Riley, 1981; Tamburino, 1980). Another kind of evidence comes from a study by Harvey (1976). He successfully trained first-graders to solve similar problems using

external partitions to distinguish the two sets. Children initially solved the problem using a single paper plate with a partition: The start set was placed on one
side of the partition and the change set was placed on the other side. The next
step involved using two paper plates. Finally, children solved the problem correctly using a single plate with no partition. The success of Harvey's training
procedure suggests that, prior to training, a main part of children's difficulty in
solving these problems was due to a failure to distinguish the start and change
sets.

MODEL (2)

The main difference between model (1) and model (2) is that model (2)
represents internally additional information about the problem situation. This
involves a schema for change problems (Figure 4.2) in which there is a mental
record kept of the structural role of each item of information. This additional
structural information enables model (2) to give the correct answer to change (3)
problems where model (1) failed.

Model (2)'s behavior in response to the first two sentences of change (3) is
identical to model (1)'s. It simply counts out three blocks in response to *Joe has*
three marbles, and represents this as a set belonging to Joe with an amount three.
The model attempts to put in additional blocks in response to *Then Tom gave him*
some more marbles, but since no amount is mentioned, it does nothing. At this
point model (1)'s and model(2)'s understanding of the problem are identical. The
difference between the two models becomes evident from the way model (2)
responds to the next input: *Now Joe has 8 marbles.* In addition to simply
increasing the existing set until there are a total of eight blocks, model (2) also
identifies the set of eight as the result set that was produced by increasing the
start set of three blocks by some unknown amount. The resulting problem representation is shown in Figure 4.7. Thus when model (2) is asked, *How many*
marbles did Tom give Joe, it can identify the separate change set in its problem
representation and determine the set's numerosity by counting all but three of the
blocks.

Although model (2) has a more complete internal representation than model
(1), it still lacks an important ability for top-down processing in its representation
of problem information. This is seen in model (2)'s performance on change (5)
problems. Recall that children at this level cannot solve this problem and give
Five (the value of the change set) as their most frequent incorrect response. The
model receives the first sentence, *Joe had some marbles,* and attempts to create a
set of blocks to represent these marbles, but realizes it does yet know exactly
how many Joe has. Therefore the model does nothing but simply remember the
fact that it heard about Joe. The second sentence results in the model putting out
five blocks for Joe. However, because the model failed to represent explicitly the

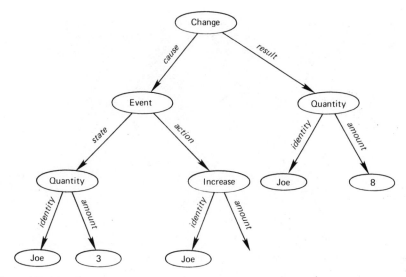

Figure 4.7. Model (2)'s representation of change (3) before determining the amount of the change set.

unknown start set, the additional five blocks are not represented as a change in the initial set, but simply as a set of five belonging to Joe. This means that when the model receives *Now Joe has 8 marbles,* the problem situation is the same as that in change (3). The model increases the set of five to eight, resulting in a representation identical to the one that was shown in Figure 4.7. Notice that the set of five blocks is identified as the start set, although it is actually the change set in the original problem. This accounts for why model (2) answers *Five* when asked *How many marbles did Joe have in the beginning?*

MODEL (3)

Model (3), like model (2), has a change schema for maintaining a structural representation of the problem situation. However, unlike model (2), model (3) can use its change schema in a top-down fashion to build a representation of the entire problem before actually solving it. This permits model (3) to operate on a quantity whose value is unknown, as required in change problems (5) and (6).

Model (3)'s ability to solve change problems (5) and (6) also involves the schema for representing part–whole relations shown in Figure 4.8. The reason for this will become clearer as we continue, but basically when the start set is unknown, the action required to solve the problem is not immediately available from the initial problem representation. We hypothesize that these problems are best understood in terms of the part–whole relations between the quantities.

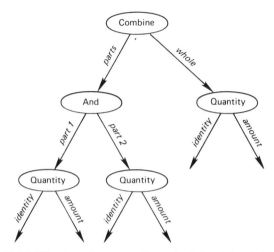

Figure 4.8. Model (3)'s schema for representing part–whole relations between quantities.

The implications of model (3)'s additional conceptual knowledge become apparent from the way it solves change (5). The model's understanding of the sentence, *Joe had some marbles,* is represented as a quantity whose identity is Joe and amount is unknown. *Then Tom gave him 5 more marbles* is understood as an increase in the initial quantity, causing the rest of the change schema to be instantiated (Figure 4.9), and the model puts out five blocks for Joe. Thus, model (3)'s representation of change (5) maintains a record of Joe's five marbles as the amount of change in the as-yet-unknown start set. This is in contrast to model (2), where no record was kept of the unknown start set, with the eventual consequence that the change set of five blocks was represented incorrectly as the start set.

Model (3) represents the third sentence of change (5), *Now Joe has 8 marbles,* as the amount of the result set and increases the existing set until it contains eight blocks. When model (3) is asked, *How many marbles did Joe have in the beginning?* it sets the goal of determining the value of the start set, but has not direct referent for this set in its blocks representation as was the case in the change (3) example (see the section on model (2)). We therefore hypothesize that identifying the appropriate action requires additional inferences about the part–whole relations in the problem, as shown in Figure 4.10.

Since the direction of change is an increase, the model infers that both the start and change sets are parts of the result set. On the basis of this inference, the model determines that the start set must consist of the additional blocks that were added to the change set to make a total of eight blocks. The model then counts these additional blocks and anwers *Three.*

Actually, the same basic solution to change (5) could just as easily have been represented by an alternative blocks procedure—one that children also frequently use (Carpenter, 1980; Riley, 1981). That is, model (3) could have delayed putting out any blocks until it inferred the part–whole relations between the quantities in the problem. Then the model could have put out the eight blocks first, used separate to remove five of the blocks, and finally identified the remaining three blocks as the answer.

Considering either blocks procedure, model (3)'s solution to change (5) suggests an alternative explanation for model (2)'s failure on change problems in which the start set is unknown. It is possible that some children did in fact use their change schema to represent correctly the problem situation with the start set unknown, but lacked the part–whole schema required to infer the appropriate operation.

The proposal that children require an understanding of the part–whole relation to solve change problems (5) and (6) is supported in Riley's study in which few children in any age group correctly solved these two problem types without first being able to solve combine problems with one of the subsets unknown. (We assume subset unknown problems also require an understanding of part–whole relations.) Furthermore, there is evidence suggesting that once children understand part–whole relations, they can use this knowledge to understand all change problems, even though our analyses have shown that knowledge of these relations is not required to solve change problems (1), (2), (3), and (4). Carpenter

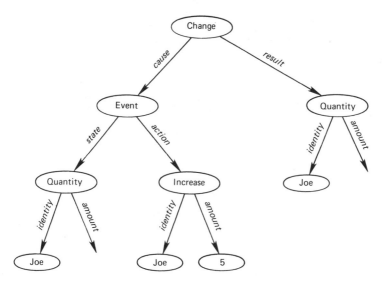

Figure 4.9. Model (3)'s representation of change (5) before determining the amount of the start and result sets.

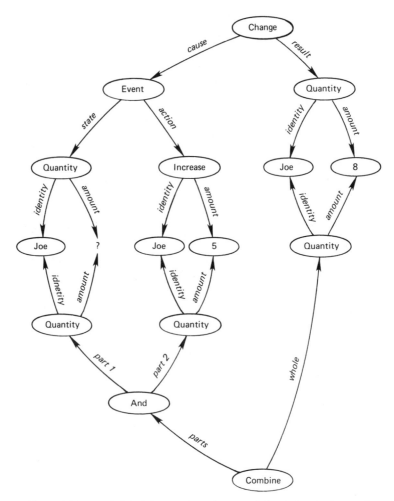

Figure 4.10. Model (3)'s inference about the part–whole relations in change (5).

(1980) reports a study of first-graders' strategies in solving word problems before and after receiving instruction in analyzing the part–whole relations of these problems. Prior to instruction, children's solution processes modeled the actions or relationships described in the problem. That is, children used separate, count-on, and match to solve change (2), change (3), and compare (1) problems, respectively, even though all of these problems involve finding the difference between two quantities. After instruction, children generally used separate for all subtraction problems. Apparently, these children were basing their solutions on the part–whole relations in the problems. A similar trend is indicated by results from a study reported by Zweng, Gerarghty, and Turner (1979) who found that the majority of third-graders and almost all the fourth-graders used subtraction to solve problems like change (3).

Models for Combine and Compare Problems

Riley's simulation models to explain children's performance on combine and compare problems are similar in their general features to those we have just described. Riley assumes that, at the lowest level, the child's representations of problems are limited to the external displays of blocks; at an intermediate level there are schemata for representing, internally, additional information about the relationships between quantities; and at the most advanced level, schemata are available that direct problem representations and solutions in a more top-down manner. The results for these models were similar to those obtained for the change problems, although the proportions of children not consistent with any of the models was somewhat greater for the compare problem set, as shown in Table 4.10. Overall, the models seem to provide a reasonable first approximation to the nature of the increased skill that children showed in solving these problems. As we will briefly describe, the models also provide a theoretical framework for integrating the various findings related to children's performance on combine and compare problems.

Recall that compare (1) problems are usually quite difficult for kindergarten and first-grade children. In Riley's compare models, failure is associated with the lack of a schema for understanding the problem situation in a way that makes

Table 4.10
Proportions of Patterns Consistent with Models

		Grade			
	Level	K	1	2	3
Combine problems	1	.52	.28	.05	0
	2	.22	.34	.10	0
	3	.13	.23	.85	.95
	Residual	.13	.17	0	.05
Compare problems	1	.52	.61	.05	0
	2	.04	.06	.40	.45
	3	0	.06	.20	.50
	Residual	.44	.27	.35	.05

contact with the model's available action schemata—in this case the match action schema. However, with instruction, first-graders apparently have no difficulty learning to apply match to solve compare (1) problems (Carpenter, 1980; Marshall, 1976). Carpenter (1980) reports a study in which children who had received such instruction were tested on a variety of problems once in February and again in May. Compare (1) problems were relatively easy for these children at both testing times; correct proportions were .67 and .74, respectively. At the same time, compare (3) problems remained difficult at both interview (.28 and .46). In February, errors on compare (3) consisted of responding with one of the givens. (Similar errors have been reported by Gibb, 1956; Hudson, 1980; Marshall, 1976; Riley, 1981; and Shores and Underhill, 1976.) In May, errors were almost equally divided between responding with one of the givens and choosing the wrong operation. This suggests that the children had learned to associate their match procedure with the *how many more than* question in the same way that children associated *giving* and *taking* with the actions put-in and take-out. But as with put-in and take-out, the children had not yet learned to represent the important relationships between the sets involved in the procedure and therefore failed to generalize the instruction to other compare problems (e.g., compare [3]). Furthermore, children did not acquire the entire compare schema at once but, as Carpenter *et al.* suggested, first focused on the difference relationships, as indicated by the increase in "wrong operation" errors for compare (3).

Riley's models propose a similar sequence of understanding to account for performance on combine problems. Even preschool children have little difficulty solving combine (1) problems. This does not, however, mean that solving these combine (1) problems involves any understanding of set inclusion or part–whole relations. Our lowest level combine model solves this problem by a simple association between the *how many altogether* question and its count-all action schema. In fact, it is this model's lack of understanding of part–whole relations that accounts for its failure on combine (2) problems. The more advanced models have a combine schema that allows them to infer the part—whole relation between, for example, the eight marbles that Joe and Tom have altogether in combine (2) and the five marbles that Joe has. These relations are not mentioned explicitly in the problem, and without this schema, children simply interpret each line of the problem separately as in change model (1) and have no way to infer the relation between the two sets. They therefore put out a set of eight blocks to represent the marbles that Joe and Tom have altogether and a separate set of five blocks to represent Tom's marbles. This leads to the incorrect response of *Eight* when asked how many marbles Tom has. The facilitation effect of rewordings like *of them* and *the rest* can be attributed to circumventing the need for a combine schema by making the source of Joe's marbles more explicit, allowing the child to remove the five from the set of eight.

Acquisition of Problem-Solving Procedures

Children's conceptual knowledge of the elements and relations in word problems also seems to be related to the acquisition of more sophisticated counting procedures. For example, model (2)'s change schema represents the model's understanding that changing a given quantity into a desired quantity involves either increasing or decreasing the given quantity by a specific amount. We therefore attribute a more sophisticated counting procedure to model (2) called count-on. Count-on allows the model to extend or count a set by beginning with the value of an existing set if it is known. Thus, if model (2) already has three blocks, but needs eight blocks, it can begin the count with three and simply keep adding in blocks until it gets to eight, instead of recounting the three as model (1) did with its count-all procedure.

Evidence for a relationship between the availability of the change schema and count-on comes from Steffe and Thompson (1981) who report a positive correlation between the ability to count-on and the ability to solve change (3) problems. Hiebert (1981) found that count-on (referred to in his study as add-on) was the procedure most frequently observed in first-graders' correct solutions to change (3) problems using blocks. The developmental relationship between the availability of count-on and the availability of the change schema will be considered in the discussion section.

Children's conceptual knowledge may also be related to the acquisition of the more efficient Min counting procedure (Groen & Resnick, 1977) in which the number counted-on changes from being the second addend given to being the smaller addend given. The mathematical property that allows this more efficient procedure is commutativity, which we believe corresponds to an implicit understanding of the part–whole relations between the addends a and b and their sum, c. That is, a and b are both parts of c, and therefore a + b and b + a are equivalent operations. It seems likely, therefore, that acquisition of the Min counting procedure would be related to children's understanding of combine (2) problems, as these problems require an understanding of part–whole relations. Fuson (1979) proposes the same basic idea by pointing out that the commutativity relation between a and b may vary with the addition problem type: Commutativity would seem to be less obvious when the roles played by the two numbers differ (as in change problems) than when the roles coincide (as in combine problems). Thus, although the sequential property of change addition problems may facilitate the transition from count-all to count-on, it may make commutativity less apparent. Fuson suggests that the transition to the more efficient Min procedure might be facilitated in the context of combine, rather than change, addition problems. It is clear that more empirical and theoretical work is required to clarify the relationship between conceptual and procedural knowledge in the development of problem-solving skill.

Conclusions

The models described in this section provide a detailed hypothesis about changes that occur in children's ability to understand relationships among quantities and to use their representations of these relationships to solve problems. We cannot claim uniqueness for the models that we have described here. Indeed, there are some redundant features in the models so that somewhat simpler accounts could be given to explain the observed improvements in skill. For example, model (2)'s ability to solve change (3) problems is related to the ability to construct a representation of the separate start and change sets and to the use of the count-on procedure. Even so, we are confident that the children's improvement in skill in these problems involves something along the lines of these models. The more skillful models are more accurate because they understand problems better. That is, their representations of problems include the relevant features of the problems more completely and in ways that lead to the choice of appropriate problem-solving actions.

RELATED ANALYSES OF CONCEPTUAL UNDERSTANDING IN PROBLEM SOLVING

We have argued that successful problem-solving performance by children depends on their understanding of certain concepts, which we have characterized as schemata used in representing information in problem situations. We want to avoid the impression that this point applies only to young children; it is equally true of adult problem solvers in domains much more complex than primary-grade arithmetic. In this section we briefly review findings from studies in two domains of intermediate complexity—algebra and physics—in which the central findings serve to emphasize the importance of conceptual knowledge in problem solving.

Word Problems in Algebra

High-school instruction in algebra usually includes solution of word problems, where the solution method presented involves translating the text into equations. The program, Student, developed by Bobrow (1968), solves algebra word problems using a method of translation into equations like the one presented in most instruction. Student uses little conceptual knowledge, focusing instead primarily on syntactic information to translate the English problem statement directly into a corresponding set of equations. It then solves the set of equations for the requested unknown. Table 4.11 presents an example of the kind of word problem Student was able to solve along with the simplified trace of

Table 4.11
Student's Direct Translation Strategy[a]

Problem

If the number of customers Tom gets is twice the number of advertisements he runs, and the number of advertisements is 45, what is the number of customers Tom gets?

Solution strategy

1. Partition the problem into phrases:
   If / the number of customers Tom gets / is twice /
   the number of advertisements he runs /, and /
   the number of advertisements he runs / is 45 /
   , what is / the number of customers Tom gets / ? /

2. Translate phrases into algebraic terms:

The number of customers Tom gets:	x
is:	$=$
twice:	2*
the number of advertisements he runs:	y
is	$=$
45:	45
what:	?

3. Organize algebraic terms into equations:
   $x = 2 * y$
   $y = 45$
   $? = x$

4. Simplify equations into single equation:
   $x = 2 \times 45$

Note. Asterisks stand for times.

[a]Adapted from Roman & Laudato, 1974.

Student's direct translation strategy. Notice how Student's solution relies almost entirely on syntactic information in the problem to guide the solution process with little understanding of the problem structure.

Although Student successfully solves many problems, a comparison of Student's performance with that of human students revealed limitations caused by its lack of conceptual knowledge. Paige and Simon (1966) and Hinsley, Hayes, and Simon (1977) found that whereas human performance was similar to Student's in important ways, humans also used a number of processes that Student did not have, but that corresponded to an understanding of the relations in the problem.

Paige and Simon (1966) noted that some human subjects proceeded, not by simply translating the verbal statements into algebraic equations (which was what they were requested to do), but by constructing a physical representation of the problem and then drawing information from the representation. Consider the following problem:

A board was sawed into two pieces. One piece was two-thirds as long as the whole board and was exceeded in length by the second piece by 4 feet. How long was the board before it was cut?

If one solves the problem by the direct translation strategy outlined in Table 4.11, an equation is obtained that yields a negative number for the length of the original board, a physically impossible result. Some students noticed this before setting up any equations, indicating that they had not proceeded by direct translation, but had instead constructed a physical representation of the situation described.

Word Problems in Physics

Important contributions regarding the influence of conceptual knowledge in problem solving have been made by comparing experts' and novices' solution procedures. Larkin, McDermott, Simon, and Simon (1979) and Simon and Simon (1978) found that skilled physics problem solvers work from elaborated representations of the problem, rather than directly from the problem description. These representations often include diagrams that make certain relationships and constraints highly salient. In effect, experts have more conceptual knowledge about problem situations than novices, and it is this conceptual knowledge that guides their more effective and efficient solutions. Although experts and novices may be observed to break problems into subparts or set subgoals to deal with difficulties, these procedures are apparently executed by experts with an emphasis on the problem "Gestalt," whereas novices tend to solve problems on the basis of more local problem features. Support for this idea comes from a study by Chi, Feltovich, and Glaser (981) in which expert and novice physics subjects were asked to sort problems into categories. The groups formed by novices contained problems with similar objects—for example, rotating objects. In contrast, the groups that experts formed contained problems related to general principles of physics, such as conservation of energy. Together these findings emphasize the importance of conceptual knowledge for constructing and transforming problem representations throughout solution, as well as the role of these representations in determining the nature and amount of procedural knowledge required to achieve a solution. Similar findings have also been obtained for the learning of early geometry proof exercises (Anderson, Greeno, Kline, & Neves, 1981; Greeno, 1980).

DISCUSSION

We have presented a theoretical analysis of both the conceptual knowledge and the cognitive procedures underlying children's performance at different stages of skill in solving word problems. In this section we summarize how these two forms of knowledge interact during problem solving and consider the role of this interaction as skill develops. Finally, we consider the general implications of this interaction for interpreting children's performance on other tasks.

Relationship between Knowledge and Procedures in Performance and Development

PERFORMANCE

We have identified three main ways conceptual knowledge and procedures interact during problem solving. One way involves the role of schemata in the selection of actions. In the models that we described on pp. 176–184, both problem schemata and action schemata are required to relate the problem statement to the actions required to solve the problem. Problem schemata are involved in interpreting the problem text. They range from model (1)'s simple schema for representing quantitative relations, to model (3)'s more complex change schema, to the schemata required for representing the complex relationships in combine (2) and complex forms of compare problems. For all the models, these schemata are associated with goals either to change the current problem situation or to obtain some information from the problem situation. Planning procedures then identify an action whose consequence matches the current goal. Sometimes there is a direct match between this goal and one of a model's action schemata—for example, the goal to increase the amount of a given set and model (1)'s put-in schema. In other cases, additional schemata are required to infer important relations in the problem situation before the appropriate action can be selected—for example, model (3)'s inferences about the part–whole relations between the quantities in change (5). In either case, the application of even simple actions in problem solving requires some mediating conceptual knowledge in the form of schemata.

The second way that conceptual and procedural knowledge interact involves the use of schemata to monitor the effects of selected actions on a problem situation. For example, whenever model (2) performs the action count-on, it uses its change schema to maintain a record of the effects of that action. This record includes information about the values of the separate start and change sets and is important for correctly answering problems like change (3). Failure to monitor the effects of actions can result in predictable errors on some problems (e.g., model [1]'s incorrect answer to change [3]).

Finally, conceptual knowledge can influence which actions get selected. For example, model (2) and model (3) solve change (3) problems correctly; however, differences in their conceptual understanding of the relationships between quantities in a change situation lead to differences in the actions chosen for solution.Model (2) understands the problem as an increase of some unknown amount in the start set and uses count-on to solve the problem. Model (3) also represents the problem as a change problem with an unknown change set, but then it infers the part–whole relations between the quantities in the problem, identifies one of the parts as the unknown, and solves the problem with reference to its understanding of part–whole relations between numbers.

We also suggested that a child's conceptual knowledge of the relations between quantities in a word problem is related to the acquisition of more efficient counting procedures. Thus, there are at least two motivations for acquiring more advanced schemata—necessity and efficiency. The question remains how more advanced schemata and procedures develop and how they interact during development.

DEVELOPMENT

Recent theories of learning and development suggest some interesting possibilities for how the acquisition of sophisticated problem-solving procedures may be related to the acquisition of conceptual knowledge. For example, Klahr and Wallace (1976) and Neches (1981) postulate some principles to constrain development that emphasize the avoidance of redundant or unnecessary processing in the developing cognitive system. They propose that once a procedure is acquired, its operation is monitored by the child by means of what is called a procedural trace—that is, a record of the procedure's functioning in some situation. This procedural trace allows "detection of consistent sequences" and eventual "elimination of redundant processing." It is feasible that procedural traces not only result in more efficient procedures, but are also the basis for the development of more advanced problem schemata. For example, consider the following mechanism to account for the transition from model (1) to model (2). If a set of a known amount—say three blocks—is already present and the child is asked to increase it to make it eight, model (1) children typically do not begin counting and adding in from the known set value, but rather use the procedure start-count-set to begin the count all over again, starting with the existing set. According to both Neches's theory and Klahr and Wallace's theory, the transition from this procedure to model (2)'s count-on would involve (a) "tracing" the operation of start-count set, (b) thereby noticing the redundancy of counting the three over again, and (c) finally eliminating this redundancy by beginning with three and counting on to the desired result set. Thus, it is not that model (1) does not form any representation of its solution procedure: It has to have some procedural trace to advance to model (2). However, the units of model (1)'s procedural traces are different from those of models (2) and (3) and do not correspond to the structural information required by some of the change problems. Anyway, at some point in transitioning between model (1) and model (2), the model's procedural trace and what we want to claim is the model's developing change schema are probably indistinguishable.

Children's available schemata may also influence what procedures get acquired. That is, levels of conceptual understanding function as intermediate steps in acquiring new procedures. Thus, it is unlikely that a child would acquire count-on until that child at least had the simple schema for representing quantitative information shown in Figure 4.6. As Kamii (1980) points out, it is impossible to put two numbers into a relationship unless the numbers themselves are solidly present in the child's mind.

In summary, we have made some general suggestions that children's procedural knowledge leads to the acquisition of schemata, and these schemata in turn are involved as intermediate steps in acquiring more advanced procedures. More work is required to explicate further the nature of this interaction.

Knowledge Underlying Problem-Solving Performance

The analyses just discussed also have some important general implications concerning the knowledge we attribute to children on the basis of their problem-solving performance. Piaget (Piaget & Szeminska, 1952) pointed out that children lack understanding of some very important concepts: conservation of number, class inclusion, seriation, and so on. Evidence for these failures of understanding came from performance that was inconsistent with the general concepts; for example, when a child sees two sets with the same number of objects and says one has more, that performance is inconsistent with the concept of number conservation. Recently, numerous investigators have shown that in other circumstances children will show performance that is consistent with those concepts. For example, Gelman and Gallistel (1978) provided considerable evidence for preschool children's understanding of number concepts involving small sets, and Trabasso et al. (1978) summarized a substantial body of evidence that under appropriate circumstances, children show that they understand the concept of class inclusion. Results of studies of word problem solving by Hudson (1980), Lindvall and Ibarra (1980a), and Carpenter et al. (1981) also fall in this category; they show that with appropriate rewordings, children are quite capable of showing that they understand concepts of quantitative comparison and set inclusion.

Implications of this are clear. Children's failure to show understanding of a concept on one kind of task should not be taken as firm evidence that they lack understanding of the concept; there may be other tasks in which their performance shows that they understand the concept quite well. At the same time, we cannot attribute the same understanding to all children who pass a simplified version of a task when these children may differ considerably in their performance on more standard versions of the task. We need to account for the differences in knowledge between children who demonstrate understanding of a concept on a single task only and children whose conceptual understanding generalizes across a range of tasks that apparently involve the same concept. The analyses discussed in this chapter have made some progress in identifying some of these differences.

Children who are more skilled have acquired schemata that act as principles for organizing the information in a problem. The schema appears to be used in a top-down fashion so that it overrides distracting features of the problem situation. Children who do not have these schemata cannot make these inferences and are dependent on modified problem situations where the relations are made explicit through rewordings or perceptual changes. Therefore we are inclined to view as very important the development of a schema to the point where it can be used to organize a problem situation and thereby override distracting, irrelevant factors. Piaget may have been wrong to assert that children lacked understanding of a schema if they failed his tests for that understanding, but it is equally misguided to assert that a schema is understood if we can find evidence for that understanding in some limited task domain. What we need is an analysis of the process of understanding in various problems situations, as well as an account of the features that are required for children at different states of development to produce an appropriate understanding of the situation and the task.

ACKNOWLEDGMENTS

We are especially grateful to Valerie Shalin for her insightful comments and careful readings of many earlier drafts of this chapter. We also thank Tom Carpenter, Karen Fuson, Maurice Lindvall, Jim Moser, Pearla Nesher, and Joe Tamburino for discussions and helpful criticisms of this chapter. Jim Mokwa collaborated in the implementation of the computer simulation models.

REFERENCES

Aiken, L. R., Jr. Verbal factors and mathematics learning: A review of research. *Journal for Research in Mathematics Education,* 1971, *2,* 304–313.

Aiken, L. R., Jr. Language factors in learning mathematics. *Review of Educational Research,* 1972, *42,* 359–385.

Anderson, J. R. *Language, memory, and thought.* Hillsdale, N.J.: Lawrence Erlbaum Associates, 1976.

Anderson, J. R., Greeno, J. G., Kline, P. J., & Neves, D. M. Acquisition of problem-solving skill. In J. R. Anderson (Ed.), *Cognitive skills and their acquisition.* Hillsdale, N.J.: Lawrence Erlbaum Associates, 1981.

Barnett, J. C., Vos, K., & Sowder, L. A review of selected literature in applied problem solving. In R. Lesh, D. Mierkiewicz, & M. Kantowski (Eds.), *Applied mathematical problem solving.* Columbus, Ohio: ERIC, November 1979.

Baylor, G. W., & Gascon, J. An information-processing theory of aspects of the development of weight seriation in children. *Cognitive Psychology,* 1974, *6,* 1–40.

Bobrow, D. G. Natural language input for a computer problem-solving system. In M. Minsky (Ed.), *Semantic information processing.* Cambridge, Mass.: MIT Press, 1968.

Bolduc, E. J., Jr. A factorial study of the effects of three variables on the ability of first-grade children to solve arithmetic addition problems (Doctoral dissertation, University of Tennessee, 1969). *Dissertation Abstracts International,* 1970, *30,* 3358A.

Brown, A. L. Knowing when, where, and how to remember: A problem of metacognition. In R. Glaser (Ed.), *Advances in instructional psychology.* Hillsdale, N.J.: Lawrence Erlbaum Associates, 1978.

Buckingham, B. R., & MacLatchy, J. The number abilities of children when they enter grade one. In *29th Yearbook of the National Society for the Study of Education.* Bloomington, Ind.: Public School Publishing, 1930.

Carpenter, T. P. *The effect of instruction on first-grade children's initial solution processes for basic addition and subtraction problems.* Paper presented at the annual meeting of the American Educational Research Association, Boston, April 1980.

Carpenter, T. P., Hiebert, J., & Moser, J. The effect of problem structure on first-grader's initial solution processes for simple addition and subtraction problems. *Journal for Research in Mathematics Education,* 1981, *12*(1), 27–39.

Carpenter, T. P., & Moser, J. M. The development of addition and subtraction problem-solving skills. In T. P. Carpenter, J. M. Moser, & T. Romberg (Eds.), *Addition and subtraction: Developmental perspective.* Hillsdale, N.J.: Lawrence Erlbaum Associates, 1981.

Chi, M. T. H. Knowledge structures and memory development. In R. S. Siegler (Ed.), *Children's thinking: What develops?.* Hillsdale, N.J.: Lawrence Erlbaum Associates, 1978.

Chi, M. T. H. Interactive roles of knowledge and strategies in development. In S. Chipman, J. Siegel, & R. Glaser (Eds.), *Thinking and learning skills: Current research and open questions* (Vol. 2). Hillsdale, N.J.: Lawrence Erlbaum Associates, 1982.

Chi, M. T. H., Feltovich, P. J., & Glaser, R. Categorization and representation of physics problems by experts and novices. *Cognitive Science,* 1981, *5,* 121–152.

Dahmus, R. M. How to teach verbal problems. *School Science and Mathematics,* 1970, *70*(2), 121–138.

Dean, A. L., Chabaud, S., & Bridges, E. Classes, collections, and distinctive features: Alternative strategies for solving inclusion problems. *Cognitive Psychology,* 1981, *13,* 84–112.

Duncker, K. On problem-solving. *Psychological Monographs,* 145, *58*(270), 1–112.

Fuson, K. C. *Counting solution procedures in addition and subtraction.* Paper presented at the Wingspread Conference on the Initial Learning of Addition and Subtraction Skills, Racine, Wisconsin, November 1979.

Gelman, R., & Gallistel, C. R. *The child's understanding of number.* Cambridge, Mass.: Harvard University Press, 1978.

Gentner, D. Evidence for the psychological reality of semantic components: The verbs of possession. In D. A. Norman & D. E. Rumelhart (Eds.), *Explorations in cognition.* San Francisco: Freeman, 1975. Pp. 241–246.

Gibb, E. G. Children's thinking in the process of subtraction. *Journal of Experimental Education,* 1956, *25,* 71–80.

Greeno, J. G. Process of understanding in problem solving. In N. J. Catellan, D. B. Pisoni, & G.R. Potts (Eds.), *Cognitive theory* (Vol. 2). Hillsdale, N.J.: Lawrence Erlbaum Associates, 1977.

Greeno, J. G. A study of problem solving. In R. Glaser (Ed.), *Advances in instructional psychology* (Vol. 1). Hillsdale, N.J.: Lawrence Erlbaum Associates, 1978.

Greeno, J. G. *Instruction for skill and understanding in mathematical problem solving.* Paper presented at 22nd International Congress of Psychology, Leipzig, July 6–12, 1980.

Greeno, J. G., Magone, M. E., & Chaiklin, S. Theory of constructions and set in problem solving. *Memory and Cognition*, 1979, *7*, 445–461.

Greeno, J. G., & Riley, M. S. *Processes and development of understanding* (LRDC Publication, 1981). Pittsburgh, Pa.: Learning Research and Development Center, University of Pittsburgh, 1981.

Groen, G. J., & Resnick, L. B. Can preschool children invent addition algorithms? *Journal of Educational Psychology*, 1977, *69*, 645–652.

Grouws, D. A. Differential performance of third-grade children in solving open sentences of four types. (Doctoral dissertation, University of Wisconsin, 1971). *Dissertation Abstracts International*, 1972, *32*, 3860A.

Harvey, C. O. *A study of the achievement and transfer effects of additive subtraction and class inclusion training*. Unpublished doctoral dissertation, University of Houston, 1976.

Hebbeler, K. Young children's addition. *The Journal of Children's Mathematical Behavior*, 1977, *1*, 108–121.

Heller, J. I., & Greeno, J. G. *Semantic processing in arithmetic word problem solving*. Paper presented at the Midwestern Psychological Association Convention, Chicago, May 1978.

Hiebert, J. *Young children's solution processes for verbal addition and subtraction problems: The effect of the position of the unknown set*. Paper presented at the 59th annual meeting of the National Council of Teachers of Mathematics, St. Louis, April 1981.

Hinsley, D., Hayes, J. R., & Simon, H. From words to equations: Meaning and representation in algebra word problems. In M. A. Just & P. A. Carpenter (Eds.), *Cognitive processes in comprehension*. Hillsdale, N.J.: Lawrence Erlbaum Associates, 1977.

Hudson, T. Young children's difficulty with "How many more _____ than _____ are there?" questions. (Doctoral dissertation, Indiana University, 1980). *Dissertation Abstracts International*, July 1980, *41*,(01).

Ibarra, C. G., & Lindvall, C. M. *An investigation of factors associated with children's comprehension of simple story problems involving addition and subtraction prior to formal instruction on these operations*. Paper presented at the annual meeting of the National Council of Teachers of Mathematics, Boston, April 1979.

Jerman, M. *Instruction in problem solving and an analysis of structural variables that contribute to problem-solving difficulty* (Tech. Rep. No. 180). Stanford: Institute for Mathematical Studies in the Social Sciences, 1971.

Jerman, M. Problem length as a structural variable in verbal arithmetic problems. *Educational Studies in Mathematics*, 1973–1974, *5*, 109–123.

Jerman, M., & Rees, R. Predicting the relative difficulty of verbal arithmetic problems. *Educational Studies in Mathematics*, 1972, *4*(3), 306–323.

Kamii, C. *Equations in first-grade arithmetic: A problem for the "disadvantaged" or for first graders in general?* Paper presented at the annual meeting of the American Educational Research Association, Boston, April 1980.

Klahr, D., & Robinson, M. Formal assessment of problem-solving and planning processes in preschool children. *Cognitive Psychology*, 1981, *13*(1), 113–148.

Klahr, D., & Wallace, J. G. *Cognitive development: An information-processing view*. New York: Halstead Press, 1976.

Kohler, W. *The mentality of apes*. New York: Harcourt Brace, 1927.

Larkin, J. H., McDermott, J., Simon, D. P., & Simon, H. A. *Expert and novice performance in solving physics problems* (CIP Working Paper 410). Pittsburgh, Pa.: Department of Psychology, Carnegie-Mellon University, December 1979.

LeBlanc, J. *The performance of first-grade children in four levels of conservation of numerousness and three IQ groups when solving arithmetic subtraction problems*. Unpublished doctoral dissertation, University of Wisconsin, 1968.

Lindvall, C. M., & Ibarra, C. G. Incorrect procedures used by primary grade pupils in solving open addition and subtraction sentences. *Journal for Research in Mathematics Education,* 1980, *11*(1), 50–62. (a)

Lindvall, C. M., & Ibarra, C. G. *A clinical investigation of the difficulties evidenced by kindergarten children in developing "models" in the solution of arithmetic story problems.* Paper presented at the annual meeting of the American Educational Research Association, Boston, April 1980. (b)

Lindvall, C. M., & Tamburino, J. L. *Information processing capabilities used by kindergarten children when solving simple arithmetic story problems.* Paper presented at the annual meeting of the American Educational Research Association, Los Angeles, April 1981.

Linville, W. J. Syntax, vocabulary, and the verbal arithmetic problem. *School Science and Mathematics,* 1976, *76*(2), 152–158.

Loftus, E. J. F. *An analysis of the structural variables that determine problem solving difficulty on a computer-based teletype* (Tech. Rep. No. 162). Stanford: Institute for Mathematical Studies in the Social Sciences, 1970.

Loftus, E. J. F., & Suppes, P. Structural variables that determine problem solving difficulty in computer-assisted instruction. *Journal of Educational Psychology,* 1972, *63*(6), 531–542.

McDermott, J., & Larkin, J. H. Re-representing textbook physics problems. In *Proceedings of the Second National Conference, Canadian Society for Computational Studies of Intelligence.* Toronto, Canada: University of Toronto, 1978.

Maltzman, I. Thinking: From a behavioristic point of view. *Psychological Review,* 1955, *62,* 275–286.

Markman, E. The facilitation of part–whole comparisons by use of the collective noun "family." *Child Development,* 1973, *44,* 837–840.

Marshall, G. *A study of the achievement and transfer effects of comparison subtraction and one-to-one correspondence training.* Unpublished doctoral dissertation, University of Houston, 1976.

Neches, R. *Models of heuristic procedure modification.* Unpublished doctoral dissertation, Carnegie-Mellon University, 1981.

Nesher, P. Levels of description in the analysis of addition and subtraction. In T. P. Carpenter, J. M. Moser, & T. Romberg (Eds.), *Addition and subtraction: Developmental perspective.* Hillsdale, N.J.: Lawrence Erlbaum Associates, 1981.

Nesher, P. S., & Katriel, T. *Two cognitive modes in arithmetic word problem solving.* Paper presented at the second annual meeting of the International Group for the Psychology of Mathematics Education, Onabruck, West Germany, September 1978.

Nesher, P., & Teubal, E. Verbal cues as an interfering factor in verbal problem solving. *Educational Studies in Mathematics,* 1974, *6,* 41–51.

Newell, A., & Simon, H. A. *Human problem solving.* Englewood Cliffs, N.J.: Prentice-Hall, 1972.

Norman, D. A., & Rumelhart, D. E. *Explorations in Cognition.* San Francisco: Freeman, 1975.

Novak, G. S. Computer understanding of physics problems stated in natural language. *American Journal of Computational Linguistics,* 1976. (Microfiche 53)

Paige, J. M., & Simon, H. A. Cognitive processes in solving algebra word problems. In B. Kleinmuntz (Ed.), *Problem solving: Research, method, and theory.* New York: Wiley, 1966. Pp. 51–119.

Piaget, J., & Szeminska, A. *The child's conception of number.* New York: Norton, 1952. (Original French edition, 1941.)

Polson, P., Atwood, M. E., Jeffries, R., & Turner, A. The processes involved in designing software. In J. R. Anderson (Ed.), *Cognitive skills and their acquisition.* Hillsdale, N.J.: Lawrence Erlbaum Associates, 1981.

Resnick, L. B. Syntax and semantics in learning to subtract. In T. P. Carpenter, J. M. Moser, & T. Romberg (Eds.), *Addition and subtraction: Developmental perspective.* Hillsdale, N.J.: Lawrence Erlbaum Associates, 1981.

Riley, M. S. *Conceptual and procedural knowledge in development*. Unpublished Master's thesis, University of Pittsburgh, 1981.

Roman, R. A., & Laudato, N.C. *Computer-assisted instruction in word problems: Rationale and design* (LRDC Publication 1974/19). Pittsburgh, Pa.: Learning Research and development Center, University of Pittsburgh, 1974.

Rosenthal, D. J. A., & Resnick, L. B. Children's solution processes in arithmetic word problems. *Journal of Educational Psychology*, 1974, *66*, 817–825.

Sacerdoti, E. D. *A structure for plans and behavior*. New York: Elsevier-North Holland Publishing, 1977.

Schank, R. C., & Abelson, R. P. *Scripts, plans, goals, and understanding: An inquiry into human knowledge structures*. Hillsdale, N.J.: Lawrence Erlbaum Associates, 1977.

Schell, L. M., & Burns, P. C. Pupil performance with three types of subtraction situations. *School Science and Mathematics*, 1962, *62*, (3, Whole No. 545), 208–214.

Shores, J., & Underhill, R. G. *An analysis of kindergarten and first-grade children's addition and subtraction problem-solving modeling and accuracy*. Paper presented at the annual meeting of the American Educational Research Association, San Francisco, April 1976.

Siegler, R. Three aspects of cognitive development. *Cognitive Psychology*, 1976, *8*, 481–520.

Siegler, R., & Klahr, D. When do children learn? The relationship between existing knowledge and the acquisition of new knowledge. In R. Glaser (Ed.), *Advances in instructional psychology* (Vol. 2). Hillsdale, N.J.: Lawrence Erlbaum Associates, 1981.

Simon, D. P., & Simon, H. A. Individual differences in solving physics problems. In R. Siegler (Ed.), *Children's thinking: What develops?* Hillsdale, N.J.: Lawrence Erlbaum Associates, 1978.

Steffe, L. P. The relationship of conservation of numerousness to problem-solving abilities of first-grade children. *Arithmetic Teacher*, 1968, *15*, 47–52.

Steffe, L. P. Differential performance of first-grade children when solving arithmetic addition problems. *Journal for Research in Mathematics Education*, 1970, *1*, 144–161.

Steffe, L. P., & Johnson, D. C. Problem-solving performances of first-grade children. *Journal for Research in Mathematics Education*, 1971, *2*, 50–64.

Steffe, L. P., & Thompson, P. W. Children's counting in arithmetical problem solving. In T. P. Carpenter, J. M. Moser, & T. Rombert (Eds.), *Addition and subtraction: Developmental perspective*. Hillsdale, N.J.: Lawrence Erlbaum Associates, 1981.

Stein, N. L., & Trabasso, T. Waht's in a story: Critical issues in comprehension and instruction. In R. Glaser (Ed.), *Advances in the psychology of instruction* (Vol. 2). Norwood, N.J.: Lawrence Erlbaum Associates, 1981.

Tamburino, J. L. *An analysis of the modelling processes used by kindergarten children in solving simple addition and subtraction story problems*. Unpublished Master's thesis, University of Pittsburgh, 1980.

Trabasso, T., Isen, A. M., Dolecki, P., McLanahan, A. G., Riley, C. A., & Tucker, T. How do children solve class-inclusion problems? In R. S. Siegler (Ed.), *Children's thinking: What develops?* Hillsdale, N.J.: Lawrence Erlbaum Associates, 1978.

Underwood, B. J., & Richardson, J. Some verbal materials for the study of concept formation. *Psychological Bulletin*, 1956, *53*, 84–95.

Vergnaud, G. A classification of cognitive tasks and operations of thought involved in addition and subtraction problems. In T. P. Carpenter, J. M. Moser, & T. Romberg (Eds.), *Addition and subtraction: Developmental perspective*. Hillsdale, N.J.: Lawrence Erlbaum Associates, 1981.

Wertheimer, M. *Productive thinking*. New York: Harper & Row, 1945. (Enlarged edition, 1959)

Young, R. M., & O'Shea, T. Errors in children's subtraction. *Cognitive Science*, 1981, *5*, 87–119.

Zweng, M. J., Gerarghty, J., & Turner, J. *Children's strategies of solving verbal problems* (Final rep. to NIE, grant no. NIE-G-78-0094). Iowa City, Iowa: University of Iowa, August 1979.

CHAPTER **5**

On the Representation of Procedures
in Repair Theory

Over the months or years that it takes students to master a procedure such as
ordinary place-value subtraction, their performance is characterized by many
systematic errors or *bugs* that indicate a flaw or incomplete understanding of the
procedure. Not only are there a very large variety of bugs across the population,
but the bugs a student exhibits often shift radically over short periods of time.
Nonetheless, there are developmental trends indicating how formal instruction
influences the student's (mis-)conceptions of the skill. Repair Theory (Brown &
VanLehn, 1980) aims to account for all these phenomena—the large variety of
bugs, their short-term instability, and their long-term relationship to instruction.

The theory draws its name from the belief that when a student has unsuc-
cessfully applied a procedure to a given problem he or she will attempt a *repair*.
Suppose that the student is missing a fragment of a procedural skill, either
because the fragment was never learned or maybe it was forgotten. Attempting to
rigorously follow the impoverished procedure will often lead to an *impasse*. That
is a situation in which some current step of the procedure dictates a primitive
action that the student believes cannot be carried out. In ordinary subtraction, an
impasse would follow from an attempt to decrement a zero, provided the student
knows (or discovers) that the decrement primitive has a precondition that is input
argument can't be zero. When a constraint or precondition gets violated, the
student, unlike a typical computer program, is not just apt to quit. Instead the
student will often be inventive, invoking problem-solving skills in an attempt to

197

THE DEVELOPMENT OF MATHEMATICAL THINKING

Copyright 1983 © by Academic Press, Inc.
All rights of reproduction in any form reserved.
ISBN 0-12-284780-6

repair the impasse so that he or she can continue to execute the procedure, albeit in a potentially erroneous way. Many bugs can be explained as "patches" derived from repairing a procedure that has encountered an impasse while solving a particular problem.

The repair concept has been incorporated into a formal computational theory. It postulates that each student has a *core procedure* that represents the student's current knowledge of the skill. The core procedure is what the student applies to solve test problems, do exercise problems, and interpret further instruction. Because applying the core procedure to solve a problem often involves reaching an impasse and repairing, it cannot be observed directly but only inferred. Although there are many bugs in the population, there are apparently few core procedures. The large variety of "surface procedures" is accounted for by "multiplying" a small set of core procedures by a small set of repairs. Not only are core procedures less variegated, but they are more stable than bugs. Much short-term shifting among bugs is due to applying various repairs to the underlying impasses of a stable core procedure.

A database of hundreds of subtraction bugs has been collected by testing thousands of students in all stages of subtraction instruction (Brown & Burton, 1978; VanLehn, 1981). In addition to providing evidence for the repair process, these data allowed inference of a set of several dozen core procedures, including details about the form or structure of each procedure. This enables much deeper questions to be investigated: What are the causes of core procedures? How are they related to instruction? Why were they acquired and not others?

The layers of Repair Theory can be graphically stated, using "→" to mean *explains.*

$$\text{instruction} \rightarrow \text{core procedures} \rightarrow \text{bugs} \rightarrow \text{errors}$$

Connecting the layers, the theory has three parts: (*a*) an acquisitional study that links instruction with core procedures, (*b*) an applicational study that explains idealized systematic errors (bugs) from repair of core procedures, and (*c*) an empirical study that abstracts bugs from actual student performances by filtering out their unintentional, "careless" mistakes (e.g., errors in recalling the basic subtraction facts). All three parts of the theory are under active development, so this chapter will not dwell on the details of any one beyond that which is necessary as background to its main purpose, which is to present part of the infrastructure of the theory.

Like many recent theories of complex human behaviors, such as problem solving or skill acquisition, that have been expressed computationally, Repair Theory uses a *knowledge representation language* to express knowledge held by the subject. This chapter shows that certain aspects of the knowledge representation language are crucial to the theory's success. These arguments are offered as one approach to understanding *mental representations* and their interface with cognitive development.

Knowledge representation languages have been the subject of great debate in the Artificial Intelligence community. The debate centers on how easily the language allows one to encode knowledge. This emphasis on representational ease is well placed. Several decades of experience in trying to construct computer programs that behave intelligently have convinced everyone that a good representation allows intelligent behavior to be captured quite simply, while a bad one makes this task complex or even impossible. By extension, the representational language used by a computational theory of cognition, such as repair theory, must have a profound impact on the simplicity, if not the overall conception, of the theory. It is an important issue to which psychologists should attend.

But psychologists have, for the most part, stayed out of the representational fray. The representation languages they choose are usually varieties of production systems or semantic nets. These are favored, I think, because they employ constructions, such as short-term memory and associations, that have enjoyed the attention of a great deal of psychological research. But just because constructions bearing names like "short-term memory" are a part of language does not mean that a theory employing the language depends in any strict way on the nature of those constructions. The theory could be a smashing success and yet its validation imputes no credit to the construction if it happens to be the case that the construction is not used *crucially* in the theory—that is, when other constructions could have been used in place of one bearing the fancy name, and the theory would succeed just as well. Worse yet is the case where alternatives to the advocated construction actually improve the theory. These are the kinds of trouble that a theory can get into if the relationships between representation language and the theory's predictions are not well understood.

Despite its importance, it is difficult to support a claim that a certain representation language is theoretically crucial. There are two reasons for this. One is that knowledge representations are quite indirectly related to observable behavior. Whereas one can make a fairly direct mapping between, say, a process model's actions and the subject's actions, such a direct mapping is unreliable in the case of knowledge. What subjects say or do is more plausibly the product of some interpretation or use of their knowledge. Consequently, the only way one can "see" the format of that knowledge is by mapping the observations backward through the interpretation/use processes. To support a claim about the knowledge representation language, one must use very complex arguments that bring together many backward-mapped observations. Moreover, in a theory as complex as Repair Theory, it is not sufficient to simply state the principles, show that the theory conforms to them, and assess the empirical adequacy. This would be treating the theory as a black box, making it impossible for anyone but its creators to change it or even fully understand it. To make a complex theory a useful and viable contribution, one must reveal the system of inferences that motivate its constructions. One should show which principles depend upon which components and which empirical observations. This is particularly important in the case of the representation language since there is a traditional tendency in psychology, even in information-processing psychology, to misunderstand (if not totally ignore) the effects of knowledge representations on theories.

The second reason for the difficulty in sustaining claims about knowledge representation languages lies in the logic required by such claims. To show that some feature of the language is crucial is to show that it is *necessary* in order for the theory to meet some criteria of adequacy. To show that it is *sufficient* is not enough. Indeed, any successful theory that uses a knowledge representation language is a sufficiency argument for that language. But when there are two theories, one claiming that *x* is sufficient and another claiming that *y* is sufficient, sufficiency itself is no longer persuasive. One must somehow show that *x* is better than *y*. Indeed, this sort of *competitive argumentation* is the only realistic alternative to necessity arguments. Such arguments form a sort of successive approximation to necessity. But to form competitive arguments requires knowing the alternatives. The more exhaustive the set of alternatives, the more closely the argument approximates a necessity argument. Although the use of formal knowledge representation languages in the cognitive sciences has been popular recently, the space of representational alternatives is not yet well understood. Instead, there is only a vague fog of jargon in which specific knowledge representation languages are embedded. There is as yet no understanding of what the important issues are that differentiate languages from each other and what range of alternatives exist for each issue. In short, this chapter is about to venture into an unknown space in search not only of the winning alternatives, but of the issues themselves.

This journey requires extensive preparation. The preparation involves motivating the major components of the theory at a certain medium level of detail. The level of detail is set high enough that the form of the mental representations plays no role. This provides a rich structure of foundational assertions for the argumentation about mental representations to work with. The arguments themselves have the form "if Repair Theory is to conform to the principles motivated at the medium level of detail and yet produce accurate empirical predictions (i.e., have the right low-level detail), then the representation language *must* have such-and-such an attribute." The first two sections of this chapter describe and motivate the theory at the medium level of detail, and the next two sections present the arguments.

More specifically, the first section discusses the empirical and applicational parts of Repair Theory, that is, how bugs are related to student performance, and how repair of core procedures generates bugs. The theory's formal expression is a *process model of application*—the interpretation and repair of core procedures. The second section presents the current thinking on how to account for the acquisition of core procedures, which differs from that presented in Brown and VanLehn (1980). It is part of a research paradigm that has emerged rather recently in developmental psychology. (For a review, see Keil, 1981.) Rather than proposing a process model for acquisition, it aims to discover *constraints on sequences of knowledge structures* that not only sharply limit the class of naturally learnable structures, but also the structural relationships that an intermediate state of knowledge can have with its predecessors and successors. The second section introduces this paradigm and applies it to a domain that has not felt its touch before: skill acquisition in formal instructional settings. The third and fourth sections present the target arguments of the chapter, concerning respectively the mental representations for the execution (or "run-time") state of core procedures and their long-term (or "schematic") form. The fifth section comments on the methodology and technology used to discover the arguments presented in this chapter.

THE REPAIR PROCESS

The initial task chosen for investigation is ordinary multidigit subtraction. Its main advantage, from a psychological point of view, is that it is a virtually meaningless procedure. Most elementary school students have only a dim conception of the underlying semantics of subtraction, which are rooted in the base ten representation of numbers. When compared to the procedures they use to operate vending machines or play games, subtraction is as dry, formal, and disconnected from everyday interests as the nonsense syllables used in early psychological investigations were different from real words. This isolation is the bane of teachers but a boon to the psychologist. It allows one to study a skill formally without bringing in a whole world's worth of associations.

In the last several years, a detailed study of thousands of student's subtraction performances has provided an extensive, precise catalog of subtraction misconceptions (Brown & Burton, 1978; Burton, 1981; VanLehn, 1981). This catalog is the major database used to develop and validate Repair Theory. It is worth a moment to discuss its nomenclature and the method of its collection before moving on to describe the repair process.

Bugs Are Precise Descriptions of Systematic Errors

It has long been known that many of the errors that students make while learning a procedural skill, such as ordinary place-value subtraction, are *system-*

atic in that the errors appear to stem from consistent application of a faulty method, algorithm, or rule (Ashlock, 1976; Brownell, 1941; Brueckner, 1930; Buswell, 1926; Cox, 1975; Lankford, 1972; Roberts, 1968). These errors occur along with the familiar unsystematic or "careless" errors that occasionally occur in expert performance as well as the learner's behavior. The common opinion is that careless errors or *slips,* as current research prefers to call them (e.g., Norman, 1981), are performance phenomena, an inherent part of the "noise" of the human information processor. Systematic errors on the other hand are taken as stemming from mistaken or missing knowledge about the skill, the product of incomplete or misguided learning. They are the results of *misconceptions.* By studying where conceptions of procedures break down, insight can be gained into the structure of knowledge about procedures.

The basic idea is that misconceptions could be formally represented and precisely described as *bugs* in a correct procedure for the skill. In brief, a bug is a slight modification or perturbation of a correct procedure. The bug-based notation is complete in the sense that it not only describes which problems the students gets wrong, but the content of each wrong answer and the steps followed by the student in producing it. The bug-based notation is the basis of the DE-BUGGY diagnostic system (Burton, 1981) and its predecessor, BUGGY (Brown & Burton, 1978). To illustrate the notion of bug, consider the following subtraction problems, which display systematic errors:

```
                                             4
    1                                 1      5
    2        7      0 17    6         2      6                    4
  3¹0¹6     8¹0   1 8¹3    7¹0 2    3 0¹0¹5  7¹0¹0¹2   3 4      2 5¹1
  -1 3 8   - 4   - 8 5    - 1 1    -    2 8  - 2 3 9  -1 4     - 4 7
    7 8     7 6     8 8    6 9 1    1 0 8 7  4 8 7 3   2 4     2 4 4
```

(The small numbers stand for the student's scratch marks.) One could vaguely describe these problems as coming from a student having trouble with borrowing, especially in the presence of zeros. More precisely, the student misses all the problems that require borrowing from zero. One could say that he or she has not mastered the subskill of borrowing across zero. This description of the systematic error is fine at one level: It is a testable prediction about what new problems the student will get wrong. It predicts for example that the student will miss 305 − 117 and will get 315 − 117 correct. Systematic errors described at this level are the data upon which several psychological and pedagogical theories have been built (e.g., Durnin & Scandura, 1977). It has become common to use testing programs based on this notion for placement, advancement, and remediation in structured curricula, such as mathematics. Such testing programs are often labeled *domain referenced* or *criterion referenced.*

Once one looks beyond what *kinds* of exercises the student misses and looks at the actual steps taken in answering it, one finds in many cases that these step sequences can be precisely *predicted* by using a procedure that is a small perturbation in the fine structure of the correct procedure. Such perturbations—called bugs—serve as a precise description of the errors.

The "student" whose work we just considered has a bug called Borrow-Across-Zero. This bug modifies the correct subtraction procedure by deleting the step wherein the zero is changed to a nine during borrowing across zero. (This bug and others like it are described thoroughly in the appendices of VanLehn, 1981 and Brown and VanLehn, 1980.) This deletion creates a procedure for answering subtraction problems. As a hypothesis, it predicts not only which new problems the students will miss, but also what each answer will be and the sequence of steps that will be used in obtaining it. Since bug-based description of systematic errors predict behavior at a finer level of detail than missing-subskill/ domain referenced testing, it has the potential to form a better basis for cognitive theories of learning and errors.

The bug description is an idealization of behavior in that it excludes slips. The tens column of the third problem shows a typical slip. The student has answered $17 - 8 = 8$. Slips are filtered from the raw data not because they are uninteresting, but because Repair Theory is probably not fine-grained enough to model them. One interesting trait of slips is "echoing." When a student makes a facts error slip, such as $17 - 8$, the digit written is often the same as one that has recently been the focus of attention, in this case an 8 (also, 8 was the answer in the preceding column). Such patterns as echoing fuel the ongoing development of theories of slips (Norman, 1981). Repair theory concerns itself more with what the student intended to do rather than what the student actually did. In a sense Repair Theory is a *competence* theory: It studies what people *can* do rather than what they *do* do. Greeno and Brown (1981) defend the importance of this kind of study. Burton (1981) and VanLehn (1981) discuss the difficulties of implementing this methodology objectively, and in particular, the algorithms used to filter out slips from the subtraction data.

It is often the case that a student has more than one bug at the same time. Indeed, the example given earlier illustrates co-occurrence of bugs. The last two problems are answered incorrectly, but the bug Borrow-Across-Zero does not predict their answers. (It predicts the two problems would be answered correctly.) A second bug called Diff-N $- N = N$ is present. When the student comes to subtract a column where the top and bottom digits are equal, instead of writing zero in the answer, he or she writes the digit that appears in the column.

Using DEBUGGY, thousands of students have been analyzed. The chart that follows summarizes the widespread extent of the phenomena by showing that about 40% of the students making errors in the two largest samples, one domestic and one foreign, were analyzed as having bugs. (These and other data following are taken from VanLehn, 1981; the 40% figure has been confirmed independently for British 10-year-olds by Young and O'Shea, 1981.):

Nicaraguan grades 5 & 6		American grades 3 & 4		Category
37		112		No errors
116	(9%)	223	(22%)	All errors due to slips
505	(39%)	417	(40%)	Most errors due to bugs
667	(52%)	386	(37%)	Cause of most errors unknown
1325		1138		Totals

It was common for a student to have more than one bug. Of the 417 American students that DEBUGGY analyzed as having bugs, 150 (36%) received a multibug diagnosis. Most of these diagnoses consisted of two or three bugs, but there were several cases of four bugs co-occurring. Overall, 77 distinct bugs occurred. (In this chapter *occurred* means that some student had the bug as his or her diagnosis, or if he or she was diagnosed as having a set of bugs, as part of this diagnosis.) The large variety of bugs and the complex patterns of relationships among them are a rich, precise database for developing and verifying a theory of how people understand and misunderstand procedures.

An Informal Introduction to Repair

When a student gets stuck while executing a flawed procedure (the product of mislearning or forgetting), he or she is unlikely to just quit as a computer does when it can't execute the next step in a procedure. Instead, the student will do a small amount of problem solving, just enough to get "unstuck" and complete the subtraction problem. These local problem-solving strategies are called *repairs* despite the fact that they rarely succeed in rectifying the broken procedure, although they do succeed in getting it "unstuck." Repairs are quite simple tactics, such as skipping the operation that can't be performed or backing up to the last branch point in the procedure and taking a different path. They do not in general result in a correct solution to the subtraction problem, but instead result in a buggy solution. For example, suppose the student has never borrowed from zero. The first time the student is asked to solve a borrow-from-zero problem, such as (a),

$$
\begin{array}{cccc}
& & & 1 \\
& & 2 & 2 \\
\text{(a)} \quad 3\ 0\ 5 & \text{(b)} \quad 3^{1}0^{1}5 & \text{(c)} \quad 3^{1}0^{1}5 \\
-\ \ 4\ 8 & -\ \ 4\ 8 & -\ \ 4\ 8 \\
\hline
& 2\ 6\ 7 & 1\ 6\ 7
\end{array}
$$

the student begins processing the units column by attempting to borrow from the tens column and immediately reaches an impasse because zero cannot be decremented. The student is stuck and so does a repair. One repair is simply to skip the decrement operation. This leads ultimately to the solution shown in (b). If the student uses this repair to the borrow-from-zero impasse throughout a whole subtraction test, he or she will be diagnosed as having a bug called Stops-Borrow-At-Zero. Suppose the student chooses a different repair, namely to relocate the decrement operation and do it instead on a nearby digit that is not zero, such as the nearest digit to the left in the top row, namely the three. This repair results in the solution shown in (c); the three has been decremented twice, once for the (repaired) borrow originating in the units column and once for the borrow originating in the (unchanged) tens column. If the student always chooses this repair to the impasse, he or she will be diagnosed as having the bug Borrow-Across-Zero.

The preceding story illustrates the repair process. It seems like plausible human behavior. What is needed next is empirical evidence, both to verify its existence and sharpen the understanding of how it works. Then we can move on to a formal model.

Empirical Motivation for Repair

A certain subset of the bugs shows a clear *cross product* or matrix-like pattern. In this pattern, concrete empirical evidence for the repair process is found. For simplicity, the pattern will be illustrated in terms of just four members of the subset. The four bugs are

1a. Smaller-From-Larger
1b. Zero-Instead-of-Borrow
2a. Smaller-From-Larger-Instead-of-Borrow-From-Zero
2b. Zero-Instead-of-Borrow-From-Zero

The first two bugs are related in that they miss problems that require borrowing. Smaller-From-Larger answers columns that require borrowing with a number that is the absolute difference of the two numbers. Zero-Instead-of-Borrow writes zero in such columns. The only problems that these bugs will answer correctly are those that do not require borrowing. The following problems illustrate these two bugs:

Smaller-From-Larger:	345	345	207
	−102	−129	−169
	243 ✓	224 ✗	162 ✗

Zero-Instead of-Borrow: | 345 | 345 | 207 |
|---|---|---|
| −102 | −129 | −169 |
| 243 √ | 220 × | 100 × |

Both bugs get the first problem correct, and miss the other two because they involve borrowing. (Correctly answered problems are marked with √, and incorrectly answered problems with ×.) This pattern of correct and incorrect answers unifies these bugs.

The other two bugs will miss only problems that require borrowing across a zero. Their answers to the the same problems would be:

		3	1
Smaller-From-Larger-Instead-of-	3 4 5	3 4¹5	2¹0 7
Borrow-From-Zero:	−1 0 2	−1 2 9	−1 6 9
	2 4 3 √	2 1 6 √	4 2 ×

		3	1
Zero-Instead-of-	3 4 5	3 4¹5	2¹0 7
Borrow-From-Zero:	−1 0 2	−1 2 9	−1 6 9
	2 4 3 √	2 1 6 √	4 0 ×

Both bugs get the second problem correct because it involves only simple borrowing. They miss the third problem because it involves borrowing from a zero. So, one dimension of the cross product pattern has been established: Bugs are grouped by the pattern of correct and incorrect answers.

The crucial second dimension of the cross product pattern is seen in the fact that the ways that the bugs in the second group miss problems *parallel* the ways that the bugs in the first group miss problems. Smaller-From-Larger-Instead-of-Borrow-From-Zero answers the units column with absolute difference, just as Smaller-From-Larger used absolute differences to answer the problems it missed. Both Zero-Instead-of-Borrow and Zero-Instead-of-Borrow-From-Zero answer the columns they miss with zero. This pattern in the answers from borrow-from-zero columns relates to the answers of the first group on borrow columns. In short, what we have here is a cross product pattern among four bugs. It could be represented informally by:

Let C = {borrow, borrow-from-zero}
Let R = {absolute-difference, max-of-zero-and-difference}
Claim: any pair in the set C × R represents a bug. Roughly speaking, the R operation is substituted for the C subprocedure.

For example, the pair ⟨borrow, absolute-difference⟩ represents Smaller-From-Larger since its hallmark is absolute differences instead of borrows.

From the standpoint of Repair Theory, the cross product pattern is just the repairs revealing themselves in the data. In Repair Theory, the set C reflects the set of *core procedures*. Core procedures are the product of mislearning or forgetting. They are named this way because, like an apple's core, they are not directly observable. Their structures are obscured by the effects of repair. The set R reflects the set of repairs. When a core procedure is executed on a problem that causes it to reach an impasse, the impasse is patched with one of the repairs, resulting in a bug.

A crucial fact about the repair process comes out clearly in the cross product pattern. It is the *independence* of repairs and impasses. Every repair is applicable to every impasse (core procedure). In principle, a bug will be found for each pairing of an applicable repair with a core procedure.

Of course, some pairs are much more popular than others, and some core procedures are more common than others. Combining an unpopular repair with an uncommon core procedure sometimes predicts a bug that has not yet been observed. This is indeed the case with the aforementioned four bugs, whose frequencies of occurrence in the 1138-student American sample are as follows:

	Borrow	Borrow-from-zero
Absolute-difference	124 (SFL)	5 (SFLIBFZ)
Max-of-zero	10 (ZIB)	0 (ZIBFZ)

The core procedure that doesn't know borrowing is more common than the one that doesn't know borrowing across zero. The repair that results in an absolute difference is more popular than the one resulting in taking the maximum of zero and the column difference. Impasse-repair independence explains why there have as yet been no occurrences of the bug resulting from pairing the unpopular repair with the unpopular core procedure. Note that this frequency distribution could not be explained in this way without reference to the repair process, which adds more evidence for the existence of that process.

Independence of repairs and impasses is one central feature of the repair process. Another relates to *when* the repair process is carried out by the subject. It could be that repair is something like forgetting or mislearning. It could happen while the student is sleeping, or watching the teacher, or explaining the procedure to a friend. All one can see in the cross product pattern is the result of repair, and not *when* it happened. However, there are data indicating that repair actually occurs during solution of problems, as the story earlier implied. These data involve a behavior labeled *bug migration*.

Bug migration is the phenomenon of a student switching among two or more bugs during a short period of time with no intervening instruction. The bugs the student is switching among are related in that they result from applying different repairs to the same impasse. That is, the student appears to have the same core procedure throughout the period of observation, but chooses to repair its impasses differently at different times. Switching repairs establishes that the repair process takes place during that period. Two periods of bug migration have been observed: intertest and intratest.

Intratest bug migration (which was called *tinkering* in earlier reports) occurs when a student answers part of the test as if he or she had one bug, and the other part as if he or she had a different bug. That is, the student appears to have the same core procedure throughout the test, but chooses to repair its impasse differently on different test problems. This establishes that repair happens during a test. Many instances of intratest bug migration have been observed (VanLehn, 1981). Figure 5.1 presents one. (Figure 5.1 is an exact reproduction of a test taken by student 22 of class 34. She misses only six problems, namely the ones that require borrowing from zero. The first two problems she misses [the second and third problems on the fourth row] are answered as if she had the bug Stops-Borrow-At-Zero. That is, she gets stuck when she attempts to decrement a zero and repairs by skipping the decrement operation. The next two problems she misses [the first two problems on the last row] are answered as if she had the bug

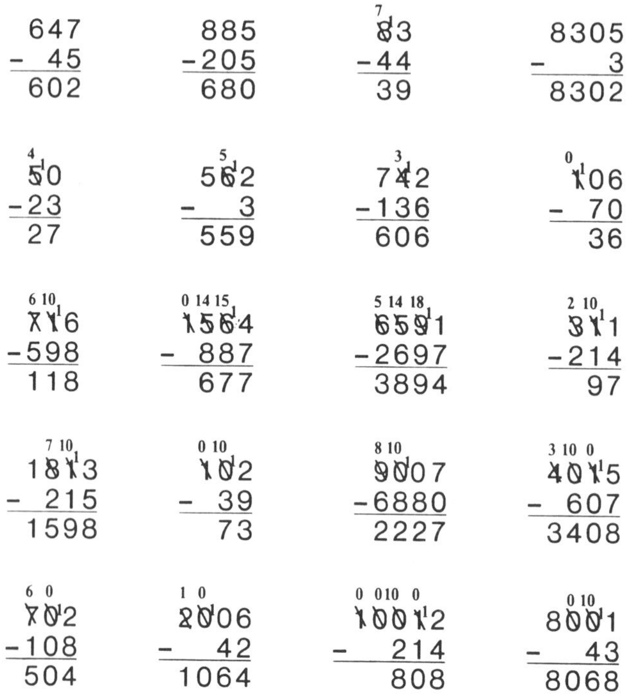

Figure 5.1. An example of intratest bug migration.

Borrow-Across-Zero. She hits the same impasse, but repairs by relocating the decrement leftward. On the third problem of the last row, she uses two repairs within the same problem. For the borrow originating in the tens column, she backs up from the decrement-zero impasse and writes a zero as the answer in the tens column [as if she had the bug Zero-Instead-of-Borrow-From-Zero]. In the hundreds column, she takes the same left-relocate repair that she used on the preceding two problems. On the last problem, she reverts to the original repair of skipping the stuck decrement of both borrows.)

The student of Figure 5.1 is typical in that her repairs occur in "runs." The first two repairs are one kind, the next two are another, and so on. This observation suggests that there can be a temporary association of an impasse with a repair. In Repair Theory, these pairs are called *patches*. Apparently, the first time the student of Figure 5.1 hit the impasse, she searched for an applicable repair and not only used it, but created a patch to remember that she used it. On the next problem, she again encounters the impasse, but instead of searching for a new repair, she just retrieves the patch and uses its repair. She completes the next problem without encountering the impasse, which is apparently enough to cause her to forget her patch, since the next time she hits the impasse, she repairs it a new way. Either the patch was forgotten during the nonimpasse problem or she chose to ignore it and try a different repair. The latter possibility is supported by her behavior at the end of the test, where she is applying different repairs for each impasse even when the impasses occur in the same problem. In short, there seems to be some flexibility in whether patches are ignored and perhaps also in how long they are retained.

Intertest bug migration is detected by testing students twice a short time apart (say, 2 days) with no intervening instruction. It is the phenomena of a student having a consistent bug on each test, but not the same bug. The bugs are related in that they can be generated by different repairs to the same impasse. It appears that the student has retained the same core procedure between the two tests, but the patch that was used on the first test was not retained. Instead, a new repair was selected, stored in a patch, and used consistently throughout the second test. Although intertest bug instability is the norm rather than the exception (only 12% of the bugs remained stable in one study (VanLehn, 1981); Bunderson (1981) reports no stable bugs at all), bug migration accounts for well over half of it (VanLehn, 1981).

Both kinds of bug migration were predicted in advance of their observation (Brown & VanLehn, 1980). They fall out as a natural consequence of viewing the repair process as modifying the execution (short-term) state of the processor that interprets the stored core procedure. An alternative is to view repair as modifying the core procedure. This "core procedure modification" view accounts for stable bugs, but not for bug migration. (Bugs are sometimes held for months or years [Brown & Burton, 1978; VanLehn, 1981].) Repairing the processor state leaves open the question of whether patches are ever abstracted and stored in long-term memory as modifications to the core procedure—it is not necessary to assume such an abstraction process, even though it is highly plausible since it suffices to explain long-standing bugs as a long-term patch retention (in fact, that is the approach used by the formal model that will be introduced in the next section). Whether patches are incorporated into the core procedure has been left unresolved for the time being.

So, from this overview of the data two important aspects of the repair process have been abstracted, as well as a variety of evidence for the existence of the process itself. First, in principle any repair can be applied to any impasse. Second, repairs are modifications to the execution state of the core procedure's interpreter. The basic patterns in the bug data were the cross-product pattern and the bug migration classes. (NB: A bug migration class is the set of bugs that switch with each other.)

A Process Model for Interpretation and
Repair of Core Procedures

The previous section introduced some of the insights that Repair Theory is based on. This section presents the theory more formally.

Repair Theory has two parts: a constraint-based account of how core procedures arise, and a process model that interprets and repairs a core procedure to generate or simulate a buggy behavior. This section discusses the second part, which is referred to as the application model since it applies the core procedure to given problems.

Figure 5.2 is a block diagram of the model's components. The model is given a core procedure and a sequence of stimuli. For subtraction, the stimuli represent subtraction problems as they appear on the test page. The model solves each problem, producing a sequence of actions that are asserted to match a subject's actions while solving the same problems. In the sense that its action sequences are expected to map onto subjects' action sequences, idealized as bugs, it is a process model. No other claims are made about the mapping. In particular, the speeds at which the actions are generated is considered irrelevant. Also, the theory is neutral about the ontological status of the model's components.

The application model is composed of the *interpreter* and the *local problem solver*. The interpreter executes the core procedure on each input, one after the other. Executing the core procedure involves constant changes to the execution state. At any given time execution state reflects not only where in the core procedure the interpreter is, but what aspects of the external state, namely the test problem, the interpreter is attending to. Reaching an impasse occurs if an attempt

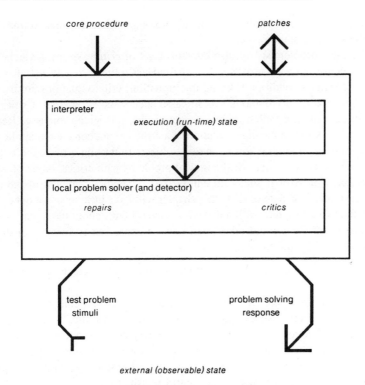

Figure 5.2. Components of the application model.

is made to execute a primitive action in a situation that violates the action's preconditions, such as attempting to decrement a zero or a blank. When this happens, control shifts from the interpreter to the local problem solver. If there is a patch for this impasse in the patch set, the patch's repair is applied. (The theorist can override this in order to match bug migrations—the default actions of the model make it produce stable bugs.) Otherwise, the local problem solver calculates which repairs are applicable, and one is selected. (The theorist can control which of the applicable repairs is selected, which in turn controls which bugs the model generates.) A patch pairing the selected repair with the impasse is created and stored in the patch set. (NB: The patch retention is not yet fully understood; although it will be shown that execution state must be a stack and not a buffer, no similar claim can be made about the architecture of patch storage.)

Regardless of whether the repair was retrieved from a patch or searched out, the local problem solver applies the repair, which modifies the interpreter's execution state. Now control switches back to the interpreter and the core procedure continues, but not of course from exactly where it left off. Some core procedures never reach an impasse on any problem. Others hit several impasses, sometimes on the same problem.

The local problem solver also contains a set of *critics*. A critic watches the core procedure as it's interpreted or repaired and objects if any action would violate the critic's condition. During interpretation, critic violations are handled just like precondition violations—an impasse occurs and is repaired. Critics also figure in determining which repairs are applicable at a given impasse. Repairs that would cause an immediate violation of a critic are not considered applicable. A typical critic for subtraction is one that objects to writing a two-digit column answer. There are a number of theoretical problems with critics, some of which are mentioned in Brown and VanLehn (1980). Since development of the critic idea is currently in a state of flux, this chapter will have little to say about critics.

In the preceding motivational section, several facts about the repair process were uncovered. These and others can now be expressed as formal constraints that are obeyed by the model:

Impasse-repair independence: Any impasse can be repaired by any applicable repair.

Local manipulation: The only information that repairs can change is the interpreter's execution state. Most commonly, they substitute an action for the "current" action, that is, the action the interpreter was about to execute but could not. They cannot execute the action themselves nor change the core procedure. All they can do is reset the execution state so that the interpreter does something different when control returns to it.

Core procedure immutability: The application model cannot change the core procedure. It does, however, change the set of patches, and this is how stable bugs and various patterns of bug migrations are captured.

Local applicability: A repair is applicable if and only if its manipulation of the execution state allows the interpreter to run. That is, repairs must "unstick" the interpreter. It may only be unstuck for a moment—a new impasse might occur immediately after the patched action is executed by the interpreter. In fact, it is rather common for two impasses to occur in a row. Another view of this constraint is that repairs cannot "look ahead" to see if their manipulation will cause ill effects later on. The local problem solver can only check preconditions and critics in the execution state that the proposed repair's manipulation would create. If none is violated, then that repair is applicable.

Domain independence: Repairs are domain independent. This is, the same set of repairs is found in other tasks besides subtraction.

One purpose of these constraints is to limit the degrees of freedom or *tailorability* of the model. The existence of a given repair can only be inferred from data, in this case by mapping backward from bug migration classes and cross-product patterns of bugs. There is no a priori exhaustive set of repairs. Since the set of bugs must be kept open, it is very important to constrain what the members of that set can do. If a repair can be arbitrarily powerful, then virtually any behavior can be "explained" by postulating a repair to fit it, which makes the theory irrefutable and vaccuous. Some of these principles directly constrain what repairs can be, as the *local manipulation* principle does. Later, more locality constraints will be proposed and play crucial roles in arguments.

Another way to limit tailorability is to make the theorist pay a heavy price for adding ad hoc repairs. That is, any change to increase the empirical coverage must make other predictions that may or may not be correct. In short, changes have *entailments*. For example, the impasse-repair independence principle and the local applicability principle work together to insure that adding a new repair in order to generate a certain observed bug will cause the theory to predict many new bugs (surface procedures), namely all those that are derived by applying the new repair to all other impasses. Some of these predicted bugs may exist, in which case the addition is good. But more often, the procedures are not bugs and in fact may be such absurd predictions that it is clear that the theory should not be making them. So, adding a repair may entail making many dubious if not incorrect predictions. By strict adherence to the principles of the theory, tailorability can be limited.

Observational, Descriptive, and Explanatory Adequacy

For the purposes of this chapter, it is sufficient to take a relatively well-known set of criteria for evaluating the theory (see Brown & VanLehn, 1980, for a different set). Pioneered by Chomsky (1965), observational, descriptive, and explanatory adequacy have become a standard in linguistics. In the context of Repair Theory, they have a particularly crisp characterization.

Observational adequacy is the ability of the theory to cover its observations. Repair theory currently uses two kinds of data. Hence, there are two instantiations of the criterion:

1. The model should be able to generate all the bugs.
2. The model should be able to generate all the bug migration classes.

Other kinds of observations could perhaps be enlisted later in testing observational adequacy. In particular, studying the co-occurrence frequencies of bugs (i.e., how often a pair of bugs occur together as part of a multiple bug diagnosis) and the overall frequencies of bugs seems a fruitful direction.

Descriptive adequacy measures the ability of the theory to cover "potential'' facts about behavior, which by their very nature cannot be observed. In Repair Theory, the most common ''potential'' facts are *star bugs*. Sometimes the model produces behavior that not only does not simulate any observed bug or bug migration, but is so absurd that it is highly doubtful that any subject will behave like that. Such absurd behaviors are called *star bugs* (after the linguistic convention of placing a star before sentences that are not a part of the language). A trivial example of a star bug is the model going into an infinite loop. A second, more subtle example is the behavior the model exhibits when it is able to borrow perfectly, even borrowing across several zeros correctly when required to, and yet despite its borrowing expertise, it is unable to complete the borrowing by subtracting the column that originated the borrow and leaves those columns' answers blank. This leads to problems answered like this:

$$
\begin{array}{r}
{\scriptstyle 8\ 9} \\
\mathbf{9\ 0^{1}5\ 7} \\
\underline{-1\ 3\ 8\ 2} \\
7\ 6\quad\ 5
\end{array}
$$

A star bug can never be an observation in the sense that bugs and bug migration classes are observations. Its assertional content is that a certain behavior will *never* be observed. That modality cannot be directly tested.

Descriptive adequacy insists that the model generate no star bugs. Generating star bugs indicates that some constraint upon it is missing; the theory is at best incomplete and at worst just plain wrong.

I will sometimes lump descriptive adequacy and observational adequacy under the nonstandard term *empirical adequacy*.

Explanatory adequacy is the ability of the theory to allow constraints on acquisition to be formulated. This is perhaps the most important measure of the three, and also the most subtle. The idea behind it is that acquisition is such a complicated process that it is best to approach it gradually by first finding out what constraints it obeys. This paradigm will be explained in detail in the next section. Put briefly here its instantiation for Repair Theory is that the bug database is mapped backward, so to speak, through repair in order to *deduce* the set of core procedures that spans the developmental period. Not only can one deduce what each core procedure does, one can infer something about its *form* as well. These inferences are some of the most interesting and difficult in the theory. They will be presented in detail in the following sections. It suffices to say here that working with the bug data allows one to discover the content and even the form of core procedures. Knowing the set of ''observed'' core procedures puts one in an excellent position to determine constraints that are obeyed by *all* core procedures, which not only makes testable predictions about core procedures that have not yet been observed, but leads perhaps to an understanding of core procedure acquisition. Changes to the theory (e.g., a new syntax for the core procedure representation language) that allow a new constraint on the "observed" core procedures to be formulated increases the explanatory adequacy of the theory.

The empirical adequacy of the original version of Repair Theory was reported by Brown and VanLehn (1980). Using a highly constrained set of nine core procedures, it generated 21 of the observed bugs (i.e., about one quarter of the observed bugs) and 1 star bug. It also predicted the eventual observation of 10 bugs and several bug migration classes. Empirical work since that first report on Repair Theory verified the existence of one of the 10 predicted bugs, as well as establishing the existence of several predicted bug migration classes (VanLehn, 1981). With a less constrained set of core procedures, the theory generated 43 of the observed bugs (i.e., about one half). Due to the way the theory was tested with this larger core procedure set, the sets of star bugs, predicted bugs, and bug migrations that it generated are not known, except that they include the corresponding sets generated by the more constrained set of core procedures.

The main difference between these versions of the theory and the one discussed here is that core procedures are being generated in a different way. This change was made mostly in order to increase the theory's explanatory adequacy, but it also improved the descriptive adequacy by eliminating the generation of the star bug.

However, the point of this chapter is not to revise the theory in order to improve its empirical adequacy. Rather, it is to present the *connection* between empirical data and existing theory. Even though the theory has been elaborated beyond even the version reported here and its predictions have improved, that development is proceeding so rapidly that it would be pointless to report it now in the middle of its flight. On the other hand, the arguments behind the theory reported here are completely stable. So far, the ongoing revisions set them in a new context but do not affect their validity. Establishing the empirical or explanatory necessity of a certain architecture or constraint seems to fix it as being a permanent feature of succeeding versions. It would perhaps not be too strong to say that adding new arguments to this theory, and perhaps any theory, seems much more important than adding more data points to its empirical coverage.

CONSTRAINTS ON ACQUISITION OF NONNATURAL KNOWLEDGE

Recent research has viewed the child's acquisition of mathematical skills as a highly structured phenomenon. Other chapters in this volume show that development of even very simple procedural skills, such as mental arthmetic (Resnick), are characterized by intermediate stages of well-formed, stable procedures. The emphasis of this research is on accurate description of the child's knowledge and how it evolves in a school setting. In the past, developmental psychology tended to avoid studying learning that is accompanied by formal instruction. On the other hand, educational research tended to describe intermediate states of skill acquisition in prescriptive terms, such as *bad habits of work, systematic errors,* or *weakness in component skills.* A recent trend is to view the misconceptions that students pass through from a developmental rather than an educational standpoint.

It has been pointed out (Keil, 1981) that there are two traditional views of cognitive development: as the acquisition of knowledge through simple and general mechanisms (learning theory views) and as a sequence of radical restructurings of knowledge (stage theory views). A new viewpoint has emerged recently. It emphasizes what remains constant in the developing cognition. It looks for constraints that are obeyed by knowledge at all stages. In a sense, this view is a natural outgrowth of the earlier views. Learning theories studied acquisition over short periods of times, often a single laboratory session. Stage theories chopped a long-term development into medium-sized lengths of time, stages. The new view studies the whole of a long-term development in order to characterize what is constant about the knowledge throughout its acquisition.

This might seem a strange thing for a developmental psychologist to do. How could knowing what is the same about knowledge at all stages be important in understanding how it changes? The crucial assumption is that the properties of knowledge that are observed to be constant are in fact *constraints* on the acquisition process. They impose real limitations on the output of whatever mechanisms implement learning. Or perhaps they are a consequence of the structure of those mechanisms or the mental representations the mechanisms work with. How these constraints are imposed by processes, mechanisms, or representations is not yet known. Although the proponents of this view tend to speak in terms of a ''generate and filter'' mechanism, where simple induction generates new knowledge from experience while the constraints filter some of it out, this is not meant, I think, as a proposal for an actual learning mechanism. The mechanisms are unknown.

It can be objected that merely to say that there are constraints on the knowledge that can be acquired is vaccuous. The mind is not a *tabula rasa.* Since there is an infinite set of hypotheses that are consistent with any finite set of observations, to explain how people converge on knowledge states that are identical (or at least very similar) requires postulating some kind of prior constraints. So, to say there are constraints is nothing new.

The rejoinder to this objection is that the constraints that have been found so far are very strong and very specific to the domain of knowledge they apply to. For example, constraints on the syntax of languages mention clauses and noun phrases—which are certainly domain specific terms—in order to sharply limit syntactic transformations of each with respect to the other (Chomsky, 1980). So the substance of the constraint theory view lies in the strength and domain specificity of constraints. One of its leading proponents, F. C. Keil (1981), puts it succinctly:

> Knowledge acquisition cannot proceed successfully unless the inductive devices that apply to various cognitive domains are constrained in their outputs. This much is a logical necessity. The important and more controversial claim to be made here is that, when one examines what sorts of constraints are present, one discovers that they are highly elaborated and domain-specific, and that such specificity is necessary to explain the ease and rapidity with which so much of human knowledge is acquired. The solution to the mystery of how children acquire so much so fast lies not in the formulation of ever more powerful and sophisticated induction procedures, but rather in the specification of how relatively simple induction procedures are limited in particular domains [pp. 199–200].

The work reported in this chapter grew out of application of the constraint theory view to the development of skill in mathematical procedures. It was originally thought that procedural skill might be a domain like language or vision in that there would be "highly elaborated and domain-specific" constraints on its knowledge structures. This led to a direct application of Chomskyan methods to its study. The results of these initial studies, which are presented next, prompted a major revision of the approach.

Two Approaches Fail: Chomsky's and Keil's

The purest Chomskyan approach is to establish such strong constraints on the output of acquisition that all procedures that meet these constraints are good core procedures. That this approach is Chomskyan can be seen by comparing it to the Chomskyan approach to grammar acquisition. Such strong constraints are placed on grammars that the learner has no choice but to learn a natural language. The language environment that the child grows up in (e.g., English, French, Walbiri) is needed only to select among the few alternatives that the constraints have left open, thereby determining which natural language is acquired. Mapping the Chomskyan approach to acquisition of procedural skills involves substituting "subtraction" or "addition" for "English" or "French." The child cannot help but learn a core procedure of a certain highly constrained form. The goal of the theory, on this approach, is to uncover what the constraints on core procedures are.

This approach gains plausibility when one considers that grammars and procedures are formally isomorphic. Mathematically, it has been shown that any context-free grammar can be converted to a push-down automation and vice versa. More realistically, transformational grammars and augmented transition nets are generally held to be notational variants at one level, and it is highly likely that any procedure can be expressed as an augmented transition net. So, it would not be surprising if the constraints-on-grammars approach could be applied to acquisition of procedures.

Strong constraints on the form of core procedures were indeed discovered with the Chomskyan approach, but they were not strong enough to restrict the set of computations that could be expressed as core procedures. The problem is that whereas there are good arguments for expressing core procedures in a certain formal language as opposed to any other, that language is nearly isomorphic to lambda calculus. It is widely accepted that lambda calculus has sufficient expressive power to represent any procedure (i.e., Church's hypothesis). To say that a core procedure can be any procedure expressible in this lambda calculus-like language is equivalent to saying a core procedure can be any procedure, which is saying nothing at all. The language constrained the *form* of the core procedures, but did not narrow the set of core procedures that could be expressed. In other words, putting constraints on the language used to express core procedures is not enough. The other approaches retain the constraints discovered with the Chomskyan approach and add more.

A second problem with the pure Chomskyan approach is that it only explains why mature knowledge has the form it does. It does not account for patterns in the development of knowledge. Keil's work on ontological knowledge is a good example of this incompleteness. He found that a certain kind of ontological knowledge was subject to the constraint that it be dendralic, that is, a tree (Keil, 1981). However, he also found that children's trees were "collapsed" versions of adult trees. He states, "The collapses are highly systematic, involving more than simple conformity to the [tree] constraint, [p. 208]." That is, a child's ontological knowledge was not just any tree, it was a certain kind of homomorphic image of an adult tree. This observation cannot in principle be captured by unary constraints on knowledge states, but they can be expressed by binary constraints: for example, a structure preserving transformation (homomorphism) relating the child's knowledge to a certain ideal or adult knowledge. Similar observations in the domain of procedural skills led Brown and VanLehn (1980) independently to adopt the same homomorphism approach as Keil.

In Brown and VanLehn (1980), all core procedures were generated by deleting parts of a mature, adult core procedure. Each deletion generated an incomplete core procedure, which like Keil's collapsed ontological trees, characterized a child's partially developed knowledge. In general, this was a rather successful approach in that it lent itself well to an exhaustive formal treatment, the precision of which led to deeper insights into the nature of acquisition. However, the deletion approach itself was abandoned due to some problems that will now be discussed.

As a step toward understanding acquisition, one needs to know which of all the possible deletions to a core procedure in fact occur. Thus, deletion was formalized as an operator that applied freely, deleting any rule or set of rules (the major syntactic unit in the language for expressing procedures is a rule). As expected, unconstrained deletion sometimes generated star bugs, that is, procedures that are so absurd that it is doubtful that they will ever be observed as bugs. The deletion operator must be constrained so that the theory will not predict that this behavior will occur as a bug. Brown and VanLehn (1980) used a set of "deletion blocking principles" to constrain deletion and avoid predicting star bugs. It is in these constraints on deletion that any understanding of acquisition lurks.

Although the details will not be presented here, this approach did not succeed empirically. The most important of the deletion blocking principles, the one dwelled upon in Brown and VanLehn (1980), had apparent implications for acquisition, but some of the others did not. Yet if the unmotivated deletion blocking principles were dropped, the theory generated star bugs.

A second empirical problem was that there are core procedures that cannot be generated in any easy way by rule deletion. As an example, there is a bug that always decrements the leftmost, top digit of a problem regardless of where the column that caused the borrow is. This bug can be explained informally by supposing that the subject who has this bug was tested at a point in his or her schooling where he or she had only practiced borrowing on two column problems. In such problems, the correct digit to decrement is indeed the leftmost, top digit. The subject has not yet had problems of the proper form to discriminate between the "leftmost" abstraction and the "left-adjacent" abstraction. It would be difficult if not impossible to generate a core procedure that used the "leftmost" abstraction in place of the "left-adjacent" abstraction using rule deletion.

Even if such a deletion operator could be formulated, it would probably not capture an important aspect of that core procedure, namely that it depends on the fact that the subtraction curriculum often involves introducing borrowing with two-column problems. Had subjects been taught borrowing some other way, the theory would still generate the "leftmost" core procedure, a prediction that would presumably be false in this hypothetical population. The deletion approach is not expressing acquisitionally relevant information in that it does not show dependence of core procedures on the teaching that subjects receive; it only shows dependence on the form of the mature skill. In cases where the same form can be acquired in the context of different teaching sequences, the deletion approach cannot make different predictions about the intermediate states of knowledge that students in the different teaching sequences can have. It is this reason more than any other that advises against applying Keil's homomorphism approach to taught knowledge. The following speculations amplify this point.

Chomsky, Keil, and others suggest that constraints are at least partially a product of human evolution (Chomsky, 1980; Keil, 1981). This implies that strong, domain-specific constraints could only be found for "natural" knowledge or "faculties," that is, knowledge that is easily imagined to be the product of evolution. Such knowledge is acquired easily and rapidly, usually without formal tutelage. It has some survival value and has been a part of the culture long enough for that survival value to cause significant natural selection. Given this, one would expect Chomskyan approaches to subtraction to fail since subtraction is clearly nonnatural knowledge (yet it is knowledge that everyone can learn). It meets none of the criteria for a "faculty." Unlike counting and simple mental arithmetic, the written multicolumn subtraction algorithm has not been observed to develop without instruction. Also, it has been common in our culture only in the last two centuries, which makes it hard to imagine it as a product of our species' evolution. In short, as nonnatural knowledge, subtraction could not be expected to exhibit strong domain-specific constraints, either of the unary or the homomorphic type. But pursuing this speculative line of reasoning a little further leads to a proposal for some contraints that it could be expected to satisfy.

Consider all the knowledge an adult has and take away natural knowledge domains: vision, language, ontology, spatial reasoning, counting, and so on. There is still a great deal of information left; call it nonnatural knowledge. Some of this nonnatural knowledge has, I assume, enough survival value that an organism that supported its transmission across generations would have a clear selectional advantage over others. Yet, by assumption, there is no direct genetic support for this knowledge per se. Indeed, this is a potential feature since it frees the knowledge to evolve at speeds far in excess of those that characterize changes in the genotype. This question is, how does this transmission work, from evolution's point of view?

If an organism is to be thrust into an information-rich world, evolution would do well to encode a special, highly restricted receptivity for knowledge with a proven survival value: natural knowledge. But restricted receptivity for nonnatural knowledge cannot be based on the structure of that knowledge in isolation since evolution cannot commit the organism in advance. To put it metaphorically, since the genes cannot decide what's useful, they must leave the decision to someone else and provide a special route for the chosen information to be written into the organism's mind. Note that it would not work to let just *any* information be absorbed by the young organisms—there is such a buzzing, blooming confusion of information in the world that information with survival value would be buried in useless information. Evolution should provide a faculty that locks out most information, but provides a "key" or "carrier wave" that allows the chosen knowledge to bypass the filter and become a part of the young organism's knowledge.

My guess is that this coded conduit is active during "teaching," an activity that sometimes goes on between teachers and pupils, masters and apprentices, or neolithic hunters and their children. Something in the structure of the instructional interaction is guiding the target knowledge (e.g., subtraction or chipping out flint arrowheads) through the barriers, allowing the student to learn it. Perhaps repetition is crucial—but nature is full of distracting, useless repeitions. More likely, it is the movement from simple cases to complex ones, regulated by the teacher's assessment of the student's comprehension, perhaps in combination with repetition. The essence of the approach described in the next section is to find formal constraints on the natural conduit that allows instruction in non-natural knowledge to succeed.

A Third Approach: Perfect Prefix Learning with Deletion

The third approach, the one taken in this chapter, is motivated by the belief that core procedures are due not only to incomplete or flawed learning, but also to incomplete traversal of the teaching sequence. In point of fact, many subjects in the subtraction database had not been taught all of the procedural skill at the time they were tested (e.g., they had not been shown how to borrow from zero, but only how to borrow from nonzero digits). In many case, subjects appear to have mastered what they had been taught, but they had only been taught the initial segments or "prefix" of the instructional sequence. The key idea is to split generation of core procedures into *perfect prefix learning* and *deletion*. That is, instead of attacking the core procedure problem with one formalism that captures all aspects, it is broken into two subproblems: flawless learning of partial instruction and the relationship between flawless core procedures and the other, flawed ones. This division is a rather natural one, given the kinds of core procedures one finds. The following two bugs illustrate why both kinds of generation are needed. The first bug can be generated by perfect prefix learning and repair:

$$
\begin{array}{lccc}
 & & \overset{3}{} & \overset{2}{} \\
\text{Stops-Borrow-At-Zero:} & 3\ 4\ 5 & 3\ 4^15 & 3^10^17 \\
 & \underline{-1\ 0\ 2} & \underline{-1\ 2\ 9} & \underline{-1\ 6\ 9} \\
 & 2\ 4\ 3\ \checkmark & 2\ 1\ 6\ \checkmark & 1\ 4\ 8\ \times
\end{array}
$$

The core procedure behind this bug does not know how to borrow across zeros. It borrows correctly from nonzero digits, as shown in the second problem. On the third problem, it attempts to decrement the zero, hits an impasse, and repairs by skipping the decrement operation entirely. The point is that this bug has a complete, flawless knowledge of borrowing from nonzero digits, but precisely at one of the known junctures in the subtraction curriculum, its under-standing stops. Now compare this knowledge state with the one implicated by the following bug:

$$
\begin{array}{lccc}
 & & \overset{3}{} & \overset{9}{} \\
\text{Borrow-From-Zero:} & 3\ 4\ 5 & 3\ 4^15 & 3\ 0^17 \\
 & \underline{-1\ 0\ 2} & \underline{-1\ 2\ 9} & \underline{-1\ 6\ 9} \\
 & 2\ 4\ 3\ \checkmark & 2\ 1\ 6\ \checkmark & 2\ 3\ 8\ \times
\end{array}
$$

This bug only does the write-nine half of borrowing across zero. It changes the zero to nine, but does not continue borrowing to the left. Because it does do half of borrowing across zero, it is likely that subjects with this bug have been taught borrowing across zero. But it is also clear that they did not acquire all of the subprocedure, or else forgot part of it. If the subtraction curriculum was such that teachers first taught one half of borrowing across zero and some weeks later taught the other half, then one would be tempted to account for this bug with perfect prefix learning. But borrowing across zero is in fact always taught as a whole (as nearly as I can tell from examining arithmetic textbooks). So some other formal technique, deletion, is implicated in this core procedure's generation.

Deletion is easily formalized by an operator that mutates a given core procedure to produce another. But formalizing perfect prefix learning requires a new technique. In order to capture the belief that perfect prefix learning depends on the sequence of instruction used with the skill, it cannot be captured as an operator on core procedures (Keil's approach and the one followed in the original version of Repair Theory), nor as constraints on just a set of core procedures (the pure Chomskyan approach). Also, it is premature to actually construct a learning algorithm that generates core procedures given something representing class-room instruction. More must be understood about the constraints on learning before such a project can be attempted. The approach used in the current version of Repair Theory is to capture perfect prefix learning with *constraints on instruction/knowledge pairs*. This is the Chomskyan approach, but applied to sets whose elements are not core procedures, but pairs consisting of a core procedure and a prefix of the instructional sequence.

The goal is to find constraints such that only the ''good'' core procedures are paired in the set. To be a ''good'' core procedure, not only must it occur in the sense that the bugs that are derived from it by repair occur or are plausible predictions, but *all of the core procedures that are derived from it by deletion must also occur*. When the output of perfect prefix learning is amplified by deletion, it must produce only correct predictions.

There is a potential confusion surrounding the deletion operator. Since it acts upon the output of perfect prefix learning, one might plausibly interpret these comments as making a chronological and almost mechanistic claim: A student first learns perfectly, only to have some process come along later and ''clobber'' part of the memory for the procedure. Such a connotation goes far beyond the claims of the theory. The intent in separating perfect prefix learning and deletion is to allow formulation of constraints that accurately express the dependence of acquisition on instruction, both perfect acquisition and incomplete acquisition.

So far, three constraints on perfect prefix learning have been identified. They all depend crucially on properties of the representation language. Indeed, the representation language has been chosen in order to allow these constraints to be chosen (cf. the discussion of explanatory adequacy on pp. 214–215). All three constraints need to mention the input to perfect prefix learning, which is supposed to represent instruction. For the sake of stating these constraints, a very simple representation will be used. It relies on the fact that instruction in procedural skills like subtraction is almost always divided into segments that introduce a new aspect of the algorithm and then drill the students on it. In a typical textbook presentation of subtraction, the segment for borrowing from zero is two pages: one page explains the new subprocedure and shows the sequence of actions needed to solve an example problem and the other is a page of exercises for the student to solve. Other typical segments are borrowing in two-column problems, borrowing in three-column problems where the borrow is initiated in the tens column, and adjacent simple borrows in three-column problems. These examples are cited to illustrate the grain-size of the formalization. In this chapter, that is all that is needed of the instruction formalization.

Given the notion of segments, the three constraints that are the targets of the arguments of the next section can be stated. The first defines what it means for a core procedure to be "perfect" vis a vis a certain prefix of the instruction:

Perfect prefix learning: If x is an example or exercise problem of segment s, and core procedure p is paired with s by perfect prefix learning, then p solves x correctly and without reaching an impasse.

The remaining two constraints describe the fact that subtraction learning is incremental. It is plausible that an individual subject moves through the sequence of core procedures as he or she moves through the curriculum. One would expect the structure of each core procedure to somehow embed the structure of its predecessor. That is, newly learned components are added on top of the existing components without changing them, or perhaps with only minimal changes. This kind of learning is usually called *assimilation,* to distinguish it from more radical kinds of learning. The hypothesis that core procedure learning is assimilation is expressed formally by the following constraint:

The assimilation constraint: If segment s' is the successor of s in the teaching sequence and pairs $\langle p,s \rangle$ and $\langle p',s' \rangle$ are generated by perfect prefix learning, then core procedure p is a proper subgraph of core procedure p'.

This constraint (and the next one as well) uses the fact that a core procedure as represented in the formal knowledge representation language happens to be a labeled, directed graph. Hence, the convenient notions of subgraph and subgraph difference can be appealed to in order to express formally the relationship between the old core procedure and the newly acquired one that is built on top of it. (Actually, a much more technical definition is necessary to be completely accurate. See the discussion of "maximal partial isomorphism" in VanLehn &

Brown, 1980.) Whereas the assimilation constraint says that learning makes only small changes to existing structure, the next constraint says that the amount of new material that can be learned is simple, in a certain sense.

The disjunction-free learning constraint: If segment s' is the successor of s in the teaching sequence and pairs $\langle p,s \rangle$ and $\langle p',s' \rangle$ are generated by perfect prefix learning, then the difference subgraph $p'-p$ contains at most one disjunctive goal, namely the one adjoining it to p.

What this constraint says is that the new material can contain no disjunctions, although a disjunction can be used to attach it to the old material. The intuition behind the disjunction-free learning constraint is that when teaching a conditional or "branch statement" in a procedure, one must first teach one side of the conditional, then the other; they can't both be taught at once. In subtraction, for example, one teaches borrowing from a nonzero digit in one segment and borrowing from a zero in another. A new disjunctive test for zero adjoins the new borrow-from-zero subprocedure, which is free of disjunctions, to the old material as an alternative to borrowing from a nonzero digit.

Stepping back to look at these three constraints as a whole, one sees that the perfect prefix learning constraint sets up a well-defined relationship between core procedures and instruction, whereas the other two, assimilation and disjunction-free learning, deliver the punch line. They begin to suggest something about potential learning mechanisms. The disjunction-free learning constraint is particularly provocative. Computational experiments with a variety of knowledge representations have shown that induction from examples is simple when the language does not allow disjunction (Dieterich & Michalski, 1980; Hayes-Roth & McDermott, 1976; Mitchell, 1978; Vere, 1978; Winston, 1975). However, when disjunction is entertained, domain-specific constraints are needed to select among the multitude of inductions that can be made—domain independent constraints, such as "simplicity," appear to be too weak (Feldman, 1972; Pinker, 1979). Combinatorial explosion continues to be a problem even when the input to the inductive algorithms contains feedback from a teacher concerning negative examples (i.e., the learner is told that the given instance is *not* an example of the concept being induced). In theory, negative examples are sufficient to guarantee convergence of induction in the limit (Gold, 1967). In practice, they have been found to be helpful, but not helpful enough at taming the combinatorics of learning when disjunction is allowed (Knobe & Knobe, 1976). In short, there is every reason to believe from an algorithmic point of view that induction must either be disjunction-free or subject to strong domain-specific constraints. By hypothesis, nonnatural knowledge does not have strong domain-specific constraints on its acquisition. Therefore, if an inductive algorithm is the underlying learning mechanism, there would have to be a disjunction-free learning constraint on acquisition. To put it differently, if no constraint like disjunction-free learning can be sustained, then doubt will be cast on any inductive explanation for learning nonnatural knowledge.

There are two comments to make with respect to testing these constraints. First, the data collection procedures for bugs did not record the instructional segment that each subject was in at the time of testing. Consequently, it has been necessary to guess which segments go with which core procedures. In one respect, this is just a mistake which has weakened the inferences that can be drawn from the data. On the other hand, it is a stroke of luck. Relying on anecdotal evidence, I suspect that the teacher's instruction to the class as a whole may be in a certain segment, but some students behave as if they are in some earlier segment. For some reason, they either did not learn or chose not to use the subskills that were taught most recently. If this phenomenon exists, it would not affect the constraints on core procedure generation. The same set of core procedure/segment pairs could be used. However, being unable to use the classroom's segment as the segment of each student in the classroom would complicate the mapping between data and predictions.

The second comment is that these constraints are highly dependent on the representation used for core procedures. Indeed, all parts of the theory depend strongly on the knowledge representation for core procedures. Not only are the acquisition constraints highly sensitive to the syntax of the expressions (procedures) of this language, but the interpreter is designed specifically to execute procedures written in the language, and the repairs (and the critics) operate on the run-time state of the interpreter, which is in turn specified by the language. The influence of the language is felt everywhere in the model. Indeed, the key point of this chapter is that: *Getting the knowledge representation language "right" allows the model to be made simple and to obey strong principles.*

My experience has been that incremental redesign of the representation language was absolutely necessary to bring the model into conformance with psychologically interesting principles. The evolution of the principles of the theory went hand in hand with the evolution of the representation language. The objective of this chapter is to analyze the crucial points in that evolution.

THE GOAL STACK

One of the earliest and most fundamental changes in computer programming languages was the move from register-oriented languages to stack-oriented languages. In register-oriented languages, one represents programs as flow charts or their equivalent. The main control structure is the conditional branch. Data flow is implemented as changes to the contents of various named registers. Stack-oriented languages added the idea of a subroutine—something that could be "called" from several places and when it was finished, control would return to the "caller" of the subroutine. Although the register-oriented languages need only a single register to keep track of the control state of the program, stack-oriented languages need a last-in-first-out stack so that the interpreter can tell not only where control is now (the top of the stack), but where it is to return to when the current subroutine gets done (the next pointer on the stack), and so on. Stack-orientation also augmented the representation of data flow. A new data flow facility was to place data on the stack, as temporary information associated with a particular invocation of a subroutine. In particular, subroutines could be called with parameters (arguments).

The fundamental distinction between register-orientation and stack-orientation has lapsed into historical obscurity in computer science, but surprisingly, psychology seems ignorant of it. When a psychologist represents a process, it is frequently a flow chart, a finite state machine, or a Markov process that is employed. Even authors of production systems, who are often computer scientists as well as psychologists, give that knowledge representation a register orientation: Working memory looks like a buffer, not a stack, and productions are not grouped into subroutines. For some reason, when psychologists think of temporary memory, whether for control or data, they think only of registers.

There are well known mathematical results concerning the relative power of finite state automata, register automata, and push-down automata. Some of these results have been applied to mental processes such as language comprehension (see Berwick, in press, for a review). However, I find myself rather unconvinced by such arguments. As Berwick and others have pointed out, these arguments must make many assumptions to get off the ground, and not all of them are explicitly mentioned, much less defended.

The project of this section is to argue for a stack-based representation of core procedures based on a rich structure of motivated assertions: the principles and architecture of Repair Theory as presented in the preceding two sections. Those sections did not make assumptions about the representation language because they dealt with the facts at a medium-high level of detail. The arguments in this section and the next show what must be assumed of the representation language in order to push the structure of the preceding sections down to a low enough level that precise predictions can be made, and made successfully. That is, they show what aspects of the mental representation are *crucial* to the theory.

The basic tools of complexity-based arguments on mental representations (cf Berwick, in press) are time and space bounds on the computation. These are risky precisely because we do not know what the processor and memory of the mind are like, or even if VanNeuman machine architecture (i.e., one processor, one memory, serial computation) is appropriate for measuring resource limitations on mental processing. In the near future, it appears that computers will be available with radically different architectures. For example, instead of one processor and a large memory, there may be thousands of processors, each with a small amount of memory. These architectures are expected to radically change the speed and memory characteristics of computations. Crucially, they are not expected to change *what* can be computed but only what can be computed within certain resource limitations. So, one can assume a VonNeuman architecture is safe from technologically stimulated changes in metaphors, as long as one only cares about what the computations are and not what resources they require.

In the following sections, resource limitations are never used as tools of argumentation. Instead, the tools are two operators that, by hypothesis, manipulate the mental representation directly. One is a certain repair that manipulates the execution state, and the other is the deletion operator that mutates core procedures. By studying the kinds of changes they make, their computations, one can understand what requirements must be placed on the mental representations that they manipulate.

The Backup Repair is Necessary

Control structure is not easily deduced by observing sequences of actions. Too much internal computation can go on invisibly between observed action steps for one to draw strong inferences about control flow. What is needed is an event that can be assumed or proven, in some sense, to be the result of an elementary, indivisible control operation. The instances of this event in the data would shed light on the basic structures of control flow. Such a tool is found in a particular repair called the *backup repair*. It bears this name since the intuition behind it is the same as the one behind a famous strategy in problem solving: backing up to the last point where a decision was made in order to try one of the other alternatives. This repair is so crucial in the remaining arguments that it is worth a few pages to defend its existence.

The existence argument begins by demonstrating that a certain four bugs should all be generated from the same core procedure. There are two arguments for this lemma, one based on explanatory adequacy, the other on observational adequacy. From the lemma, it is argued that the backup repair is the most observationally adequate way to generate the four bugs.

For easy reference, the four bugs will be broken into two sets called big-borrow-from-zero and little-borrow-from-zero. Big-borrow-from-zero bugs seem to result from replacing the whole column processing subprocedure when the column requires borrowing from a zero. Its bugs are

		^{3}	^{1}
Smaller-From-Larger-Instead-of-	3 4 5	3 4^{1}5	2^{1}0 7
Borrow-From-Zero:	−1 0 2	−1 2 9	−1 6 9
	2 4 3 √	2 1 6 √	4 2 ×

$$
\begin{array}{lccc}
 & & \overset{3}{} & \overset{1}{} \\
\text{Zero-Instead-of-} & 3\ 4\ 5 & 3\ 4^1 5 & 2^1 0\ 7 \\
\text{Borrow-From-Zero:} & -1\ 0\ 2 & -1\ 2\ 9 & -1\ 6\ 9 \\
\hline
 & 2\ 4\ 3\ \checkmark & 2\ 1\ 6\ \checkmark & 4\ 0\ \times
\end{array}
$$

When a column requires borrowing from zero, as the units column does in the last problem, the first bug takes the absolute difference instead of borrowing and taking a regular difference, whereas the second bug just answers the column with the maximum of zero and the difference, namely zero.

The little-borrow-from-zero bugs have a smaller substitution target. Only the operations that normally implement borrowing across the zero are replaced, namely the operations of changing the zero to nine and borrowing from the next digit to the left. Its bugs are

$$
\begin{array}{lccc}
 & & \overset{3}{} & \overset{1\ \ 11}{} \\
\text{Borrow-Add-Decrement-} & 3\ 4\ 5 & 3\ 4^1 5 & 2\ 0^1 7 \\
\text{Instead-of-Zero:} & -1\ 0\ 2 & -1\ 2\ 9 & -1\ 6\ 9 \\
\hline
 & 2\ 4\ 3\ \checkmark & 2\ 1\ 6\ \checkmark & 5\ 8\ \times
\end{array}
$$

$$
\begin{array}{lccc}
 & & \overset{3}{} & \overset{1\ \ 10}{} \\
\text{Stops-Borrow-At-Zero:} & 3\ 4\ 5 & 3\ 4^1 5 & 2\ 0^1 7 \\
 & -1\ 0\ 2 & -1\ 2\ 9 & 1\ 6\ 9 \\
\hline
 & 2\ 4\ 3\ \checkmark & 2\ 1\ 6\ \checkmark & 4\ 8\ \times
\end{array}
$$

In the first case absolute difference has been substituted for decrementing. Hence, the zero in the third problem is changed to the absolute difference of zero and one, namely one, during borrowing. The second bug, Stops-Borrow-At-Zero, is generated by substituting the max-of-zero-and-difference operation for decrement. This causes the bug to cross out the zero of the last problem, and write a zero over it. (Both these bugs, by the way, have other derivations than the ones discussed here.)

The cross product relationship between these four bugs is exactly the kind of pattern that the repair process captures. The most straight forward way to formalize it would be to postulate two core procedures, one for big-borrow-from-zero and another for little-borrow-from-zero. However, all four bugs miss the same kind of problems, namely just those problems that require borrowing from a zero. Intuitively, they seem to have the same cause: The subskill of borrowing across zero is missing from the subject's knowledge. It is a fact that subtraction curricula generally contain a segment that teaches borrowing from nonzero digits. Now the perfect prefix learning constraint implies that all core procedures associated with a certain segment of instruction will miss just the problems that lie outside the set of examples and exercises of that segment. If two core procedures were used and the perfect prefix learning constraint is to be obeyed, then both core procedures would have to be paired with the instructional segment that teachers borrowing from nonzero digits. This means that whatever the mechanism is that implements perfect prefix learning, it must explain how it could generate two core procedures that differed *only by how much of the procedure was repaired*. Imposing this added task on the learner could make it more complex. So, in order to simplify the learner and maintain the perfect prefix learning constraint, the four bugs should be generated from the same core procedure.

The previous argument was based on explanatory adequacy. There is a second based upon observational adequacy. It stems from the empirical claim that all bugs in a bug migration class are generable from the same core procedure. Figure 5.3 shows the first six problems of a subtraction test taken by subject 19 of classroom 20. This third-grader gets the first four problems right, which involve only simple borrowing. He misses the next two, which require borrowing from zero. Crucially, these two problems are solved as if the subject had two different bugs from the cross product pattern. This is an instance of intratest bug migration. The fifth problem is solved by a little-borrow-from-zero bug: He hits the impasse (note the scratch mark through the zero) and repairs it by skipping the decrement, a repair that generates the bug stops-borrow-at-zero. He finishes up the rest of the problem without borrowing—Apparently he wants to "cut his losses" on that problem. On the next problem, he again hits the decrement zero impasse, but repairs it this time by backing up and taking the absolute difference in the column that originated the borrow, the units column. These repairs generate the bug Smaller-From-Larger-Instead-of-Borrow-From-Zero. Since both bugs are in the same bug migration class, both are somehow derived from the same core procedure via repair.

Two arguments have forced the conclusion that all four bugs come from the same core procedure. Now the problem is to find repairs that will generate all of them, given that they all stem from the same impasse. One way to do that would be to use four separate repairs. However, that would not capture a fact about these bugs that was highlighted in their description: They fall into a cross product pattern whose dimensions are the "size" of the patch (i.e., just the decrement versus the whole borrowing operation) and its "function" (i.e., absolute difference versus maximum of zero and difference). It will be shown that this pattern can be captured by postulating a backup repair, and hence that approach is more descriptively adequate.

As mentioned earlier, the backup repair resets the execution state of the interpreter back to a previous decision point in such a way that when interpreta-

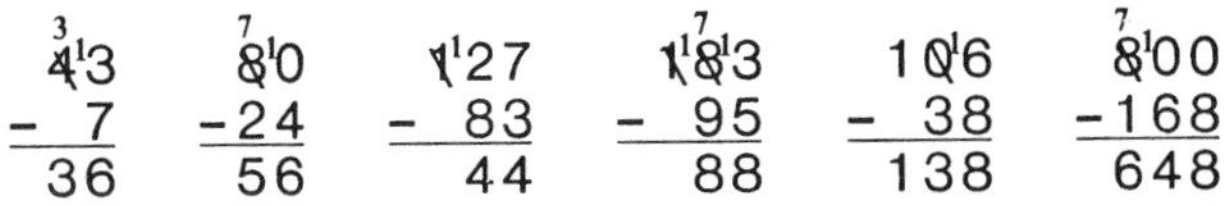

$$
\begin{array}{cccccc}
\overset{3}{\cancel{4}}\text{'}3 & \overset{7}{\cancel{8}}\text{'}0 & \cancel{1}\text{'}27 & \overset{7}{1}\cancel{8}\text{'}3 & 1\,\cancel{0}\text{'}6 & \overset{7}{\cancel{8}}\text{'}00 \\
-\ \ 7 & -24 & -\ 83 & -\ 95 & -\ 38 & -168 \\
\hline
36 & 56 & 44 & 88 & 138 & 648
\end{array}
$$

Figure 5.3. The first six problems show bug migration.

tion continues, it will choose a different alternative than the one that led to the impasse that backup repaired. The backup repair is used for the big-borrow-from-zero bugs but not the little-borrow-from-zero bugs. Using backup in those cases causes a secondary impasse. The secondary impasse is repaired with the same two repairs that are used for the little-borrow-from-zero bugs. This is perhaps a little confusing, so it is worth a moment to step through a specific example.

Figure 5.4 is an idealized protocol of a subject who has the bug Smaller-From-Larger-Instead-of-Borrow-From-Zero. The (idealized) subject does not know about borrowing from zero. When he tackles the problem 305 − 167, he begins by comparing the two digits in the units column. Since 5 is less than 7, he makes a decision to borrow (episode *a* in the figure), a decision that he will later come back to. He begins to tackle the first of borrowing's two subgoals, namely borrowing-from (episode *b*). At this point, he gets stuck since the digit to be borrowed from is a zero and he knows that it is impossible to subtract a one from a zero. He's reached an impasse. The backup repair gets past the decrement-zero impasse by "backing up," in the problem-solving sense, to the last decision

a.
$$\begin{array}{r} 305 \\ -\,167 \end{array}$$
In the units column, I can't take 7 from 5, so I'll have to borrow.

b.
$$\begin{array}{r} 305 \\ -\,167 \end{array}$$
To borrow, I first have to decrement the next column's top digit. But I can't take 1 from 0!

c.
$$\begin{array}{r} 305 \\ -\,167 \\ \hline 2 \end{array}$$
So I'll go back to doing the units column. I still can't take 7 from 5, so I'll take 5 from 7 instead.

d.
$$\begin{array}{r} {}^{2}\!\not{3}05 \\ -\,167 \\ \hline 2 \end{array}$$
In the tens column, I can't take 6 from 0, so I'll have to borrow. I decrement 3 to 2 and add 10 to 0. That's no problem.

e.
$$\begin{array}{r} {}^{2}\!\not{3}05 \\ -\,167 \\ \hline 142 \end{array}$$
Six from 10 is 4. That finishes the tens. The hundreds is easy, there's no need to borrow, and 1 from 2 is 1.

Figure 5.4. Pseudo-protocol of a bug generated with backup repair.

which has some alternatives open. The backing up occurs in episode c where the subject says, "So I'll go back to doing the units column." In the units column he hits a second impasse, saying, "I still can't take 7 from 5," which he repairs ("so I'll take 5 from 7 instead"). He finishes up the rest of the problem without difficulty. His behavior is that of Smaller-From-Larger-Instead-of-Borrow-From-Zero. The other big-borrow-from-zero bug would be generated if he had used a different repair in episode c. (e.g, He might say, "I still can't take 7 from 5, but if I could, I certainly wouldn't have anything left, so I'll write 0 as the answer.")

It has been shown that the backup repair is the best of several alternatives that generate the four bugs. Needless to say, it plays an equally crucial role in the generation of many other bugs, but the argument stuck with just four bugs for the sake of simplicity. Backup is the tool that will be used to reveal constraints on core procedures.

Flexible Execution State is Necessary

There are several important aspects of the preceding example that indicate something about the kind of execution environment that backup operates within. The decrement-zero impasse that backup repaired occurred when the focus of attention was on the tens column. After backup had returned control back to an earlier decision (the decision about whether to borrow), the focus of attention was on the ones column. Somehow, backup knew to shift not only the flow of control, but the data flow as well. (I am assuming that the control and data flow are distinct, and that the column that is the focus of attention is held in a data flow construct—a register, variable, message, local binding, etc. There is an argument for this assumption, which is based on the ability of a competent subtractor to answer problems requiring borrowing across arbitrarily many zeros, but it will not be presented here.)

An even more impressive example of the power of backup to shift data and control flow occurs with a common core procedure that forgets to change the zero when borrowing across zero. This core procedure produces answers like (a):

$$
\begin{array}{llll}
\;\;\;2 & \;\;\;1 & & \quad\;\;\;0 \\
\;\;\;3 & \;\;\;0 & \quad\;\;\;0 & \quad\;\;\;0 \\
(a)\;\; 4^{10}12 & (b)\;\; 1^{10}12 & (c)\;\; 1\;\,0^{12} & (d)\;\; 1^{10}12 \\
\;\;\;-1\;3\;9 & \;\;\;-\;\;3\;9 & \;\;\;-\;\;3\;9 & \;\;\;-\;\;3\;9 \\
\hline
\;\;\;\;1\;7\;3 & \;\;\;\;1\;7\;3 & \;\;\;\;\;\;\;\;3 & \;\;\;\;\;\;7\;3
\end{array}
$$

The 4 was decremented once due to the borrow originating in the units column and then again due to a borrow originating from the tens column because the tens column was not changed during the first borrow as it should have been.

The crucial fact is seen in (b). Because this example is such a critical one

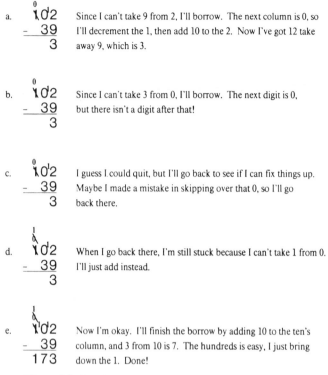

a. $\overset{0}{\cancel{1}}\overset{\cdot}{0}2$ Since I can't take 9 from 2, I'll borrow. The next column is 0, so
   $-\ \ 39$ I'll decrement the 1, then add 10 to the 2. Now I've got 12 take
   $\ \ \ \ \ 3$ away 9, which is 3.

b. $\overset{0}{\cancel{1}}\overset{\cdot}{0}2$ Since I can't take 3 from 0, I'll borrow. The next digit is 0,
   $-\ \ 39$ but there isn't a digit after that!
   $\ \ \ \ \ 3$

c. $\overset{0}{\cancel{1}}0\overset{\cdot}{2}$ I guess I could quit, but I'll go back to see if I can fix things up.
   $-\ \ 39$ Maybe I made a mistake in skipping over that 0, so I'll go
   $\ \ \ \ \ 3$ back there.

d. $\overset{1}{\underset{\cancel{0}}{\cancel{1}}}\overset{\cdot}{0}2$ When I go back there, I'm still stuck because I can't take 1 from 0.
   $-\ \ 39$ I'll just add instead.
   $\ \ \ \ \ 3$

e. $\overset{1}{\cancel{1}}\overset{\cdot}{0}2$ Now I'm okay. I'll finish the borrow by adding 10 to the ten's
   $-\ \ 39$ column, and 3 from 10 is 7. The hundreds is easy, I just bring
   $1\ 7\ 3$ down the 1. Done!

Figure 5.5. Pseudo-protocol of Borrow-Across-Zero with backup.

throughout this section, it is illustrated in Figure 5.5 with by a pseudo-protocol. The procedure decrements the one to zero during the first borrow (episode *a* in the figure). Thus, when it comes to borrow a second time, it finds a zero where the one was and borrows across it. This causes an attempt to decrement in the thousands columns, which is blank. An impasse occurs (episode *b*). The most common repair to this impasse is *quit*—a repair that just gives up on the problem. This would give the answer shown in (c). The answer shown in (b) and in Figure 5.5 is generated by assuming the impasse is repaired with backup. Backup returns to the decision made in the hundred's column concerning the zero-ness of its top digit (episode *c* in the figure). It takes the open alternative to this decision, which causes an attempt to decrement the top digit in the hundred's column. This causes a secondary impasse (episode *d*). If it is repaired one way, one sees the decrement replaced by an absolute difference as in (b) and the figure. If it is repaired a different way, one sees the familiar maximum of zero and difference patch, as in (d). The existence of several repairs to this secondary impasse confirms its existence. So, the assertion that backup returns to the decision at the hundreds column is well-motivated. The crucial fact is that the backup repair shifted the focus even though both the source and the destination of the backing up were highly similar: They both concerned borrowing-from (as opposed to borrowing-into) and they both involved single digits rather than columns.

The stage has been set to uncover what kind of execution state the interpreter and backup are using. The execution or "run time" state of an interpreter holds whatever temporary information the interpreter needs to execute the procedure. Three alternatives will be contrasted: (*a*) minimal state, (*b*) fixed state, and (*c*) flexible state.

At a minimum, the execution state requires a register to indicate the current locus of control in the procedure. Since data flow is assumed to be independent of control flow, at least one more register would be needed to hold a pointer to the column or digit being operated on. However, this minimal control state puts an enormous burden on backup if it is to account for the facts. Since the only thing that backup has at the time that an impasse occurs is a control pointer to the action that is stuck and a data flow pointer to the column or digit that was being focused on when the impasse occurred, it can only use these as a starting point in a search through the core procedure to locate a decision point to go back to. That is, it must analyze the core procedure enough to "walk control backward" through it, and thus locate a decision to return to. Moreover, it must pay special attention to any data flow manipulations that it passes since these will have to be undone. Only by reversing the data flow can backup account for the shift in focus of attention that was observed in the examples above. Giving backup this much analytic ability means repairs cannot be considered weak and local. If one were to give repairs as much power as this version of backup requires, then it would be possible to account for virtually any subject behavior by postulating a complex enough repair. This much tailorability would render the theory vacuous.

An alternative to minimal state is to have a certain fixed number of registers, say one per "subprocedure." The idea is that each suprocedure would have its own control and data registers. Thus, instead of analyzing the core procedure, backup would search through the registers in the execution state. It doesn't matter much what a subprocedure is, but to account for the facts just stated, borrow-from and process-column would have to be distinct subprocedures. To account for the first example (Figure 5.4), the impasse happens at borrow-from, and backup shifts back to process-column. Because they have separate data flow registers, this effects a shift from the tens column to the units column. Although multiple registers suffice to account for the first fact, they fail on the second (Figure 5.5). There the impasse occurs at borrow-from in the thousands column, but backup shifts to borrow-from in the hundreds column. Since borrow-from has only one data flow register, backup would have to do some clever analysis of the core procedure (which should not be allowed in the theory).

Other possible finite execution states are to assign a register to each data type, which in this case means roughly one register for columns and one for digits. But this fails to account for the shift of the second example since both the impasse (decrementing a blank) and the decision point returned to (testing a digit for zero-ness) involve the digit data type. Hence, there can be only one register involved, and one is again forced to give backup the intelligence necessary to change it in order to account for the facts.

A further alternative still is to assign one control register to each column. The basic idea (which is inspired by object-oriented programming languages such as Smalltalk and Simula) is that when an impasse occurs in one column, backup moves rightward and resumes with whatever action that column's control register indicates. However, this could not account for cases where the impasse and the decision that backup returns to are in the same column. There are examples of this, but for the sake of brevity, they will not be presented here. Anyway, one could go on in this spirit, using a fixed set of registers with various semantics. However, I think the point has been made that the whole fixed state approach is flawed observationally. If it could be made to work at all, its register semantics would probably be quite implausible.

A third option for the execution state involves a capacity to hold an arbitrary amount of state that is not determined in advance of the procedure's execution. Given this faculty, one can stipulate that each time a decision is made, the control flow and the data flow are saved in a safe place in the execution state. By "safe," I mean that as control flow and data flow change later, these saved values will not be affected. This allows backup to be very simple. It searches among the saved decision points for one meeting its criterion (e.g., the most recent one). Not only does this allow backup to account for the facts just exhibited, but it makes the strong prediction that whenever backup restores control to a past decision, the data flow must be reset to whatever it was at the time of that decision. Whereas the other schemes allowed the possibility that control could be shifted independently of data—backup could choose not to reverse the data flow as it walked control backward through the core procedure—this one forces them to be shifted together, if at all. That is, instead of just accounting for the fact that control and data flow are shifted together, the flexible-state architecture *explains* it.

This kind of argument is a familiar one in computer science. It trades off increased storage of information against decreased processing on the part of backup. The more information is saved in the run time state, the less information backup has to compute. Since there are good methodological reasons for making repairs as simple as possible, this tilts the tradeoff in favor of increasing execution state. The fact that this end of the tradeoff netted us an explanation rather than a mere description of the facts of backing up indicates that this tradeoff is more than a free parameter of the theory, but integral to it.

There is a second argument for flexible execution state, which turns on explanatory adequacy. The argument shows that recursive control flow is necessary for disjunction-free learning. That is, if execution state is not flexible, thus allowing the language to use recursion, then the disjunction-free learning constraint cannot be imposed on the perfect prefix learning. The argument involves learning a certain way to borrow across zero, which is exemplified in the following problem:

$$2 \ 9$$
$$3^1 0^1 5$$
$$\underline{-1 \ 2 \ 9}$$
$$1 \ 7 \ 6$$

The zero has ten added to it, then the three is decremented, then the newly created ten is decremented. The claim is that the only way to learn this way of borrowing in a nonrecursive language violates the disjunction-free learning constraint. To make the argument crisp, a particular nonrecursive language, namely flowcharts, is used. Figure 5.6a shows borrowing from a core procedure that only knows how to borrow from nonzero digits. Figure 5.6c shows borrowing after borrowing across zero in the fashion shown earlier has been learned. Clearly, there are two branches to learn. One moves control leftward across a row of zeros, and the other moves back across them until the column originating the borrow is found (i.e., the "Home?" predicate is true of the column B). There are many other ways that borrowing could be implemented, but if recursive control is not available, they would all have to have two loops—one for searching leftward, one for searching rightward.

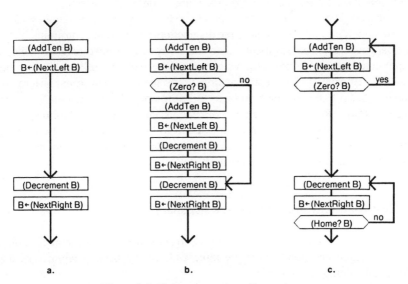

Figure 5.6. Finite state version of borrowing.

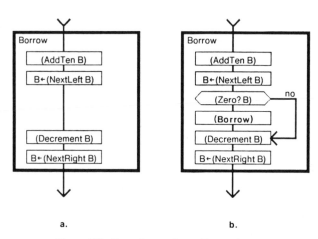

Figure 5.7. Recursive versions of borrowing.

Now disjunction-free learning sanctions the acquisition of at most one branch, and this must be one that adjoins the newly learned steps to the older material. A formal definition of branches and adjunction depends on the syntax of the language, but the essence of it should be clear by examining the difference between Figure 5.6a and 5.6b. Functionally, the difference is that the core procedure of Figure 5.6b has learned how to borrow from one zero. Syntactically, there is one branch, and it is an adjoining branch because one arm of the branch skirts the new material. The essence of adjunction is that one arm of the new conditional replicates the old procedure's control pathway.

It should be clear that the transition from Figure 5.6a to 5.6c requires adding two disjunctions and thus violates the disjunction-free learning constraint. Yet, if the language allowed recursion, then the borrowing-from-zero goal could be represented as in Figure 5.7b, with a recursive call to itself (the heavy box labeled "Borrow"). This representation allows the transition from nonzero borrowing (5.7a) to borrowing-from-zero (5.7b) to obey the disjunction-free learning constraint. Recursion requires a flexible execution state because the control state must be saved across the recursive call to borrow. In short, the language must have a flexible-state control structure so that a certain acquisitional transition obeys the disjunction-free learning constraint.

A Goal Hierarchy and a Stack Are Necessary

The backup repair sends control back to some previous decision. The question is, which decision? There are three well-known backup regimes used in Artificial Intelligence (AI):

Chronological backup: The decision that is returned to is the one made most recently, regardless of what part of the procedure made the decision.

Dependency-directed backup: A special data structure is used to record which actions depend on which other actions. When it is necessary to backup, the dependencies are traced back to find an action that doesn't depend on any other action. This means a decision was made (an "assumption" in the jargon of dependency-directed backtracking) whose first effect was this action. That decision is the one returned to.

Hierarchical backup: To support hierarchical backup, the procedure representation language must be hierarchical in that it supports the notion of goals with subgoals. In order to find a decision to return to, it searches up the hierarchy starting from the current goal, going up from goal to supergoal. The first (lowest) goal that can "change its mind" is the one returned to. (NB: In A.I., this is not usually thought of as a form of backup. It is usually referred to by the LISP primitives used to implement it, e.g., Catch and Throw in Maclisp.)

The argument that follows shows that the first two backup regimes have observational and descriptive problems. Hierarchical backup is the only one of the three that can generate the crucial bugs while avoiding the generation of star bugs. This conclusion has consequences for the representation language, namely that the language must force a modular or goal-subgoal structure on procedures, and furthermore, that this kind of backup forces the execution state to be a stack.

Chronological backup is able to generate the bugs mentioned earlier. The walk-through of Figures 5.4 and 5.5 should be evidence enough of that. However, by the impasse/repair independence principle, it can be used to repair any impasse, and here it has problems. When it is applied to the impasses of certain independently motivated core procedures, it generates star bugs. This point will be made with an example of one of them. It is a core procedure that is needed to generate bugs like:

		3	1 9
Smaller-From-Larger-	345	345	207
With-Borrow:	−102	−129	−169
	243 √	214 ×	32 ×

		3	1 9
Zero-After-Borrow:	345	345	207
	−102	−129	−169
	243 √	210 ×	30 ×

After completing the borrow-from half of borrowing these bugs fail to add ten to the top of the column borrowed into. Instead, Smaller-From-Larger-With-Borrow answers with the absolute difference of the column, and Zero-After-

Borrow answers the column with zero. Apparently, and quite reasonably, the core procedure is hitting an impasse when it returns to the original column after borrowing. Since the column has not had ten added to the top digit as it should, the bottom digit is still larger than the top. This causes the impasse, which is seen being repaired two different ways in the two bugs.

When the core procedure thus independently motivated, one should find that a bug is generated when chronological backup is applied to the impasse. But since the core procedure knows how to borrow across zero, at the time it reaches the impasse, its most recent decision was that the digit that it was to borrow from was nonzero and hence could be decremented, thereby finishing up the borrow (I'm making assumptions about the order of steps during borrowing, but these are not essential to the argument—somewhere in the middle of borrowing there will be a decision point). Since this is the most recent decision, chronological backup causes control to go back and "continue" the borrowing, even though it already has borrowed. Worse, when it gets done with this superfluous borrowing, it comes right back to the backup patch, which sends it back to borrow again! The chronological backup patch results in an infinite loop, and a rather bizarre one at that. Clearly, this is a star bug and should not be predicted to occur by the theory. Chronological backup is ruled out on grounds of descriptive inadequacy.

Dependency-direct backup is really not a precise proposal for this domain until the meaning of *actions depending on other actions* is defined. Consideration of the star bug that was just described leads to a plausible definition. Part of what makes the star bug absurd is its going back to change columns to the left of the one where the impasse occurred. It seems clear that difficulties in one column don't depend causally on actions in a different column. If dependency between actions is defined to mean that the actions operate on the same column (or more generally, have the same locative arguments), then dependency-directed backup would not make the mistake that chronological backup did. It would never go to another column to fix an impasse that occurs in this column. However, even this rather vague definition limits dependency-directed backup too strongly. Several examples of backup were presented earlier (Figures 5.4 and 5.5) where the location was shifted, and hence dependency-directed backup cannot generate these bugs. As it is presently defined, it is observationally inadequate.

In essence, these two approaches to backing up show that neither time nor space suffice. That is, such natural concepts as chronology or location will not support the kind of backing up that subjects apparently use. That leaves one to infer that they must be using some knowledge about the procedure itself. The issue that remains is what this extra knowledge is. A goal hierarchy is one sort of knowledge that will do the job, as the following demonstration shows.

The basic definition of hierarchical backup is that it can only resume decisions which are supergoals of the impasse that it is repairing. With this stipulation, any one of a number of goal structures will suffice to block the star bug that chronological backup generated (as well as to generate all the backup bugs that have been presented so far). One such goal structure is shown in Figure 5.8. This figure shows the goal–subgoal relationships with arrows. In this goal structure, the borrow-from goal (the goal that tests whether the digit to be borrowed from is zero) is not a supergoal of the take-difference goal (the operation that reaches an impasse in the star bug's generation). There would be a chain of arrows from borrow-from to take-difference if it were. Hence, when backup occurs at the take difference impasse, it cannot go to the borrow-from decision even though that decision is chronologically the most recent.

In the previous section, it was shown that decisions must be saved in the execution state so that backup doesn't have to analyze the core procedure. But now a new constraint has been added to backup, namely one that it only return to decisions that are supergoals of the impasse. To maintain the constraint that the control information that repairs need is in the execution state, the set of saved decision points must be structured to reflect the goal structure. One way to do this is to remove decision points from the execution state when control ascends through their goal in the goal structure. This guarantees that the only decisions that are left in the execution state are ones that are supergoals of whatever goal is currently executing. This amounts to keeping goals in a last-in-first-out stack.

Of course, there are many more powerful control regimes than stack-based ones. As an example of the trouble more powerful control regimes cause, consid-

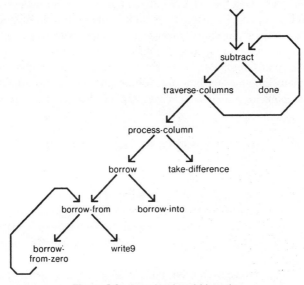

Figure 5.8. A goal–subgoal hierarchy.

er a simple one: Coroutines are a control structure that allows independent processes, each with their own stack. But this control structure increases the expressiveness of the language, which allows perfect prefix learning to generate absurd core procedures. To demonstrate this point, suppose the language allowed coroutines, and consider perfect prefix learning of simple borrowing. The instructional sequence would perhaps have pictures or blackboard demonstrations presenting the actions of borrowing in a sequence, such as:

$$
\begin{array}{cccc}
 & {}^3 & {}^3 & {}^3 \\
4\,{}^1 5 & 4\,{}^1 5 & 4\,{}^1 5 & 4\,{}^1 5 \\
\underline{-2\ 9} & \underline{-2\ 9} & \underline{-2\ 9} & \underline{-2\ 9} \\
 & & 6 & 1\ 6
\end{array}
$$

Given that coroutines are allowed, one way to construe the first two actions, the new ones, is that they are a new coroutine. It happens that the example has this coroutine executing before the old one, but the learner need not take that as necessary. The core procedure could execute the coroutines interleaved, as in:

$$
\begin{array}{cccc}
 & & {}^3 & {}^3 \\
4\,{}^1 5 & 4\,{}^1 5 & 4\,{}^1 5 & 4\,{}^1 5 \\
\underline{-2\ 9} & \underline{-2\ 9} & \underline{-2\ 9} & \underline{-2\ 9} \\
 & 6 & 6 & 1\ 6
\end{array}
$$

Because this core procedure can execute either interleaved or correctly, it seems an absurd prediction to make, a star bug. Yet with the addition of few minor constraints on sequencing the two coroutines, it meets the perfect prefix learning constraint and, I suppose, the others as well if the notion of subgraph can be defined for coroutines. In short, the use of a more powerful control regime allows perfect prefix learning to generate core procedures that it should not. Restricting the control regime to be a stack improves the descriptive adequacy of the theory by blocking the star bug. In addition, the stack constraint is in itself a restriction on acquisition and so increases explanatory adequacy as well.

SATISFACTION CONDITIONS, DELETION, AND THE APPLICATIVE CONSTRAINT

In the previous section, it was shown that the execution state should be a stack, which entails that the knowledge representation for core procedures has a goal–subgoal calling hierarchy. That is, a goal should have explicit subgoals that it can call. But this constraint leaves many issues unsettled. One open question is, when should a goal be popped from the stack? Certainly a goal is finished when all the subgoals have been tried, but are there occasions when it should exit before trying all its subgoals? If there are, how should such information be represented? There are three reasonable possibilities:

1. Pop after *one* subgoal has been successfully executed.
2. Pop only after *all* applicable subgoals have been tried.
3. When to pop is controlled by a special information associated with each goal.

The first possibility cannot be the only exit convention used by the control structure because it will not allow expression of conjunctive goals, such as borrow, which has two subgoals that must both be executed. Consequently, it must be used along with some other exit convention, and goals must be typed to indicate which one applies. It is simplest to use it together with the second exit convention, a control regime equivalent to AND/OR graphs (Nilsson, 1971). So this control structure really amounts to giving goals either an AND or an OR type.

On the other hand, the second convention for when to pop goals can be used alone. Indeed, it is the one used by most production systems (i.e., those that use the recency and refractoriness conflict resolution principles (McDermott & Forgy, 1978) and do not erase goals from working memory). Goals do not need to be typed. Because it is simpler in that it does not require typing of goals, it is preferable to the AND/OR control regime.

The third option is a generalization of the other two. As it stands, it is just a catchall category and would have to be fully defined before it can be evaluated empirically. However, it is clearly more powerful and less constrained than the other two and should be considered only if the data force abandonment of the others (which they do, as it turns out). So, the three options, ranked in order of simplicity, are

1. AND: pop when all applicable subgoals have been tried.
2. AND/OR: goals have a binary type.
3. Otherwise.

The theme of this section is determining which of the three control structures optimizes the theory's adequacy. However, the route to resolving this question is in many ways more interesting than its answer. The preceding section uncovered some of the structure of the internal execution state by examining how repairs manipulated it. The details of how backup changed "short-term memory" revealed its structure. Here the task is similar. The exit conventions for goals are, in a sense, part of the "long-term memory" structure of goals. Repairs have been excluded from changing this structure by the core procedure immutability principle. To see into this "long-term" structure, one needs something that manipulates it or in some way depends on its form. Two will be used: perfect prefix learning and deletion. So, part of the argumentation that resolves the exit convention controversy involves taking a stand on certain aspects of two formal devices for describing acquisition.

The AND Exit Convention
Is Not Compatible with Assimilation

As a result of adopting a stack architecture in the previous section, a subset of the subtraction core procedures directly mirrors the structure of the teaching of subtraction. The segments of the subtraction curriculum are linearly ordered. It is plausible that an individual subject moves through some sequence of core procedures as he or she moves through the curriculum, assimilating new material on top of the old, with relatively little change to the old material. In order to capture this hypothesis formally as the assimilation constraint, the representation of knowledge must have a certain format. The argument that follows shows that capturing assimilation formally is incompatible and with the AND exit convention, the one used by production systems. It forces too much of the existing core procedure to be changed in order to assimilate the new material.

The problem with the AND exit convention is due to the cumbersome way that disjunctive goals must be expressed. To express the fact that two subgoals are mutually exclusive, one must put mutually exclusive "applicability" conditions on them. For example, to express the fact that there are two mutually exclusive ways to process column, depending on whether it is a two-digit column or a one-digit column, one would write

To process-column (X):
1. (blank? (bottom X)) $\Rightarrow$ (bring-down-top X)
2. (not (blank? (bottom X))) $\Rightarrow$ (take-difference X)

(In this example and the ones following it, a rule-oriented syntax has been adopted. Each subgoal is a rule, with its name and arguments on the right side. The conditions determining when the subgoal is applicable are separated from the rest of the subgoal to come a set of "applicability conditions," which are shown on the left side of the rule. The locally bound data flow that was argued for in the preceding section is implemented by giving goals arguments; in this case, X is a variable (argument) standing for the column being processed. Lisp function/argument syntax is used instead of the usual mathematical notation (i.e., $f(x)$ is written $(f\ x)$.). Nothing in the following argument depends on the adoption of a rule-oriented syntax as opposed to, for example, networks or schemata.)

Both rules must have applicability conditions in order that they by mutually exclusive. In the following problem:

$$\begin{array}{r} 37 \\ -\ 4 \\ \hline \end{array}$$

the first rule must be prevented from applying to the units column, so its applicability condition is necessary. The second rule's applicability condition is necessary to prevent it from applying to the tens column. Because the AND exit convention tries to execute all subgoals, one can only get mutual exclusion by using mutually exclusive applicability conditions.

This implies that assimilating a new alternative method of accomplishing a goal involves rewriting the applicability conditions of the existing subgoals. If the applicability conditions are not changed, then the new subgoal will not turn out to be mutually exclusive of the old subgoal. For example, to assimilate a new method of processing columns, say one that handles columns whose top and bottom digits are equal, one would have to modify the previous goal to become

To process-column (X):
1. $(= (\text{top } X)(\text{bottom } X)) \Rightarrow (\text{write-zero-in-answer } X)$
2. $(\text{blank? } (\text{bottom } X)) \Rightarrow (\text{bring-down-top } X)$
3. *(and* $(\text{not } (\text{blank? } (\text{bottom } X))$
   (not $(= (top\ X)(bottom\ X)) \Rightarrow$ $(\text{take-difference } X)$

Adding the new subgoal forced the applicability conditions of one of the existing subgoals to be changed (the italicized material was added). The essential point here is that the AND convention forces mutually exclusive alternatives to be highly interdependent. This lack of modularity defeats assimilation since assimilation is forced to modify existing material even though that material's function has not changed.

Despite the elegance of having just one exit convention and the simplicity of leaving goals unannotated, the AND exit condition must be abandoned. This implies that goals must be annotated with at least a single bit, discriminating between AND and OR, and perhaps with something rather complex.

This annotation is welcome in that it simplifies the backup repair. The function of the repair is to pop the goal stack back to the first goal that has some alternatives left to try. When goals are typed, it is trivial to tell whether a supergoal has any alternatives left. If it is an AND goal, by definition it does not. If it is an OR goal, then only one of its alternatives has been tried (because it normally pops after trying one subgoal), so all the rest must be open. If the representation language adds the syntactic constraint that all OR goals must have more than one subgoal, then backup's search becomes trivial: Pop the stack back to the first OR goal. In short, typing goals allows us to strengthen the locality constraints on repairs even more, reducing tailorability and leading perhaps to insight into the processes underlying local problem solving.

Finally, the AND/OR distinction allows the disjunction-free learning constraint to be formalized as a constraint on syntax rather than domain-dependent predicates, which is how disjunction must be represented under the AND exit convention.

Conjoined Rule Deletion

Perfect prefix learning is characterized by the constraint that all core procedures generated by it from a given prefix of the instructional sequence can answer all the exercises of that prefix correctly, without hitting impasses. If core procedures that are apparently the products of forgetting or partial learning were included in the output of perfect prefix learning, the constraint would be false. So, in order to increase the explanatory adequacy of the theory, they are generated by a separated operator. The issue is, what should that operator be? To motivate it empirically, consider the form of the following goal:

To borrow-from-zero (X), do all of:
  1. (borrow-from (next-column X))
  2. (write9 (top X))

This goal is an AND goal. When the first rule is deleted, a core procedure is generated that happens to survive application without reaching any impasses to become the bug Borrow-From-Zero:

$$
\text{Borrow-From-Zero:} \quad
\begin{array}{r} 3\ 4\ 5 \\ -1\ 0\ 2 \\ \hline 2\ 4\ 3\ \checkmark \end{array}
\qquad
\begin{array}{r} {}^{3}\\ 3\ 4^{1}5 \\ -1\ 2\ 9 \\ \hline 2\ 1\ 6\ \checkmark \end{array}
\qquad
\begin{array}{r} {}^{9}\\ 2\ 0^{1}7 \\ -1\ 6\ 9 \\ \hline 1\ 3\ 8\ \times \end{array}
$$

This bug only changes the zero to nine during borrowing across zero. It doesn't borrow recursively to the left since rule 1 has been deleted. On the other hand, if rule 2 were deleted instead, it would be the writing of nine that is omitted, generating the bug borrow-across-zero

$$
\text{Borrow-Across-Zero:} \quad
\begin{array}{r} 3\ 4\ 5 \\ -1\ 0\ 2 \\ \hline 2\ 4\ 3\ \checkmark \end{array}
\qquad
\begin{array}{r} {}^{3}\\ 3\ 4^{1}5 \\ -1\ 2\ 9 \\ \hline 2\ 1\ 6\ \checkmark \end{array}
\qquad
\begin{array}{r} {}^{0}\ \ {}^{1}\\ 2^{1}0^{1}7 \\ -\ \ 6\ 9 \\ \hline 4\ 8\ \times \end{array}
$$

The basic idea is that rule deletion of any of the rules under the AND goal seems to generate core procedures that not only lead to bugs, but are core procedures that one cannot generate with perfect prefix learning because they do not reflect perfect learning of any known segment in the subtraction curriculum.

So, deleting rules under AND goals is good. But look what happens when rule deletion is applied under the following OR goal

To borrow-from (X), do one of:
1. (zero? (top X)) $\Rightarrow$ (borrow-from-zero X)
2. ( ) $\Rightarrow$ (decrement (top X))

Deleting the first rule generates a familiar core procedure, namely one that impasses whenever it is asked to borrow from a zero. Because the rule that handles borrowing from zero is missing, it tries to decrement the zero. (This happens despite the fact that rule deletion has left the goal borrow-from-zero intact; only the call to it has been deleted.) But there is no need to generate this core procedure via deletion since it is already generated by perfect prefix learning. Even worse things than redundant generation happen when the second rule or the OR goal is deleted. The following star bug is generated:

$$
\begin{array}{lccc}
 & & 9 & \\
\text{*Only-Borrow-From-Zero:} & 3\ 4\ 5 & 3\ 4^1 5 & 2\ 0^1 7 \\
 & -1\ 0\ 2 & -1\ 2\ 9 & -1\ 6\ 9 \\
 & \overline{2\ 4\ 3}\ \checkmark & \overline{2\ 2\ 6}\ \times & \overline{1\ 3\ 8}\ \times
\end{array}
$$

This star bug misses all problems requiring borrowing because it never performs a decrement, despite the fact that it shows some sophistication in borrowing across zero in that it changes zeros to nines. The juxtaposition of this competency in borrowing across zero with missing knowledge about the simple case makes the behavior highly unlikely.

One possible deletion operator is to delete any rule (subgoal) of an AND goal. The rules of OR goals are not to be deleted. As the examples just given implied, this operator fits the data rather well. Note that it cannot be formulated without appealing crucially to the AND/OR distinction. Hence, this deletion operator cannot be used in a control structure that uses only the AND exit condition (i.e., it won't work with production systems). Its name reflects this dependence: conjoined rule deletion.

Satisfaction Conditions

This deletion operator is a little too unconstrained. Some of its deletions lead to star bugs. For example, the main loop of subtraction, the one that traverses columns can have the following goal structure:

To process-all-columns (X), do one of:
1. (not (blank? (next-column X))) $\Rightarrow$ (column-traverse X)
2. ( ) $\Rightarrow$ (process-column X)

To column-traverse *(X)*, do all of:
1. (process-column *X*)
2. (process-all-columns (next-column *X*))

The second goal is an AND goal, so either of its subgoals can be deleted. Deleting rule 1 creates a core procedure which only answers the leftmost column in a problem. This is a star bug. Deleting rule 2 creates another stag bug, one that only does the units column.

Similar problems occur with other conjunctive goals. If the only goal types are AND and OR, borrowing must be represented with an explicit subgoal for taking the column difference after the modifications to the top row are made. This causes problems with the deletion operator. Specifically, when borrowing is represented with AND/OR control structure, as in:

To process-column *(X)*, do one of:
1. (blank? (bottom *X*)) $\Rightarrow$ (bring-down-top *X*)
2. (less? (top *X*) (bottom *X*)) $\Rightarrow$ (borrow&take-difference *X*)
3. ( ) $\Rightarrow$ (take-difference *X*)
To borrow&take-difference *(X)*, do all of:
1. (borrow-from (next-column *X*))
2. (add10 (top *X*))
3. (take-difference *X*)

deleting the third rule of the AND goal means borrow&take-difference will do all the setting up necessary to take the column difference, but will forget to actually take it. This leads to the following star bug:

		3	₁ 9
*Blank-With-Borrow:	3 4 5	3 4¹5	2 0¹7
	−1 0 2	−1 2 9	−1 6 9
	2 4 3 √	2 1 ×	3 ×

It is perhaps possible to put explicit constraints on conjoined rule deletion in the same way that deletion blocking principles were used in Brown and VanLehn (1980). This approach would probably collapse into a nest of unmotivated constraints. However, a second way to prevent overgeneration is to make the operator inapplicable by changing the type of the goal so that it is not an AND. That is, one changes the knowledge representation rather than the operator.

The proposed change is to generalize the binary AND/OR type to become "satisfaction conditions." The basic idea of an AND goal is to pop when *all* subgoals have been executed, whereas an OR goal pops when *one* subgoal has been executed. The idea of satisfaction conditions is to have a goal pop *when its satisfaction condition is true*. Subgoals of a goal are executed until either the goal's satisfaction condition becomes true or all the applicable subgoals have been tried. (Note that this is not an iteration construct—an ''until'' loop—since a rule can only be executed once.) AND goals become goals with FALSE satisfaction conditions. Since subgoals are executed until the satisfaction condition becomes true (which it never does for the AND) or all the subgoals have been tried, giving a goal FALSE as its satisfaction condition means that it will always execute all its subgoals. Conversely, OR goals are given the satisfaction condition TRUE. The goal exits after just one subgoal is executed.

With this construction in the knowledge representation language, one is free to represent borrowing in the following way:

To process-column (X), do until (not (blank? (answer X))):
  1. (blank? (bottom X)) $\Rightarrow$ (bring-down-top X)
  2. (less? (top X) (bottom X)) $\Rightarrow$ (borrow X)
  3. ( ) $\Rightarrow$ (take-difference X)
To borrow (X), do until FALSE:
  1. (add10 (top X))
  2. (borrow-from (next-column X))

The AND goal borrow now consists of two subgoals. After they are both executed, control returns to process-column. Because process-column's satisfaction condition is not yet true—the column's answer is still blank—another subgoal is tried. Take-difference is chosen and executed, which fills in the column answer. Now the satisfaction condition is true, so the goal pops.

Given this encoding of borrowing, conjoined rule deletion does exactly the right thing when applied to borrow. Deleting the first rule generates a core procedure that hits an impasse which repair amplifies to generate several bugs, such as:

		3	19
Smaller-From-Larger-	345	345	207
With-Borrow:	-102	-129	-169
	243 $\checkmark$	214 $\times$	32 $\times$

		3	19
Zero-After-Borrow:	345	345	207
	-102	-129	-169
	243 $\checkmark$	210 $\times$	30 $\times$

(These bugs are generated because skipping the add10 operation causes an impasse when control returns to process the column that initiated the borrow. The top digit of that column is still larger than the bottom digit, so a repair is necessary. Different bugs result from different repairs.) Deleting the second rule generates the bug Borrow-No-Decrement:

Borrow-No-Decrement:	3 4 5	3 4^{1}5	2^{1}0^{1}7
	$-1\ 0\ 2$	$-1\ 2\ 9$	$-1\ 6\ 9$
	2 4 3 $\checkmark$	2 2 6 $\times$	1 4 8 $\times$

The point is that it is no longer possible to generate the star bug. Similarly, the star bugs associated with column traversal can be avoided by structuring the loop across columns as:

To process-all-columns (X), do until (blank? (next-column X)):
1. ( ) $\Rightarrow$ (process-column X)
2. ( ) $\Rightarrow$ (process-all-columns (next-column X))

By using a satisfaction condition formulation here, generation of the star bugs is avoided. From this example and the preceding ones, it is clear that augmenting the representation with satisfaction condition and using conjoined rule deletion creates an observationally and descriptively adequate treatment of perfect prefix learning and deletion.

Satisfaction conditions also play a crucial role in the formulation of empirically adequate critics. It turns out that one of the problems with critics, the so-called "blank answer critic" problem mentioned in Brown and VanLehn (1980), can be solved using satisfaction conditions. The details of this argument will be omitted here in favor of some comments of a different kind.

The Applicative Constraint

Having an operator that mutates the knowledge representation allows one to "see" the structure of the representation. An important use of this tool is to uncover one of the tacit constraints on data flow. One of the prominent facts about bugs is that none of them requires deletion of locative focus shifting functions. For example, if one knows about borrowing-from, one knows to borrow from a column to the left. No bug has been observed that forgets to move over before borrowing-from. This fact deserves explanation.

In all the illustrations so far, focus shifting functions such as (next-column X) have been embedded inside calls to actions, predicates, or subgoals. This is no accident. Suppose one did not embed them, but instead made them separate subgoals themselves, as in:

To borrow (X), do all of:
1. (add10 (top X))
2. $(X \leftarrow (\text{next } X))$
3. (borrow-from X)

where "$\leftarrow$" means to change the binding (value) of the local variable X. A star bug could be generated by deleting rule 2. This star bug would borrow from the column that originates the borrow:

$$
\begin{array}{llll}
 & & \overset{14}{} & \overset{9}{}\ \overset{16}{} \\
\text{*Borrow-From-Self:} & 3\ 4\ 5 & 3\ 4\vert 5 & 2\vert 0\vert 7 \\
 & \underline{-1\ 0\ 2} & \underline{-1\ 2\ 9} & \underline{-1\ 6\ 9} \\
 & 2\ 4\ 3\ \checkmark & 2\ 2\ 5\ \times & 1\ 3\ 7\ \times
\end{array}
$$

In order to avoid such star bugs, focus shifting functions must be embedded, so a constraint upon the knowledge representation is needed to make this explicit. About the strongest constraint one can impose is to stipulate that the language be applicative. That is, data flows by binding variables rather than by assignment. There are no side effects: A goal cannot change the values of another goal's variables, nor even it's own variables. The only way that information can flow "sideways" is by making observable changes to the external state, that is, by writing on the test page.

The applicative constraint is extremely strong, forcing data to flow vertically only in the goal hierarchy. The procedure can pass information down from goal to subgoal through binding the subgoal's arguments. Information flows upward from subgoal to goal by returning results. No counterexamples to the applicative constraint have been found in the subtraction domain.

The applicative constraint could have a profound effect on learning. It seems to make learning context-free. That is, learning a procedure becomes roughly equivalent to inducing a context-free grammar. The basic idea is that the applicative constraint together with the stack constraint force data flow and control flow to exactly parallel each other. To put in terms of grammars, the data flow *subcategorizes* the goals. This in turn makes it possible to induce the goal hierarchy from examples. Inducing hierarchy for procedures from examples has been an unsolved problem. Neves (1981) had to use hierarchical examples to get his procedure learner to build hierarchy. However, subtraction teachers do not always use such examples. Badre (1972) recovers hierarchy by assuming examples are accompanied by a written commentary; each instance of the same goal is assumed to be accompanied by the same verb (e.g., *borrow*). This is a somewhat better approximation to the kind of input that students actually receive, but again it rests on delicate and often violated assumptions. The applicative constraint cracks the problem by structuring the language in such a way that hierarchy can be learned via a context-free grammar induction algorithm (subject to the disjunction-free learning constraint, of course).

VARYING THE REPRESENTATION LANGUAGE IS THE KEY

As mentioned previously, the main methodological point of this chapter is that: *Getting the knowledge representation language "right" allows the model to be made simple and to obey strong principles.*

There have been five major changes in Repair Theory's representation language (so far). Each one had far-reaching effects on the simplicity and principledness of the model. In contrast, there has been only one change on the basic processes of the model, when perfect prefix learning was separated from rule deletion. There are several personal observations to make about this development.

Although it was intuitively obvious all along that "perfect learning" and "forgetting" could be separated, the need for changes in the representation language was never immediately obvious. The impact of knowledge representation was very subtle. Its structure was not at all apparent from the data, and intuition was no guide either. It was only by reimplementing the model to use the new representation that its effect could be evaluated.

The changes in the representation language were absolutely necessary for bringing the model into conformance with psychologically interesting principles. A typical scenario was to note a particularly ugly part of the model that could be simplified and made more elegant by changing the representation language; after the change was made, reflection upon the new elegance of the model uncovered a principle that the model now conformed to. Hard on the heels of that insight would often come several more ideas, usually of the form that changing a certain component process of the model to take advantage of the new construct in the representation language would allow it also to become simpler and to obey a new principle. In short, the evolution of the principles of the theory went hand in hand with the evolution of the representation language.

The process model and the deletion operator were implemented in a computer program—called the "workbench"—in such a way that, given a core procedure, all possible deletions and all possible repairs could be applied and evaluated automatically. The evaluation was carried out by having the parameterized model "take" a highly diagnostic subtraction test and submit its answers to DEBUGGY (Burton, 1981). DEBUGGY would diagnose the pseudo-student's answers to determine which bugs, if any, were exhibited. This program could be given a set of core procedures and left to run overnight. The morning's results revealed the observational adequacy of that set of core procedures. Although the initial investment in constructing this automatic theory evaluation system was high, it more than repaid that investment in facilitating the evaluation of subtle changes in the representation language as well as varying expressions of individual core procedures.

Although changing the representation language allows discovery of its impact on the theory, quite a bit more work was required to separate these effects from the context of their discovery in order to make well-formed arguments. To put it graphically, the journey that the model and I took through the space of representations was a single twisted path. A great deal of thought was required not only to uncover the critical junctions along that path, but to understand the *dimensions of the space.* The analysis sought to free the choices in alternative representational structures from their discovery context and present them in a neutral context, so they could be evaluated independent of each other. Once put in a neutral context, use of the medium-level of detail of Repair Theory as a rich structure of starting assumptions enabled competitive argumentation in most cases to settle the choice, establishing the particular set of choices that constitutes the representation language at the end of the path as a global maximum, rather than as a local one, in the space of mental representations.

REFERENCES

Ashlock, R. B. *Error patterns in computation.* Columbus, Ohio: Bell and Howell, 1976.

Badre, N. A. *Computer learning from English* (memo ERL-M372). Berkeley: University of California, Berkeley, Electronic Research Laboratory, 1972.

Berwick, R. Cognitive efficiency, computational complexity and the evaluation of grammatical theories. *Linguistic Inquiry,* in press, *13.*

Brown, J. S., & Burton, R. B. Diagnostic models for procedural bugs in basic mathematical skills. *Cognitive Science,* 1978, *2,* 155–192.

Brown, J. S., & VanLehn, K. Repair Theory: A generative theory of bugs in procedural skills. *Cognitive Science,* 1980, *4,* 379–426.

Brownell, W. A. The evaluation of learning in arithmetic. In *Arithmetic in general education.* 16th Yearbook of the National Council of Teachers of Mathematics. Washington, D.C.: N.C.T.M., 1941.

Brueckner, L. J. *Diagnostic and remedial teaching in arithmetic.* Philadelphia, Pa.: John C. Winston, 1930.

Bunderson, C. V. *Cognitive bugs and arithmetic skills: Their diagnosis and remediation* (interim tech. rep.). Provo, Utah: Wicat, April 1981.

Burton, R. B. DEBUGGY: Diagnosis of errors in basic mathematical skills. In D. H. Sleeman & J. S. Brown (Eds.), *Intelligent tutoring systems.* London: Academic Pres, 1981.

Buswell, G. T. *Diagnostic studies in arithmetic.* Chicago: University of Chicago Pres, 1926.

Chomsky, N. *Aspects of the theory of syntax.* Cambridge, Mass.: MIT Press, 1965.

Chomsky, N. Rules and representations. *The Behavioral and Brain Sciences,* 1980, *3,* 1–63.

Cox, L. S. Diagnosing and remediating systematic errors in addition and subtraction computations. *The Arithmetic Teacher,* 1975, *22,* 151–157.

Dietterich, T. G., & Michalski, R. S. *Learning and generalization of structured descriptions: Evaluation criteria and comparative review of selected methods.* (Rep. 1007). Urbana: University of Illinois, Department of Computer Science, 1980.

Durnin, J. H., & Scandura, J. M. Algorithmic approach to assessing behavior potential: Comparison with item forms. In J. M. Scandura (Ed.), *Problem Solving: A structural/process approach with instructional implications.* New York: Academic Press, 1977.

Feldman, J. Some decidability results on grammatical inference and complexity. *Information and Control,* 1972, *20,* 244–262.

Gold, E. M. Language identification in the limit. *Information and Control,* 1967, *10,* 447–474.

Greeno, J., & Brown, J. S. *Theories of competence.* Paper presented at the Sloane conference in Boulder, Colo., 1981.

Hayes-Roth, F., & McDermott, J. Learning structured patterns from examples. *Proceedings of the Third International Joint Conference on Pattern Recognition,* Stanford, Calif., 1976, 419–423.

Keil, F. C. Constraints on knowledge and cognitive development. *Psychological Review,* 1981, *88,* 197–227.

Knobe, B., & Knobe, K. A method for inferring context-free grammars. *Information and Control,* 1976, *31,* 129–146.

Lankford, F. G. *Some computational strategies of seventh grade pupils.* Charlottesville: University of Virginia, 1972. (ERIC Document)

McDermott, J., & Forgy, C. L. Production system conflict resolution strategies. In D. A. Waterman & F. Hayes-Roth (Eds.), *Pattern-directed inference systems*. New York: Academic Press, 1978.

Mitchell, T. M. *Version Spaces: An approach to concept learning* (Tech. Rep. 78–711). Stanford, Calif.: Stanford University, Department of Computer Science, 1978.

Neves, D. M. *Learning procedures from examples*. Unpublished doctoral dissertation, Department of Psychology, Carnegie-Mellon University, Pittsburgh, Pa., 1981.

Nilsson, N. J. *Problem-solving methods in Artificial Intelligence*. New York: McGraw-Hill, 1971.

Norman, D. A. Categorization of Action Slips. *Psychological Review*, 1981, *88*, 1–15.

Pinker, S. Formal models of language learning, *Cognition*, 1979, *7*, 217–283.

Roberts, G. H. The failure strategies of third grade arithmetic pupils. *The Arithmetic Teacher*, 1968, *15*, 442–446.

VanLehn, K. *Bugs are not enough: Empirical studies of bugs, impasses and repairs in procedural skills* (Tech. Rep. CIS-11). Palo Alto, Calif.: Xerox Palo Alto Research Centers, 1981.

VanLehn, K., & Brown, J. S. Planning Nets: A representation for formalizing analogies and semantic models of procedural skills. In R. E. Snow, P. A. Federico, and W. E. Montague (Eds.), *Aptitude, learning and instruction: Cognitive process analyses*. Hillsdale, N.J.: Lawrence Erlbaum Associates, 1980.

Vere, S. A. Inductive learning of relational productions. In D. A. Waterman & F. Hayes-Roth (Eds.), *Pattern-directed inference systems*. New York: Academic Press, 1978.

Winston, P. H. Learning structural descriptions from examples. In P. H. Winston (Ed.), *The Psychology of Computer Vision*. New York: McGraw-Hill, 1975.

Young, R. M., & O'Shea, T. Errors in children's subtraction. *Cognitive Science*, 1981, *5*, 153–177.

Complex Mathematical Cognition[1]

Considering the great complexity of educational phenomena and the substantial areas of real disagreement, it is not surprising that educators have sought a "scientific" basis for the study of teaching and learning. This search has led educators to look at the physical sciences—seeming models of simplicity, agreement, and objectivity—and in some fashion to try to imitate them. The imitation has in many cases failed, and perhaps nowhere more than in the study of complex mathematical cognition—how do humans *think about mathematical problems,* how do they *develop the ability* to think about mathematical problems, and *what obstacles impede this development?*

But perhaps the difficulty lies in an incorrect understanding of physical science. Within the physical sciences, it is *not* true that objective data are generally available—more commonly the researcher has to begin by making decisions on how and where to seek data, and these decisions are not themselves objective; nor is it true that data lead directly to theories. Theories are, indeed, the main product of science, but they are created somewhat as Euclidean geometry must have been, largely by *proclamation,* by the explication of someone's intuitions. And this, of course, opens the door to quite substantial disagreements between different people who are building on different intuitions. We need to pay more heed to the nature of such disagreements.

[1]The research at the University of Illinois Curriculum Laboratory, which is reported here, was supported by grant NSF SED77-18047 from the National Science Foundation.

THE DEVELOPMENT OF MATHEMATICAL THINKING

THE EMERGENCE OF AN ALTERNATIVE PARADIGM

Thomas Kuhn's well-known volume, *The Structure of Scientific Revolutions* (Kuhn, 1970), has given currency to the idea of a *paradigm shift* in science—the process of coming to view phenomena in a new way, as when Priestley's "phlogiston" view came to be replaced by Lavoisier's lists of *elements* and *compounds*.

Something of this sort is now happening in the study of the teaching and learning of mathematics, or in the study of the processes of thinking about mathematical topics. Indeed, part of the alternative paradigm is concerned with just which words we choose here: An earlier paradigm focused on *teaching* and *learning*, whereas the emerging paradigm focuses on *the processes of thinking about mathematical problems*. In the new paradigm, *learning* is regarded as definable in terms of *transitions* between one form of mathematical thinking and another and is in this sense not fundamental, but is a kind of "derived" or "second-order" concept. The *fundamental* task is to describe the various forms of mathematical thought themselves, that form the "before" and the "after" of any change.

But perhaps Priestley and Lavoisier are not the best parallels to what is now occurring in mathematics education. A closer parallel would probably be the earlier eighteenth century study of electricity, because the simultaneously coexisting views not only focused on different phenomena, but saw them in terms of a different set of basic concepts. What phenomena should properly be regarded as fundamental aspects of "electricity?" Frictional generation? Attraction? Repulsion? The "fluid" ability to "run through conductors?" What *concepts* should be used in thinking about electricity? Mechanical-corpuscular particles? The idea of a "fluid" or "effluvium?" "[Although] most experimenters read each other's works, their theories had no more than a family resemblance [Kuhn, 1970, p. 14]."

Kuhn's description sounds very similar to what one might say about mathematics education in the United States today. The "standard" approach focuses on external observable behavior, uses multiple-choice tests and questionaires as data sources, avoids discussion of thought processes, employs statistical tests of significance, commonly works with moderately large numbers of students (because common statistical methods require this), often compares alternative treatments, regards teaching and learning as fundamental processes, and typically deals with "events" (often alternative teaching approaches) that involve 30 or more students for a period of perhaps 6 weeks, or something on that order of magnitude.

By contrast, the "alternative" paradigm that this chapter discusses regards mathematical thought processes as fundamental (and, deriving "learning" from the observation of *changes* in these thought processes, regards learning as NOT usually fundamental), focuses on the *observation* and *description* of mathematical behavior, recognizes the necessity of relating observations to a *postulated* theory of "metaphoric" processes for dealing with information (how the individual *thinks about* some mathematical problem), gets data most often from task-based interviews, deals with "events" as small as why a single student gave a single response to some specific question, often takes *one student* as the "universe of discussion" (and is concerned with how this single student thinks about certain tasks within mathematics), may follow the same student for as long as 5 years (appropriate when a student is the "universe" you are studying!), and borrows its basic concepts from the field of Artificial Intelligence (or cognitive science).[2] As this list suggests, the alternative paradigm is closer to physics, chemistry, or even history, than it is to statistical quality control or statistical hypothesis testing. Yet it is also close to the thinking of many experienced teachers, who give serious consideration to the way individual students are analyzing mathematical problems of various types—and to the task of improving the way these students approach mathematical tasks.

The alternative paradigm that we describe in this chapter is based especially upon the work of Minsky, Papert, Simon, Newell, Larkin, Clement, diSessa, Matz, Brown, VanLehn, Greeno, Neves, Rissland, Ginsburg, Easley, Winston, Winograd, Karplus, Lenat, Groen, Carry, von Glasersfeld, Erlwanger, Friend, Norman, Rumelhart, McDermott, Krutetskii, and a research group at the Curriculum Laboratory at the University of Illinois. A partial listing of relevant works is included in the reference list. (There is also an unmistakable debt to Polya and Piaget, and there are clear antecedents in the work of George Miller and Kurt Lewin.)

Task-Based Interviews

For the alternative paradigm, probably the most common source of data is the *task-based interview* (see Chapter 1, this volume; see also Ginsburg, 1981; Opper, 1977). Different investigators use the technique somewhat differently, but the central idea is usually recognizable and usually follows this pattern: A student is seated at a table and is given paper and pencil (or pen); his or her actions and words are recorded (often via videotape or audiotape), a mathematical task is presented to the student, and an interviewer talks with the student. There may also be one or more additional observers who makes notes of actions, phrases, gestures, inflections, impressions, and so on.

---

[2]Obviously, these brief descriptions cannot completely describe the work of a considerable number of investigators. They are intended as generally suggestive sketches to focus the present discussion.

The interviewer may play a severely restrained role, perhaps saying nothing at all for an extended time interval; at the opposite extreme, the interviewer may participate rather actively by perhaps handing the student useful implements (e.g., rulers, compasses, Dienes MAB blocks, etc.), posing additional questions, making helpful hints, and so on.

Some interviewers employ direct questions, such as: "How did you decide to write a 7 there?" Others prefer to seek the same kind of information indirectly, for example, by asking, "If this number were a 2, would you solve the problem in the same way?" In any case, at the end of the interview, the investigator has (*a*) the papers on which the student has written, (*b*) an audiotape or videotape record of the interview, (*c*) notes made by observers (and by the interviewer), and (*d*) recollections of what happened (which can be very valuable in analyzing the other three items).

Before one dismisses this as a simple procedure, it is worth noting that few interviewers do a good job, and even the best interviewers nearly always find, in analyzing an interview afterward, that they wish they had conducted the interview differently. For example, it very often turns out that a crucial uncertainty that shows up in postinterview analysis could have been resolved if the proper question had been posed at the right time during the interview.

Another difference among investigations is the extent to which an advance protocol for an interview is prepared beforehand and how rigidly or flexibly it is used. Tradition, imported from the earlier research paradigm, favors considerable rigidity, but most alternate paradigm[3] investigators believe that unexpected situations arise so often during interviews that one must be prepared to deal with them, and that one must therefore retain considerable flexibility.

The mathematical content of the task may be drawn from arithmetic, algebra, geometry, or calculus, or it may be a specific ad hoc task of a mathematical nature.

The rest of this chapter deals with the analysis of records obtained in this way, with some specific results that have been found by such methods, and with the point of view that is emerging as a result of this research.

---

[3]Clearly, as Kuhn's remarks imply, it is important for coexisting paradigms to communicate and to make sympathetic and determined efforts to understand one another and even to coalesce into a larger, all-embracing synthesis of their various partial views. (Indeed, it is only to this "synthesis" that Kuhn applies the word *paradigm*.) Consequently, the terms chosen here are not intended to generate animosity where there needs to be cooperation. Our most common terms will be *traditional paradigm* or *earlier paradigm* for that approach which one has commonly encountered in the 1950s and 1960s (and is still the prevalent approach today) and to use the phrase *alternative paradigm* for the approach that takes its data mainly from task-based interviews and relates the data to a postulated theory cast in cognitive science terms.

Basic Conceptualizations

The Greeks—perhaps the true parents of modern science—might have stared at islands and oceans and mountains for quite a long time, with little "scientific" result, had they not taken the decisive step of *postulating* some entities that made "space" more amenable to effective human thought: specifically, entities such as points, planes, lines, distances, etc.

A similar approach is a key to the alternative paradigm for the study of mathematical thought: the postulation of structures and procedures that represent mathematical knowledge and mathematical information processing. What should be postulated? Three dualities need to be considered: (*a*) the static representation of knowledge versus the dynamic processing of information, (*b*) "gestalt" or "aggregate" or "chunk" entities versus sequential procedures, and (*c*) storage and retrieval of entities versus the real-time ad hoc construction of representational entities. What each of these dualities means will become clearer in a few pages.

Presumably, before postulating things like "points" and "planes," the Greeks had thought quite a bit about physical space and about possible intellectual tools that might make it easier to discuss space. In analyzing intellectual thought before postulating comparable mathematical tools, one needs to observe a large number of instances of human mathematical behavior. Such observations have been carried out (though vastly more are needed) (cf. Davis, Jockusch, & McKnight, 1978), and the consideration of instances has suggested the postulation of several devices for processing mathematical information, which we now list.

SEQUENTIAL PROCESSES

Procedures.   By "procedure" we mean an algorithmic, step-by-step activity, such as the cognitive sequence for adding 11 + 3 by starting with "eleven," then saying counting words to "count onward" from 11 by counting the points of the symbol "3": "twelve" [⌐³3]; "thirteen" [→3]; "fourteen" [⌐→3]. "So eleven plus three is fourteen."

At least two kinds of procedures exist and produce observably different behavior: *visually moderated sequences* and *integrated sequences*.

Visually moderated sequences have the form of an input (usually visual) that cues the retrieval (from memory) of a procedure; execution of the procedure modifies the visual input; the modified visual input cues the retrieval of a new procedure; and the cycle continues until some process (possibly completing the solution) triggers termination. A very typical instance would be long division, being performed by someone who is not the full master of algorithm:

A visual cue

$$21\overline{)7329}$$

triggers the retrieval of a   [''Uh, yes! How many times
  procedure   does 2 go into 7?'']

which produces a new visual   3
  cue[4] $21\overline{)7329}$

which triggers retrieval from   [''Oh, yes! Now I multiply 3
  memory of another procedure   times 21.'']

which produces a new visual   3
  input $21\overline{)7329}$
                                  63

. . . and the process continues.

Factoring quadratic polynomials is another example—again, if the student doing it is not yet the complete master of the topic:

The visual input $x^2 - 5x + 6$

triggers the retrieval of a
  procedure

that produces a new visual input   (   ) (   )

which triggers the retrieval of a
  procedure that produces

a new visual input $(x\quad)\,(x\quad)$

. . . and so on.

With sufficient practice, a visually moderated sequence can become independent of visual cues to trigger retrieval of the smaller component sequences that need to be strung together. Those who know long division well can describe the entire process without dependence on written cues. (They may, however, require paper as temporary storage for interim numerical results.) Sequences which, through sufficient practice, have become independent of visual cues for program guidance are called integrated sequences.

Relations among Procedures.   Within computer programming there is an important relationship among procedures: One procedure, A, may ''call upon'' or *transfer control to* a second procedure, B. When B has completed its assigned task, it returns control to procedure A. In such a relationship, procedure B is said to be a *subprocedure* of procedure A, and A is called the *superprocedure*.

It seems appropriate to postulate a similar relationship among procedures in the information processing that is part of human mathematical thinking. Indeed, observational data collected by Erlwanger (1973, 1974) indicate that, for the sixth-grade students observed, errors were entirely in superprocedures calling for

[4]Clearly, this composite ''protocol'' is presented here in a considerably simplified form.

wrongly chosen subprocedures. The subprocedures themselves functioned correctly (cf. Davis, 1977). This, of course, is partly a comment on the school curriculum; over-learning of antecedent tasks had occurred satisfactorily, but the new tasks had not yet been mastered. Erlwanger's evidence from remedial tutoring suggested that, for most of these sixth-graders, the fifth- and sixth-grade tasks probably never would be mastered.

As one example, one student answered .3 + .4 = ? with the answer:

$$.3 + .4 = .07$$

The subprocedure that accepted, as inputs, the numbers 3 and 4, and returned 7, operated correctly; so did the subprocedure that counted "1 decimal place" and "1 decimal place," and returned the format "2 decimal places" (i.e., .07). Of course, this second subprocedure *should not have been called upon* in the process of solving the given addition problem; it should have been used only for multiplication.

In cases where a superprocedure called upon the wrong subprocedure, Erlwanger's data showed a persistent relationship between the subprocedure A (say) that *should* have been chosen and the subprocedure (call it B) that actually *was* chosen. Almost without exception, the visual stimuli to elicit retrieval of A and B were extremely similar—for example, "3 + 3" versus "3 × 3", or

$$\text{versus} \quad \begin{array}{c} 10 \\ \underline{17} \end{array} \quad \begin{array}{c} 10 \\ \underline{17} \end{array}$$

A further pattern has virtually no exceptions: Within Erlwanger's data, it is very nearly always the case that subprocedure B (which *was* chosen) is something that was learned *earlier* in the school curriculum. In other words, some recently encountered *new* subprocedure has erroneously been ignored, and its place has been taken by some more familiar "old friend."

An interesting explanation of this phenomenon can be given in terms of Minsky's theory of K-lines, but the details are complex, and quite beyond the scope of this article (cf. Minsky, 1980).

THE GENERAL PROBLEM OF FLEXIBILITY

Thinking of offices, bureaucracies, and other human organizations, we all have some experience with the limits of flexibility within an organization; at some point, one reaches the boundary, and office procedure cannot deal appropriately with some specific instance because that instance lies outside of the original design for office procedures, and no further provisions for adaptation have been made.

Clearly, an analogous phenomenon bedevils human thought; one can reach a point where person A cannot cope because no *explicit* procedures learned by A will suffice, and the creation of a new (and appropriate) procedure lies outside of A's capability.

To understand this phenomenon, most researchers attempt to postulate some definite body of procedures and knowledge representation structures (which we shall temporarily call *the system*) and then to distinguish "operations within a system" from "operations that involve stepping outside of the system." This distinction is made with exceptional clarity by Hofstadter (1980). When a procedure orders up some subprocedure, all of the activity is "within the same level of the system"), rather as if a carpenter asks a fellow carpenter to hold a board in place while he nails it there. But clearly there are other kinds of operations that are needed.

A computer, asked to find the phone number of George Washington, first President of the United States, might call on a "Philadelphia" subprocedure to scan the Philadelphia listings, or a "Virginia" subprocedure to scan listings for Virginia, or even a "D.C." subprocedure to scan the D.C. listings. That sort of thing could go on for a long time, unless there were some information-processing operators of a different type that were able to deal with different aspects of the task—for example, a "plausibility" operator that could make a historical check, perhaps calling on a "historical" subroutine that could query when Washington lived and when the telephone was invented. The original "phone directory" procedure and its "Philadelphia" and "Virginia" subprocedures are on the same level (in this classification), a level that might be described as "finding the phone numbers." The "plausibility" operator, and its "historical" subroutine, are on a "higher" level (and on the *same* higher level) since they do not perform "phone-number-look-up" tasks, but "reflect" on the nature of such tasks. (To return to our carpenters, it is as if there were a "higher level" of operations carried out by efficiency experts, architects, economists, etc., who do not cut boards and drive nails, but *who study the process* of cutting boards and building houses.)

This is an old issue in artificial intelligence and cognitive science; very often when a computer performs "stupidly" (as in spending vast resources in the search for George Washington's phone number), it is because the machine has been programmed with "task-performing" procedures, but *without* any higher level procedures to step back from, as it were, the assembly-line, and to look at what is going on.

There are various ways to provide for these higher level operators, including at least these three (the first of which is NOT actually higher level):

1. Checks may be inserted *at the original level*—for example, before looking up any phone number. The look-up procedure can check dates, locations, and other reasons for believing that a phone number probably exists. (This, of course, is not a real solution because there remains the possible involvement of relevant attributes that have *not* been provided for in the checking procedure, as in the case of looking for the phone number of Hans Solo, or Lieutenant Uhuru, or KAOS).
2. Procedures can be created that do not perform tasks on the original level, but exist on a higher level and *operate on* lower level procedures (a mechanism postulated by Skemp, 1979, and by Hofstadter, 1980, and others).
3. Operations can take place in two separate areas: a "task-performance space" and a separate and distinct "planning space" (a solution implied by Simon, Minsky, Papert, and others).

All three solutions are possible for computers, although the second is, for computers, usually the most difficult; presumably all three are possible also for humans (who appear to make extensive use of the second method, which is one of the ways humans differ from today's computer programs.) It consequently seems desirable to postulate all three possibilities. We deal with the first method here (since it is really on the lowest, or task-performing, level) and also postulate some mechanisms to provide for the second method. We defer discussion of the third method until later, when we deal with *heuristics* and *planning*.

Critics.   A *critic* is an information-processing operator that is capable of detecting certain kinds of errors.

Example 1. A beginning calculus student wrote

$y = \sec x^2$
$dy = (\sec x^2)(\tan x^2)\, 2x$

The teacher instantly recognized that there MUST be some error here, because a differential dy could not be equal to an expression which did not involve differentials. The teacher's collection of information processors included a *critic* that was not contained in the student's collection.

Example 2. For reasons we shall consider below, a third-grade student wrote

```
  7002
-   28
  5084
```

Many adults possess a critic, related to the *size* of numbers, that should come into play here. After all, if I have about $7000 and I spend $28, I should NOT end up with about $5000. Something is wrong! The third-grade student, however, believed her work to be correct.

One particular kind of information-processing arrangement, the so-called *production system,* provides an especially straight-forward method for dealing with critics. For a discussion of production systems, refer to Newell and Simon (1972), or Davis and King (1977).

In both of the preceding examples we see differences in mathematical behavior that would be attributed to the presence, or absence, of certain specific critics.

Operations on Procedures.   It is commonly postulated that a memory record is kept of procedures that have been used (Davis, Jockusch, & McKnight, 1978; Minsky, 1980; Winograd, 1971). It is also usually postulated that there are procedures that use the *sequence of active* (lower level) *procedures* as their *input* and that *output* modifications of either *the collection of lower level procedures,* or else *the control structure.* John Seely Brown, for example, postulates a higher level operator that recognizes when the operational sequence is in a loop and that intervenes in the control structure so as to terminate the loop. Other higher level operators that have been postulated include a "look-ahead" operator that, with repetition, makes possible the prediction of which operator (or which input data) will be encountered next (Davis, Jockusch, & McKnight, 1978), and a "recognition" operator that can detect repetitions (idem). There is also evidence for a "simulate a run and observe" operator, as when a student, confronted with

$$x + 2 \overline{)x^4 + 2x^3 - 2x^2 - x + 6}$$

can "run through" in his or her head the algorithm for the long division of integers, as in

$$31 \overline{)6851}$$

"observe mentally" what happens, and thus solve the division-of-polynomials problem.

METAPHOR AND ISOMORPHISM

Of course, underlying the ability to recognize the precise parallel between an algorithm for dividing integers and an algorithm for dividing polynomials, there is something far more fundamental: an ability to match up input data with some kind of knowledge representation structure that has been stored in memory. In typical information-processing explanations, four steps are postulated:

1. use of some cues to trigger the retrieval from memory of some specific knowledge representation structure.

2. mapping information from the specific present input into "slots" or "variables" that exist within the knowledge representation structure.
3. making some evaluative judgments on the suitability of the preceding two steps (and cycling back where necessary).
4. If the judgment is that steps 1 and 2 have been successful, then the result is used for the next stage in the information processing.

One could illustrate these four steps as follows: If the task were to solve the following equation:

$$x^2 - 5x + 2 = 0,$$

then step 1 consists of observing some visual cues in the equation that cause us to say (in effect): "Aha! It's a quadratic equation!", with the result that we retrieve from memory the quadratic formula:

> For the equation
>
> $$ax^2 + bx + c = 0,$$
>
> the solutions are given by
>
> $$x = \frac{-b \pm \sqrt{b^2 - 4ac}}{2a}$$

Step 2 now involves looking at our specific present input—namely, the equation $x^2 - 5x + 2 = 0$—and taking from it certain specific information to enter into the "slots" or "variables" of our memorized rule. We see that "1" should be used as a replacement for the variable a, that "-5" should be used as the replacement for the variable b, and that "2" should be used as the replacement for the variable c.

Step 3 involves whatever checks we carry out in order to convince ourselves that this is all correct, after which use of the quadratic formula (step 4) easily produces the answers.

In one respect this example is too simple and might therefore be misleading. In most examples of human information processing, the "knowledge" that is involved is more complex, and hence the knowledge representation structures are more complex. Suppose the content dealt not with solving simple equations, but instead with reading and understanding a story about two people taking a trip of some sort:

Leslie and Dana knew that they had several hours to travel, so they decided to seize the opportunity of having lunch.

The "variables" in this case deal not with simple numbers like 1, -5, and 2, but rather with such matters as: the sex of Leslie and of Dana (either could, after all, be either male or female); their mode of travel (horseback? bicycles? commercial airliner? a car?); in what sense they are "seizing the opportunity" (stopping at an inn? stopping beside the road at a point where there is a good view? telling the stewardess that they do want lunch?); and so on.

By considering Leslie and Dana, we have not left the domain of mathematics, which after all requires us to read words, sentences, paragraphs, equations, tables, and so on. In every case, however, the basic four-step operation is usually postulated as a fundamental part of the human information processing; in this, at least, information-processing theories do not treat reading and mathematics as being very different. The convenient accident that, within mathematics, the "slots" in knowledge representation structures *may* be labeled as mathematical variables—*a, b, x, y,* and so on—is merely that: a convenience, but not an essential difference. (And, of course, a great deal of mathematical knowledge is stored in memory in other forms that do not make use of literal variables.)

One is dealing here with one of the most fundamental matters in human information processing. Presumably a successful selection, retrieval, and matching is what is meant by the word *recognize*—an instance of remarkable "folk" insight built into a common English verb. Hofstadter (1980) suggests this is what is commonly meant by *meaning,* and Minsky and Papert (1972) remark that, when a situation on a chess board has been analyzed correctly and (say) a "pin" has been recognized, it seems almost as if the pieces in question had suddenly changed color. The small pieces of input data are suddenly linked up with an important memorized data representation structure—the small pieces have suddenly become a large "chunk" (in Miller's phrase). Instead of tiny "meaningless" bits of information, we now have a chunk to deal with, and it is this chunk which gives meaning to this aggregation.

These chunks are sometimes called *assimilation paradigms,* and the teaching strategy that consists of *first* establishing such assimilation paradigms (or metaphors) and then subsequently exploiting them is sometimes called the *paradigm teaching strategy* (Davis, Jockusch, & McKnight, 1978). (Note that this educational use of the word *paradigm* is unrelated to Kuhn's historical use of this word!) This paradigm teaching strategy is used with stunning effectiveness by Hofstadter (1980) to teach Gödel's theorem and the theorem that recursively enumerable sets are not necessarily recursive. For his assimilation paradigm metaphors, which he establishes beforehand, Hofstadter uses drawings and lithographs by M. C. Escher and some Lewis Carroll style dialogues he created himself. The paradigm teaching strategy for precollege mathematics is used in Davis (1980). In one example, a bag is partially filled with pebbles. By adding pebbles to the bag, by removing pebbles from the bag, and by interpreting the result as more or fewer pebbles than when one started, it is possible to give a meaningful interpretation of mathematical statements such as:

$$4 - 5 = -1$$

and

$$7 - 3 = +4$$

(Davis, 1967, pp. 57–61). Thus, the effect of the combined acts of putting 4 pebbles *into* the bag and removing 5 pebbles *from* the bag is to leave the bag holding 1 less pebble than it held beforehand. In the case of the second equation, the combined acts of putting 7 pebbles into the bag and taking 3 out produce the net effect of leaving the bag holding 4 *more* pebbles than it held before. The mental imagery of "putting pebbles into the bag" and "taking pebbles out of the bag" serves as a paradigm that guides the process of calculation; it is easy to demonstrate that this imagery serves this purpose considerably better than an explicit set of verbal rules can. (Notice that, if this "pebbles-in-the-bag" model is used, then the *knowledge representation structure* that is created within the student's memory is *not* like the quadratic formula, with explicit literal variables, but instead more closely resembles the kind of episode-based memory trace that one ordinarily associates with reading comprehension, rather than with mathematics.)

We turn now to knowledge representation structures in general.

KNOWLEDGE REPRESENTATION

Several arguments establish the need to postulate, within knowledge representation structures, some form of aggregates or chunks. One of the most telling arguments is arithmetical: The number of possibilities that would need to be discriminated if, say, every symbol in a book were independent of all others, can be estimated as something like

$$70^{585000} = (70^6)^{97500} \doteq 10^{1072500},$$

if one assumes 70 possible characters (a, b, c, . . . ; A, B, C, . . . ; 0, 1, 2, 3, . . . , 9; plus punctuation and a "space"), 65 characters per line, 45 lines per page, and 200 pages. It is inconceivable that a human could discriminate $10^{1072500}$ different things; and one gets a hint of the difficulty if one imagines a 200-page book, where every symbol in every line on every page was a decimal digit, so that the entire book contained one huge number, 585,000 digits long. One could not "read" such a book because the human mind cannot process so much information

Clearly, then, a book that contains 585,000 characters does *not* contain this much information. Rather, it contains far less because it is highly redundant. (As one trivial example, a symbol "q" in a word will necessarily be followed by a "u"; relatively few capital letters will appear; in the symbol string:

$$\text{space, t, h, } \underline{\hspace{1cm}} \text{, space}$$

the blank can *only* contain "e" or "o" (with a few rare exceptions).

But our assembling little bits of data into larger chunks goes much further than this; we normally deal in words, sentences, or even larger aggregates— which, incidentally, is what makes proofreading so difficult: we see what we are prepared to see, and this may not coincide with what is actually there.

Frames.   But even more forceful arguments can be given that demonstrate that the information in one's mind must typically be organized into quite large aggregates (cf. Davis & McKnight, 1979; Minsky, 1975). For some of these larger aggregates, Minsky has used the word *frame* (although Rumelhart and Ortony use *schema,* and Schank uses *script*). A *frame,* then, is an abstract formal structure, stored in memory, that somehow encodes and represents a sizeable amount of knowledge.

A frame differs from a *procedure* in (among other things) the fact that a frame is not ordinarily sequential—it allows multiple points of entry and provides some flexibility in its use.

Retrieval and Matching.   Consider what needs to occur when an eleventh-grade student is asked to solve the equation

$$e^{2t} + 6 = 5\,e^t \ ,$$

or a calculus student is asked to integrate

$$\int \frac{e^{\sqrt{x}}}{\sqrt{x}}\,dx$$

The first student presumably knows how to solve

$$ax^2 + bx + c = 0,$$

and the second knows how to integrate

$$\int e^u du,$$

but each, of course, knows many other things that might be relevant to these tasks. A complicated process must take place, leading ultimately to the realization that

$$e^{2t} + 6 = 5\,e^t$$

matches exactly the pattern

$$ax^2 + bx + c = 0$$

if you write

$$e^{2t} - 5\,e^t + 6 = 0,$$

and make the correspondences

$$a \leftrightarrow 1$$
$$b \leftrightarrow -5$$
$$c \leftrightarrow 6$$
$$e^t \leftrightarrow x$$

This matching process succeeds only because the correspondence $e^t \leftrightarrow x$ necessarily implies the correspondence $e^{2t} \leftrightarrow x^2$, as a result of the addition law for exponents.

A similar analysis shows what must occur in the case of the integration problem.

How correct retrieval can occur—and often occurs almost instantaneously—is a considerable mystery. Minsky's recent "K-Lines" theory may provide an answer (Minsky, 1980), but we do not pursue the fundamental retrieval question further in this chapter. (Next, however, we shall consider some heuristics that students can learn that can improve student performance on retrieval problems of this general type.)

A *frame*, then, is a formal data representation structure that is stored in memory, hopefully to be retrieved when needed. This retrieval often occurs almost instantly. In a well-known experiment, Hinsley, Hayes, and Simon (1977) found that, for some subjects, merely the first three words ("A river steamer . . .") in the statement of a problem were sufficient to trigger the retrieval of an appropriate frame (indicated by the subject interrupting after the third word to say something like: "It's going to be one of those river things with upstream, downstream, and still water. You are going to compare times upstream and downstream—or, if the time is constant, it will be the distance.")

Frames possess considerable internal organization. Especially important are the variables (or "slots") for which the frame will seek specific values from input data. When the present input does not provide enough information to permit certain slots to be filled, the frame may insert some tentative "guess," based on past experience. When slots are filled in this way, we say they contain "default evaluations"—data inserted (from past experience) to make up for gaps in the present input. Thus, in the earlier story about Leslie and Dana, if we knew they were on horseback, heading for a cattle drive in the old west, we might *assume* they were both male—but, of course, we could be wrong. Default evaluations are not guaranteed!

Default evaluations are important primarily in nonmathematical frames, where the matching of input data into slots is usually approximate; *it is a peculiarity of mathematics to require that every matching must be complete and precise!*

Pointers. Following computer practice, it is usually postulated that one mechanism by which one data representation structure can be related to another is the device of *pointers;* in effect, when a certain structure has been retrieved from memory and rendered active, and when certain definite conditions are met, then a pointer causes the retrieval of some other data representation structure and/ or causes some specific change in control. Pointers can also be used *within* a data representation structure to "weld" the whole unit together.

PLANNING LANGUAGE, PLANNING SPACE, AND META-LANGUAGE

It seems clear that complicated problems are often dealt with by carrying on two somewhat separate activities: actual calculations and the process of *planning* what calculations are to be performed and how. To provide for this, it is common to postulate (*a*) descriptors that identify the possible uses of "action-level" processors, and (*b*) mechanisms for dealing with descriptors (which can include tree searches, backward-chaining, etc.)

Example: Suppose a student encounters the problem: The plane P passes through the points A (a, O, O), B (O, b, O), and C (O, O, c). Find the distance from the origin (O, O, O) to the plane P. Suppose also that this is, for the student, a novel problem, one that is not already familiar.

Presumably the student has the requisite knowledge to solve the problem, but this knowledge is scattered among the many techniques that the student knows. The task, then, is to select the correct techniques and to relate them correctly to one another. Hopefully, the student possesses, for example, a technique that might be described as:

How, given a nonzero vector V, one can find a unit vector ü that is parallel to V;

and another technique that might be described as:

How, given the equation
$$e\,x + f\,y + g\,z = k,$$
one can find a vector that is perpendicular to the plane represented by this equation;

and so on. By sequencing these techniques correctly, the student can solve the problem, even though it is an entirely new problem that the student has never seen before (for details, cf. Davis, Jockusch, & McKnight, 1978).

APPLICATIONS

How can such conceptualizations, imported from cognitive science or Artificial Intelligence, be useful in studying the process of carrying out mathematical tasks, or learning to do so? We consider here some specific studies of this type.

Some Specific Frames

Can we identify some specific frames (i.e., knowledge representation structures) that most students build up and store in memory? The answer is yes. From some general rules about how frames are created, one can deduce some probable frames; from the existence of certain frames, one can deduce observable behaviors. One can then check actual student performance protocols, to see if these behaviors do in fact occur.

THE UNDIFFERENTIATED ADDITION FRAME

A common law of frame creation, used by Feigenbaum and others, is that discrimination procedures are no finer than they need to be. In the first year of elementary school, children typically learn (at least at first) only one arithmetic law, namely, addition. Hence they presumably synthesize a frame that will input 3 and 5 and output 8. When this frame is invoked, it will demand its two numerical inputs; it will, however, ignore the operation sign "+" because it has no need to consider this sign. There being only one arithmetic operation, discrimination among operations is not necessary.

When, in later months (or years), students encounter, say, 4×4, one should expect the wrong answer "8," and this is by far the most common wrong answer (cf. Davis, Jockusch, & McKnight, 1978).

There is further evidence of the operation of this frame: Friend (1979) reports the seemingly curious fact that, of the three addition problems,

$$
\begin{array}{ccc}
(A) & (B) & (C) \\
235 & 235 & 235 \\
14 & 45 & 114 \\
\underline{12} & \underline{42} & \underline{12}
\end{array}
$$

problem (A) is the most difficult for elementary school students in Nicaragua, whereas, naively, one would expect (A) to be the easiest. After all, problem (B) involves one additional "carry" from one column to another, and problem (C) involves one more addition (the "2 + 1" in the leftmost column).

In terms of frame operations, however, one would assume that a "column-addition frame," now being learned and tested, will make repeated subprocedure calls on the primary-grade addition frame, *which demands two numerical inputs*. When, in problem (A), this demand is frustrated in the leftmost column, a general law postulated by John Seely Brown predicts some intervention in the control sequence so that the program can be executed (see also Matz, 1980); this intervention distorts the column-addition pattern so as to obtain the required second input for the primary-grade addition frame, often by picking up a numeral from the tens column, so as to get the wrong answer 361 (taking "1" from the "12") (for details, see Friend, 1979).

Consider another arithmetic operation learned in the primary grades: subtraction. At first, subtraction problems are of the form $5 - 3$, but are *never of the form $3 - 5$. Hence, once again following the Law of Minimum Necessary Discriminations, students synthesize a frame that inputs the two numbers "3" and "5" and outputs "2." The frame ignores order since a consideration of order has never been important.*

*In later years, of course, the student will need to deal with both $7 - 3$ and $3 - 7$, and will need to discriminate between them. Such discrimination capability has not been built into the frame (which is why it is called symmetric). Conse-*quently, in later years certain specific errors are easily predicted and are, in fact, precisely what one observes (cf. Davis & McKnight, 1979).

(In a similar way, many adults are confused between "dividing into halves" and "dividing by one-half." The most common answer to $6 \div \frac{1}{2}$ is the wrong answer "3.")

Karplus, in some elegant (forthcoming) studies of ratio and proportion, reports protocols such as the following:

> In a story presented to the student, there is a boy, John, who is making lemonade with 3 teaspoonfuls of (sweet) sugar and 9 teaspoonfuls of (sour) lemon; a girl, Mary, is also making lemonade, but using 5 teaspoonfuls of (sweet) sugar and 13 teaspoonfuls of (sour) lemon. Appropriate illustrative pictures accompany the story. The interviewer asks, "Whose lemonade will be sweeter?"

1. Student: Let me see. Mary's would be sweeter.
2. Interviewer: Mary's would be sweeter? Um-hum [thoughtful tone invites further explanation . . .]
3. Student: Because Mary's has two less lemons in contrast with this [pointing to pictures], with John's.
4. Interviewer: Actually, she has 4 *more*.
5. Student: She *does*? [tone of great surprise]
6. Interviewer: [explaining his preceding remark] Well, she has 13 compared to 9.
7. Student: Yeah, but in relation to the *sugars*.

8. Interviewer: Could you explain to me how you figured that she has 2 less in relation to sugar?
9. Student: Well, O.K. . . . There's 3 and 9.
10. Interviewer: Uh-huh.
11. Student: And 3 goes into 9 3 times, and then you go 5 and 13 . . .
12. Interviewer: Yes . . .
13. Student: 15 goes into 5 3 times [sic!], so it's really too much . . .
14. Interviewer: Uh-huh. So it's 2 less. And so if Mary wanted to make it come out the same sweetness as John's . . .
15. Student: It would have to be 15.
16. Interviewer: She'd have to use 15. So, I see she has 2 less. O.K.

This interview excerpt can be split into three sections: utterances 1–7 show frame-like behavior, utterances 8–13 show sequential behavior under frame control, and utterances 14–16 are the interviewer's attempt to restate the student's idea in more explicit language.

The student's comparison of the table

	Sugar	Lemon
John	3	9
Mary	5	13

with a different table, *which is nowhere in evidence except in her own imagination,* is stunning! The alternative table, namely,

	Sugar	Lemon
John	3	9
Mary	5	15

is so real to her that she at first rejects the interviewer's "common sense" numbers—which are the numbers that are actually in evidence—in favor of her "ideal" table.

This, and other evidence in various Karplus interviews, suggests strongly that many students have a "recipe" frame, which allows them great facility in doubling recipes, halving recipes, and so on.

A "UNITS" OR "LABELS" FRAME

Clement, Kaput, and their colleagues (Clement & Kaput, 1979; Clement & Rosnick, 1980; Lochhead, 1980) have carried out an important series of experiments dealing with student responses to this question: "At a certain university, there are six times as many students as there are professors. Please write an algebraic version of this statement, using S for the number of students and P for the number of professors.''

The correct answer, of course, is $6P = S$; but an exceedingly common wrong answer, even among engineering students, is $6S = P$.

By itself this might mean very little. After all, humans are fallible, and $6S = P$ is almost the only wrong answer that any reasonable person would invent. Furthermore, one could explain the error as a manifestation of internal information processing that simply follows the time-sequential order (or left–right order) ''there are SIX times as many STUDENTS as there are PROFESSORS'' (or a possible abbreviated version: ''SIX STUDENTS for each PROFESSOR'').

But the phenomenon is far deeper than this: Clement and his co-workers have varied word order, studied different populations, used different degrees of ''meaningfulness'' (after all, there are almost always more students than professors—but how about the ratio of sheep to cows in a certain farm?), and varied the ratio (e.g., ''there are five professors for every two students''). Most importantly they have tape-recorded interviews in which a tutor attempted to correct the error in students who were writing the wrong equation. No simple ''accidental slip'' is involved here, as one sees from four facts:

1. Students who are initially wrong protest vigorously against the change. (''I can't think about it that way!'' ''You're getting me all mixed up!'' ''That's weird!'')
2. Students who are initially wrong are very reluctant to change, and if they do change to writing correct equations, they soon slip back to the wrong versions.
3. In order to preserve their wrong equations, students would make egregious variations in the definition of variables, even concluding that S must stand for the number of professors (sic!) and P must stand for the number of students (or S must stand for the number of cows, and C must stand for the number of sheep).
4. Students who wrote the wrong equations tended to *verbalize* the problem differently from students who wrote the correct equation: Students writing the correct equation tended to say ''Six times *the number of* professors equals *the number of* students,'' whereas students writing the wrong equation tended to say ''There are six students for each professor.''

In any situation of very common and very persistent errors of this type, one who believes in frames will quickly suspect the presence of a frame that is itself perfectly useful (and in that sense ''correct''), but which is being retrieved when it should not be and put to use for a task where it is not appropriate. The fourth characterization listed gives a strong confirmation of this conjecture and even indicates what this alternative frame probably is: It is a frame that has been developed for dealing with *units* and with *labels*. We have all seen "equations" such as:

$$12\text{in.} = 1 \text{ ft}$$
$$3\text{ft} = 1 \text{ yd}$$
$$5280\text{ft} = 1 \text{ m.}$$
$$2.54\text{cm} = 1 \text{ in.}$$

and so on. The labels, "inches", "foot", etc., are *not* variables and do not behave like variables. For contrast, let I be Mohammed Ali's height in *inches,* and let F be his height in *feet.* What equation can you write between I and F? (cf. Davis, 1980).

The "Greater Than" Relation

Richard J. Shumway, of Ohio State University, has pointed out the discrepancy between formal definitions of $a < b$ versus what students actually do to decide whether $a < b$. For example, one may define: "$a < b$ if and only if there exists a positive number N such that $a + N = b$."

Now, ask a student which is smaller, 31 or 2986. What thought processes will the student employ? Does the student start adding numbers to 2986 to see if he or she can get 31 as a result? If the student *did* start such a search, how would he or she know when to give up?

1. For *positive integers,* the first step in deciding probably looks like this (where ℓ (A) denotes the number of digits in the decimal representation of the integer A, etc.).

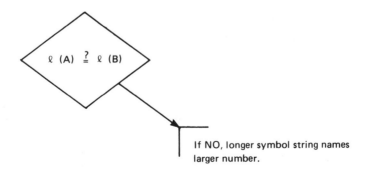

$$\ell \text{ (A)} \stackrel{?}{=} \ell \text{ (B)}$$

If NO, longer symbol string names larger number.

2. But suppose ℓ (A) = ℓ (B); what then? Typical students presumably use some stepwise decision procedure—or else a production system—*roughly* equivalent to this:

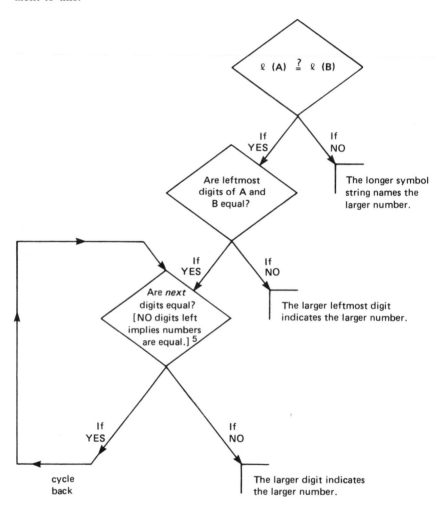

This is all very well for positive integers, but something more is needed to deal with negative numbers, rational numbers, etc. For example:

[5] In fact, two identical numbers would almost certainly be recognized in the first step of the comparison. We have not bothered to indicate this test for identity, which probably occurs at the start of the procedure. The fact that humans probably *begin* with checks for identity, whereas typical computer programs perhaps do not, is food for thought!

1. Which is larger, -1239 or -37?
2. Which is larger, .039 or 1.15?
3. Which is larger, .0395 or .00953?

We do not pursue this example further in this chapter. The main point of Shumway's example is that older analyses—in terms of formal definitions or, cognitively, "having a concept"—although these may be valuable for certain purposes, *are NOT as useful for understanding a student's thinking as the "procedure-and-frame" kind of analysis is. Clearly, a student does not find a positive number to add to .039 so as to get 1.15 in order to answer the question of whether 1.15 is larger than .039.*

Deeper-Level Procedures

It is a well-known fact (cf. Davis, Jockusch, & McKnight, 1978) that students who have learned to solve quadratic equations by factoring

$$x^2 - 5x + 6 = 0$$
$$(x - 3)(x - 2) = 0;$$

so, either $x - 3 = 0$, or else $x - 2 = 0$; hence, either $x = 3$ or else $x = 2$, tend to make the following mistake:

$$x^2 - 10x + 21 = 12$$
$$(x - 7)(x - 3) = 12;$$

so either $x - 7 = 12$ or $x - 3 = 12$; therefore, either $x = 19$, or else $x = 15$. This error is very difficult to eradicate—or is, at least, very difficult to eradicate *permanently*. Even when classes of able students, using a seemingly excellent textbook, receive careful instruction—with emphasis on the special role of zero in the *zero product principle*—it is still the case that this error will continue to crop up in student work. Despite careful explanations of why it is an error and despite short-term elimination of the error, it keeps coming back.

Matz (1980) presents a theory of cognitive processing that explains the persistence of this error. She postulates two levels of procedures (stateable as "rules"). The surface level rules are ordinary rules of algebra; the deeper level rules serve the purpose of *creating* superficial-level rules, *modifying* superficial-level rules, or *changing the control structure*.

Now one of the deeper level rules must surely generalize over number; that is, it must say, in effect,

$$[P(a), P(b), P(c)] \Rightarrow [\forall_x P(x)].$$

In other words, if I show you how to add

$$\begin{array}{r} 23 \\ +14 \\ \hline 37 \end{array}$$

you will never master arithmetic if you persist in believing that this works *only* for 23 and 14. You must believe that this same procedure works also for

$$\begin{array}{r} 34 \\ +15 \\ \hline \end{array}$$

or

$$\begin{array}{r} 34 \\ +25 \\ \hline \end{array}$$

and so on. In short, in order to learn arithmetic you *must* possess a deeper level rule of the type that Matz postulates.

Now, it is a known property of rules that, in their less mature forms, they tend to be applied too widely; the appropriate constraints on their applicability have not yet become attached to the rules.

The *zero product principle*

$$[A \cdot B = 0] \Rightarrow [(A = 0) \vee (B = 0)] \tag{1}$$

is very nearly the first law students have encountered where some specific number (zero) must not be generalized. Predictably, then, the lower level "generalizing" rule will be used to extend Equation (1) to read

$$[A \cdot B = 0] \Rightarrow [(A = C) \vee (B = C)] \tag{2}$$

Equation (2) would be a correct generalization of Equation (1) if generalizing were appropriate in this case. Unfortunately it is not.

Then this error is hard to eradicate for the same reason that dandelions are—something is creating "new" dandelions, so even when you eliminate dandelions, the job is not permanently accomplished. Even when you achieve enough remediation that a student ceases—temporarily—to commit the error, your work may not be finished, for the student has, in his or her own mind, the deeper level rules that are capable of creating anew the incorrect factoring pattern, in accordance with commonly accepted laws of cognitive processes.

John Seely Brown has used this same notion of deeper level rules to explain the observed fact that multiple errors occur in arithmetic problems more often than ordinary probabilities would predict. Brown postulates that a first *bug* (i.e., systematic error) may produce a control error (such as an infinite loop); an "observer" procedure notes this control error and intervenes; the intervention takes the form of modifying the control procedure so as to exit from the infinite loop; but this intervention will itself tend to create further performance errors. Consequently, multiple errors occur disproportionately often in student work.

Knowledge in Other Forms:
The Semantic Meaning of Symbols

One who is skillful in the performance of mathematical tasks must be able to work with mathematical symbols in a relatively "meaningless" way, guided only by patterns and formal rules, but they must also be able to deal with the meanings of the symbols—or at least, with some of the meanings of the symbols. We have seen earlier that third-graders will often subtract

$$7002$$
$$-\underline{28}$$

incorrectly, following the standard algorithm, but using a version of that standard algorithm that contains a bug:

$$
\begin{array}{r}
6 \\
\mathrm{\nabla}00\,{}^1 2 \\
-\underline{2\ 8} \\
4
\end{array}
$$

$$
\begin{array}{r}
5 \\
6 \\
\mathrm{\nabla}0\,{}^1 0\,{}^1 2 \\
-\underline{2\ 8} \\
50\ 8\ 4
\end{array}
$$

In a recent study, Davis and McKnight (1980) used this subtraction problem as an interview task in order to compare children's *algorithmic* knowledge (which could, of course, be rote) with the knowledge that these same children had concerning the *meaning* of the various symbols and operational steps.

The interviews sought to study each child's possession of five kinds of knowledge that might be labeled "meaningful." The interest in this question arises from a question of *knowledge representation;* the different kinds of knowledge would require coding into different representational forms.

The *size* of the numbers should immediately signal an error—"about seven thousand" minus "a few" ought not to turn out to be "about five thousand." This would be coded in "critic" form. Interview data showed that no third-grader interviewed (from four different schools) had this level of understanding of approximate sizes of numbers. Since the students did not possess this kind of understanding, it could not be used to help identify and correct the error in the algorithm.

The use of simpler numbers is another kind of knowledge. Perhaps numbers like "seven thousand" are essentially meaningless to these students; perhaps, then, smaller numbers would be more meaningful and might thus allow *meaning* to guide the algorithm. This kind of knowledge would be coded as a heuristic strategy. Unfortunately, since the error in question involves "jumping borrowed ones over zeros," it is not possible to use truly small and familiar numbers—but one can, at least, use *smaller* numbers. For example,

$$
\begin{array}{r}
702 \\
-\ 25 \\
\hline
\end{array}
$$

No improvement in student performance was achieved by switching to smaller numbers.

Adults who regularly use "mental arithmetic" to solve such problems *without writing* do not usually do so by visualizing the standard algorithm. On the contrary, they take advantage of the special properties of specific numbers. For example, for $7002 - 28$, one can say: $7000 - 25$ would be 6975. If I subtract 3 more (because $28 = 25 + 3$), I will end up with three less, so $7000 - 28$ must be 6972. If I now *start with two more*, I must end up with two more, so $7002 - 28$ must be 6974. This kind of knowledge would be procedural. In interviews, this general method was taught to students. The results were

1. Those students who learned it well enough to get *correct* answers to $7002 - 28$, etc., nonetheless had more confidence in the correctness of their (wrong) algorithmic answers.

2. Students more often attempted to visualize the usual algorithm—in short, they persisted in unchanged algorithmic behavior, with only the modification that they attempted to visualize the algorithm instead of actually writing it down on paper.

3. In no case was this kind of knowledge used to correct the bug in the algorithm.

Borrowing, as in

$$
\begin{array}{r}
6 \\
\nabla^1 592 \\
-\ \ \ 621 \\
\hline
6\ 971 \\
\end{array}
$$

can be interpreted as "giving the cashier a thousand-dollar bill and receiving in exchange 10 hundred-dollar bills." This kind of knowledge would be coded as a frame based on past experience. The interviews revealed that every third-grader could deal correctly with such "cashier transactions" when they were presented directly (and not by implication, in a subtraction problem); that is, if asked "How many hundred-dollar bills could you get in exchange for a thousand-dollar bill?," every third-grader answered such questions correctly. However, no third-grader saw the relevance of this to the subtraction algorithm, and none sought to correct the algorithmic error as a result of the discussion of "cashier exchanges." (The interviewers carefully avoided suggesting how the two might be related since it was the goal of the study to see if the students would spontaneously see the relevance of cashier exchanges to the subtraction algorithm.)

Dienes' multibase arithmetic blocks provide a physical embodiment for place-value numerals and allow a physical "subtraction" that is precisely analogous to the subtraction algorithms (cf. Davis & McKnight, 1980). This would also be coded as a frame. In one school included in this study, students did learn how to represent 7002 correctly in terms of MAB blocks ("7 blocks and 2 units"), and similarly for 28 ("2 longs and 8 units"), and could interpret the task 7002 − 28 in MAB terms ("you have 7 blocks and 2 units, and you are asked to give someone 2 longs and 8 units"). Nonetheless, no third-grader saw (a) that this indicated an error in their algorithmic calculation, or (b) that this MAB task showed how to *correct* the error in the algorithmic calculation.

The over-all result was that the students were entirely wedded to an algorithmic performance of this subtraction task, preferring their algorithmic answers (even when wrong) to answers obtained in other ways (even when these answers were in fact correct). Asked to check their answers, they merely repeated the algorithm. No knowledge of possible meanings of the symbols was brought to bear on the algorithmic task. None!

It is interesting to compare this result to a persistent theme that emerges from many studies by Ginsburg, who reports that students commonly possess important mathematical competences of "nonschool" origin which they do not relate to school tasks (cf. Ginsburg, 1977; Ginsburg, Posner, & Russell, 1981).

IMPLICATIONS FOR TEACHING

Does this mean that, say, MAB blocks are no help in learning algorithms? No, surely not. Many experienced teachers believe that the MAB blocks can be quite helpful. Furthermore, in general, mathematically experienced people frequently report being guided in their calculations by a knowledge of the meaning of the symbols. Presumably what this unexpected outcome does indicate is that the school in question was not relating MAB blocks to algorithmic calculations, so that, even though the students were getting a good knowledge of how to set up MAB representations, they were not *using* these representations to guide them through the algorithm.

Adults can easily see the students' point of view here: Anyone who is following a very unfamiliar and complicated recipe for the first time may find themselves checking up primarily by checking through the recipe, one line at a time, to see if it seems to have been followed correctly—and this is exactly what the students did. One doesn't "think about the task in other terms" because one lacks the tools to be able to do so. With experience, all of this can change, and knowledge of other sorts can be brought to bear on the task at hand.

HOW DO "OTHER MEANINGS" RELATE TO THE THEORY?

The study of third-graders has been presented primarily in terms of observable behaviors. How could these phenomena be formulated in theoretical terms?

The algorithmic performance is the easiest to conceptualize—a programmable hand-held calculator can exhibit this kind of performance, and it is readily conceptualized as a procedure consisting of a sequence of simple "unit" steps.

The knowledge of MAB blocks which the students demonstrated is probably contained in a collection of procedures, any one of which performs some specific task—such as trading ten units for one long, or recognizing that "7396" calls for three flats (and also 7 blocks, 9 longs, and 6 units). (Of course, this could be in the form of relatively powerful, relatively general procedures that can deal with, say, any "10-for-1" trading situation, or it could be in the form of a larger number of more specific procedures, such as "trading ten longs for one flat.") Because such behavior shows little of the sequential rigidity of a procedure, it would be classified as "frame-like."

But for the students who had both the algorithm procedures and the MAB frame and who nonetheless failed to relate the two, what was lacking?

There are several possibilities, including at least these:

1. The students may lack *pointers* in the algorithm procedure that would invoke the MAB block frame.
2. The students may lack a goal-oriented control mechanism that would relate to the MAB blocks frame to establish goals for block exchanges and to sequence exchanges so as to achieve these goals.
3. The students may never have *reflected* on the MAB frame and may never have discovered pattern similarities between the MAB frame and the algorithmic procedures.

Concepts

In the 1950s it seemed that mathematicians meant one thing by the word *concept*, whereas psychologists and educators meant something else. Mathematicians spoke of the concept of *function*[6] of the concept of *limit*. By contrast,

---

[6]It is important to note that *function* is a special term in mathematics and is a noun, not a verb. Within mathematics, there are disagreements in the best way to define *function*, and the historical evolution of the idea may or may not be well represented in certain current definitions. But in any case, function is a special term, as in "y is a function of x", or $y = f(x)$.

psychological studies of concepts seemed to deal only with rules for inclusion in a certain class of things. (To be sure, it may appear that anything whatsoever *can* be formulated as a class inclusion problem, but this often distorts the reality so badly as to be positively harmful. On October 6, 1980, the Los Angeles Dodgers batters would have loved to have known the defining properties for the class of swings that would get hits off of Joe Niekcro's pitches—but it probably would not have helped them much if they had tried to analyze the problem in such terms.)

Artificial Intelligence, cognitive science, or even what is nowadays called *knowledge engineering*, provides a way to express something that is closer to the mathematician's notion of a concept. If you have mastered, say, the concept of *limit of an infinite sequence*, you possess adequate knowledge representation structures of certain specific types and you possess an adequate array of pointers to guide certain appropriate associations. You also possess a collection of useful examples (or the means of creating new examples) and an ability to relate examples to general statements.

For the concept of limit of an infinite sequence, you would need at least knowledge structures representing:

1. The graph of $u_1, u_2, u_3, \ldots$ showing u_n versus n, with a horizontal line for the Limit L, a strip representing $L - \epsilon < u_n < L + \epsilon$, and a representation of the "cut point" N (for $n > N$).

2. The interpretation of

$$|u_n - L| < \epsilon$$

in terms of the *distance* from u_n to L.

3. An ability to convert between

$$|u_n - L| < \epsilon$$

and

$$L - \epsilon < u_n < L + \epsilon$$

and the graphical representation (as in 2).

4. "Metaphoric" language, describing ϵ as an "allowed tolerance" and N as a "cut point."

5. Knowledge of the consequences of choosing ϵ first, and N second versus choosing N first, and ϵ second.

6. Metaphoric language to describe (5) intuitively;

7. Even more intuitive formulations, such as: "L is the limit of the sequence $u_1, u_2, \ldots$ if every term u_n in the sequence is equal to L—except that when I say *equal* I will forgive an "allowed error" ϵ, and when I say *every term* I mean "except for a finite number at the beginning.""

8. The usual ϵ, N definition.

9. An ability to relate all of the preceding structures.

10. Knowledge of how "arbitrarily close" works, or can be used.

11. Knowledge of how *indirect proofs* are employed, especially by using (10).

12. Knowledge of how (11) uses the Law of Trichotomy.

13. The axiom or theorem that every Cauchy sequence converges.

14. A classification of sequences as: montonic, increasing, nondecreasing, oscillating, convergent, divergent, bounded, and unbounded.

15. Either a collection of sequences that are examples of the categories in (14), or else an ability to generate examples as needed.

16. Knowledge of various common errors, and precisely *why* they are errors. (For example, the error of claiming that "the sequence 1, 1, 1, . . . is not convergent because the terms are not getting nearer to any number"; the error of claiming that "the limit of .9, .99, .999, . . . must really be less than one, because the terms of the sequence are always less than one"; the inadequacy of defining *limit* by saying "the limit is the number that the terms are getting nearer to"; the error in defining *limit* by saying "given any $\epsilon > 0$, there must be an integer N such that there exists a term u_n, with $n > N$, and with $|u_n - L| < \epsilon$"; the error in assuming that "the limit of a sequence is either an upper bound for the terms of the sequence, or else a lower bound.")

The relation between *general statements* and *examples* is so important that it deserves special attention (cf. Rissland, 1978). Student errors reveal something of this relationship. One twelfth-grade calculus student defined the *limit of a sequence* by writing: The limit of a sequence is a number that the terms approach but never reach.

This student had seen, in class, sequences such as

$$.9, .99, .999, . . .$$

and

$$1, 1.4, 1.41, 1.414, . . .$$

Even for these sequences, the student's answer is inadequate, but it fails flagrantly for sequences such as

$$1, 1, 1, . . .$$

or

$$1, 0, \tfrac{1}{2}, 0, \tfrac{1}{3}, 0, \tfrac{1}{4}, . . .$$

Subsequent interviews showed, unsurprisingly, that the student was not bringing to mind examples of this type to test the suitability of his definition.

The other main error in the student's answer can also be revealed by testing his statement against appropriate "test case" examples. The first error, which we have seen, involved his use of the phrase "but never reach," and is revealed by considering examples of possible *sequences*. The second error is revealed by testing his definition against examples of possible *limits*. Consider the sequence

$$.9, .99, .999, \ldots$$

Clearly, the terms of this sequence approach—that is, get nearer to—the number 1, which is what the student had in mind. But the terms also get nearer to the number 1.01; they get nearer to the number 1.5; they get nearer to the number 2; and, in fact, they get nearer to the number 1 million. To be sure, they never get very near to 1 million, but .999 is closer to 1 million than .9 is!

The relation between known examples and general statements is so important in mathematics (Rissland, 1978) that it seems necessary to postulate two information-processing capabilities:

1. Given a general statement, one can retrieve from memory, or construct, examples by which to test the statement. (The postulate does not assert how successfully this will be done in any particular case, only that in principle it is something that *can* be done—just as humans possess, in general, the ability to move from one place to another, whereas most plants do not.)
2. Given a collection of examples stored in memory, one can make general statements that describe common attributes of these examples.

Mathematical thinking is heavily dependent upon these two capabilities.

When a physicist or mathematician says "I must educate my intuition" about certain matters, it seems likely that part (at least) of this process means synthesizing data representations of appropriate examples and establishing relations among them—as, for instance, when the phenomenon of a returning space capsule hitting the earth's atmosphere can be better understood by relating it to skipping flat stones across the surface of a lake. Both are useful examples—one familiar, and the other not—of the surprising ability of a solid to glance off of a liquid if its velocity is adjusted in a certain way.

Planning Space

Earlier we considered the problem: Plane P passes through the three points A (a, 0, 0), B (0, b, 0), and C (0, 0, c). If a, b, c are all nonzero, find the distance from the plane P to the origin O (0, 0, 0).

This problem involves *planning* only if, as we suppose to be the case, it is a novel problem, which the student has not previously encountered. As a novel problem, it is rather difficult. We further suppose, of course, that the student has learned the separate techniques (in vector form) needed for a solution. Thus, the student's real task is to select (and retrieve from memory) the correct techniques, sequence them correctly, and establish the proper relations between them.

This task is fairly easy, however, if and only if the student has the correct descriptors attached to each technique and has developed the procedures needed for searching among these descriptors. One can think of this, informally, as if each procedure is a specific tool, and attached to each tool is a tag that describes what the tool can be used for. Execution, of course, requires the use of the tools themselves, but planning is carried out merely by reading the "tags" or "labels." One tag, for example, says: *If you have a vector $\vec{V}$ (of any nonzero length) and a unit vector $\vec{u}$, you can find the component of $\vec{V}$ in the direction $\vec{u}$ by computing the "dot product" or "inner product":*

$$\vec{u} \cdot \vec{V}$$

Another tag says: *If you have any nonzero vector $\vec{W}$, you can get a unit vector $\vec{u}$ by dividing $\vec{W}$ by its length.*

Yet another tag says: *If you have the equation of a plane P in the form*

$$a\,x + by + cz = d,$$

where $a^2 + b^2 + c^2 > 0$, you can immediately write down a vector $\vec{N}$ that is normal to plane P.

Carrying out such advance planning of how to attack a problem depends upon (*a*) knowing the necessary techniques, (*b*) possessing appropriate descriptors (tags or labels), for each technique, which specify what the technique can accomplish, (*c*) (probably) possessing a definite collection of recognizable subgoal *candidates* (that is to say, a "menu" of *possible* subgoals from which appropriate subgoals for a given problem can be selected), (*d*) using mechanisms for identifying appropriate subgoals and retrieving the appropriate tags or labels, (*e*) given a tag or label, using a mechanism for retrieving its associated tool, and (*f*) using mechanisms for assigning correct inputs for each "tool" or subprocedure.

It is sometimes assumed that a problem solver has laid out in his or her mind a complete "tree" of possibilities—of *all* the possibilities, that is. There seems to be no observational data in support of such an assumption. On the contrary, people commonly "see" (or "bring to mind") very *few* possibilities, and may, indeed, omit the most promising ones. (This happens, for example, in the puzzle: Draw a sequence of connected straight line segments without lifting pen from paper, so that each dot lies on at least one of the segments; do this with the smallest possible number of segments:

The smallest number turns out to be four, as in this solution which most people do not consider.)

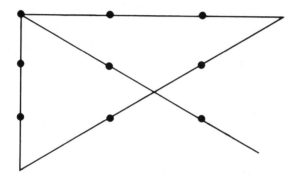

The search problem may seem simple, if one unconsciously assumes the student knows just where to look; but—for the "distance to the plane" problem—consider all of the things the student *could* do, such as:

1. Form the vector $\overrightarrow{AB} = (-a, b, 0)$ from the point (A (a, o, o) to the point B (o, b, o)
2. Form seven times the vector $\overrightarrow{AB}$:
$$7\,\overrightarrow{AB} = (-7a, 7b, 0)$$

3. Form the cross product
$$\overrightarrow{AB} \times \overrightarrow{BC}$$

4. Find the unit vector
$$\frac{\overrightarrow{AB}}{\|\overrightarrow{AB}\|}$$

5. Form the triple scalar product
$$(\vec{A} \times \vec{B}) \cdot \vec{C}$$

6. Find the distance from point A to point B

The number of possibilities is clearly infinite; but even when the number of possibilities is finite but large, people do not typically recognize most of them, and may omit some of the most important. What guides good problem solvers to "grow" the search tree in the most useful directions?

Another example of planning in advance of calculation is given by Suzuki (1979), a student in eleventh-grade calculus, who decided to solve the following problem by as many different methods as she could devise:

The curve C is defined by the equation

$$5x^2 - 6xy + 5y^2 = 4.$$

Find those points on C which are nearest to the origin.

Among the methods Suzuki concocted were

1. Recognize this as the equation of an ellipse whose axes are rotated in relation to the coordinate axes; therefore rotate the coordinate axes to coincide with the axes of the ellipse and read off the semimajor and semiminor axes by inspection.

2. Let $\ell^2 = x^2 + y^2$ be the square of the distance from the origin to the point P (x, y). Minimize ℓ^2, subject to the constraint that the point P must lie on the curve C, by getting a pair of simultaneous equations in the differentials dx and dy, and use the Cramer's rule requirement that the determinant of the coefficients must be zero.

3. The vector $\vec{T} = (1, y')$ is tangent to the curve C; it must be perpendicular to the vector $\vec{R} = (x, y)$ from the origin to the point of tangency; therefore set the dot product equal to zero: $\vec{R} \cdot \vec{T} = 0$.

4. Introduce a mythical "temperature" T, defined as $T = 5x^2 - 6xy + 5y^2 - 4$. The gradient ∇T points in the direction in which T increases most rapidly and is normal to a curve of constant temperature. At the nearest point, the vector $- \nabla T$ must point directly toward the origin and hence be parallel to (and therefore a scalar multiple of) the vector $\vec{R} = (x, y)$.

5. Convert to polar coordinates and minimize r.

The specific planning by which Suzuki created these strategies was not reported; nonetheless, it seems clear that she possesses a very powerful capability for planning novel ways of solving problems. (More recently, she has made excellent showings in various problem-solving contests.)

Yet one more method for solving this same problem has been employed by Kumar (1980), while a student in twelfth-grade calculus: Examination shows that C is a smooth curve contained in an annular ring centered at the origin. Therefore, a very small circle

$$x^2 + y^2 = r^2,$$

where r is small, will not intersect C. But as r becomes larger, the circle will intersect C. Finally, for still larger r, the circle will be outside the annular ring and will not intersect C. Hence, find the smallest positive value of r such that the system of simultaneous equations

$$\begin{cases} 5x^2 - 6xy + 5y^2 = 4 \\ x^2 + y^2 = r^2 \end{cases}$$

has a solution in real values of x and y. One can set out to solve this system; if the result is a quadratic equation, the criterion $b^2 - 4ac = 0$ will identify the desired value of r.

So much for *strategic* planning; when this strategy is implemented, the result is not a quadratic equation, but rather a fourth-degree equation. With a little ingenuity, a tactical step can be inserted that transforms the fourth-degree equation to a quadratic, and the problem is then easily solved.

Finally, one can look at planning as it is carried out by more experienced problem solvers. R, an experienced calculus teacher, was planning a lesson on writing the equations of tangents and normals to various curves. He attempted to sketch out his a priori plans to the extent that he was aware of them.[7]

I thought of two things: (i) distinguish the point (x, y) [the "general" point that moves along a curve, or tangent line] from the point (x_1, y_1) [the fixed (or "constant") point of tangency]; and (ii) use two geometric criteria to get two equations—the tangent and curve must *intersect* at (x_1, y_1), and they must *have the same slope* at (x_1, y_1).

Later on, as I began to think more seriously about the problem, I wanted to write equations for the pencil of lines through the point P $(3/2, 0)$, and to do this I selected the form

$$\frac{y - 0}{x - 3/2} = \frac{y_1 - 0}{x_1 - 3/2}$$

Still later, when I saw C, [another mathematics teacher in the same school] use the point–slope form of the equation, that struck me as more natural, and so I switched to using that form [interviews with students and teachers over past 10 years].

Notice especially the use of an appropriate and rather well-developed *meta-language*, with terms such as *get two equations, the general point, the fixed point of tangency, the pencil of lines, the point–slope form of the equation, struck me as more natural*, and so on. (To see a remarkably sophisticated meta-language used by a 17-year-old high school student, cf. Parker, 1980).

[7]Insertions in square brackets have been added.

SUMMARY

For most adults in past years, *mathematics* has probably meant memorizing certain specific techniques, and thereafter recalling them and using them as necessary. This is probably still true for many people even today. For a minority of people—including mathematicians, engineers, scientists, computer specialists, and a growing number of people in the health care field, in psychology, in education, in art, and elsewhere—the use of mathematics has nowadays a different quality. For these people, mathematics is an expandable tool for solving novel problems, for which no previously learned algorithm will be entirely sufficient. An engineer designing a more fuel-efficient automobile engine or a composer using a computer to generate a new piece of music is not merely plodding along by stepping in someone else's footprints. He or she is exploring new territory—and this often means exploring new *mathematical* territory, as well. Hence thoughtfulness, the rethinking of old assumptions, intelligent guessing, insight, and shrewd planning are all integral parts of the task.

Unfortunately, this "creative" aspect of mathematics has typically been ignored in most school curricula, and the whole idea is foreign to most of the general public. Something very important thus gets lost. In order to correct this situation, we must devote more effort in school to teaching the creative aspects of mathematics, and in our research and development work as well we need a greater emphasis on creativity, understanding, and problem solving.

In recent years, an alternative research paradigm has appeared in the world of mathematics education. This alternative paradigm shows considerable promise for increasing the emphasis on creativity, vision, understanding, insight, and the like, while at the same time paying proper attention to rote drill, routine practice, and "meaningless" algorithmic performance. This alternative paradigm gets data especially from task-based interviews, but also uses other data-collection methods, including the analysis of error patterns, and even the precise measurement of response times. In typical cases, the analysis of this data is based upon information-processing conceptualizations, often drawn from the fields of cognitive science and artificial intelligence.

This approach is beginning to demonstrate its ability to create a serious theory for the analysis of those processes of thinking that are required in dealing with tasks of a mathematical nature.

REFERENCES

Brown, J. S. *Towards a theory of semantics of procedural skills in mathematics.* Paper presented at Session 26.14, San Francisco Meeting of the American Education Research Association, 1979.

Brown, J. S., & Burton, R. B. Diagnostic models for procedural bugs in basic mathematical skills. *Cognitive Science,* 1978, *2,* 155–192.

Brown, J. S., & Van Lehn, K. *Repair theory: A generative theory of "bugs."* Palo Alto, Calif.: Xerox Palo Alto Science Center, 1980.

Carry, L. R., Lewis, C., & Bernard, J. E. *Psychology of equation solving—an information-processing study.* Austin: University of Texas, undated.

Clement, J., & Kaput, J. J. Letter to the editor. *Journal of Children's Mathematical Behavior,* 1979, *2*, 208.

Clement, J., & Rosnick, P. Learning without understanding: The effect of tutoring strategies on algebra misconceptions. *Journal of Mathematical Behavior,* 1980, *3*, 3–27.

Davis, R., & King, J. An overview of production systems. In E. W. Elcock & Donald Michie (Eds.), *Machine intelligence 8. Machine representations of knowledge.* Ellis Horwood Ltd. (Dist. by Halstead Press; John Wiley and Sons, New York), 1977.

Davis, R. B. *Explorations in mathematics. A text for teachers.* Palo Alto, Calif.: Addison-Wesley, 1967.

Davis, R. B. Representing knowledge about mathematics for computer-aided teaching, Part I.— educational applications of conceptualizations from artificial intelligence, In E. W. Elcock & D. Michie (Eds.), *Machine intelligence 8. Machine representations of knowledge.* Ellis Horwood Ltd. (Dist. by Halstead Press; John Wiley and Sons, New York), 1977, 363–386.

Davis, R. B. (with T. Romberg, S. Rachlin, & M. G. Kantowski). *An analysis of mathematics education in the Union of Soviet Socialist Republics.* Columbus, Ohio: ERIC Clearinghouse for Science, Mathematics, and Environmental Education, 1979.

Davis, R. B. *Discovery in mathematics: A text for teachers.* New Rochelle, N.Y.: Cuisenaire Co. of America, 1980. (a)

Davis, R. B. *Some new directions for research in mathematics education: Cognitive science conceptualizations.* Paper presented at ICME IV, Berkeley, Calif., 1980. (b)

Davis, R. B. The postulation of certain specific, explicit, commonly shared frames. *Journal of Mathematical Behavior,* 1980, *3*, 167–201. (c)

Davis, R. B., & Jockusch, E., & McKnight, C. Cognitive processes in learning algebra. *Journal of Children's Mathematical Behavior,* 1978, *2*, 10–320.

Davis, R. B., & McKnight, C. The influence of semantic content on algorithmic behavior. *Journal of Mathematical Behavior,* 1980, *3*, 39–79.

Davis, R. B., & McKnight, C. Modeling the processes of mathematical thinking. *Journal of Children's Mathematical Behavior,* 1979, *2*, 91–113.

diSessa, A. A. *Unlearning Aristotelian physics: A study of knowledge-based learning.* Cambridge, Mass.: Division for Study and Research in Education, Massachusetts Institute of Technology, 1980.

Erlwanger, S. H. Benny's conception of rules and answers in IPI mathematics. *Journal of Children's Mathematical Behavior,* 1973, *1*, 7–26.

Erlwanger, S. H. Case studies of children's conceptions of mathematics. Doctoral dissertation, University of Illinois, Urbana, Illinois, 1974.

Friend, J. Column addition skills. *Journal of Children's Mathematical Behavior,* 1979, *2*, 29–54.

Ginsburg, H. P. *Children's arithmetic: The learning process.* New York: D. Van Nostrand, 1977.

Ginsburg, H. P. The clinical interview in psychological research on mathematical thinking: Aims, rationales, techniques. *For the Learning of Mathematics,* 1981, *1*, 4–11.

Ginsburg, H. P., Posner, J. K., & Russell, R. L. Mathematics learning difficulties in African children. *The Quarterly Newsletter of the Laboratory of Comparative Human Development,* 1981, *3*, 8–11.

Hinsley, D., Hayes, J., & Simon, H. From words to equations: Meaning and representation in algebra word problems. In M. Just & P. Carpenter (Eds.), *Cognitive processes in comprehension.* Hillsdale, N.J.: Lawrence Erlbaum Associates, 1977.

Hofstadter, D. R. *Gödel, Escher and Bach: An eternal golden braid.* New York: Vintage Books, 1980.

Karplus, R., Karplus, E., Formisano, M., & Paulsen, A. C. *Proportional reasoning and control of variables in seven countries.* Berkeley, Calif.: Lawrence Hall of Science, 1975.

Karplus, R., Pulos, S., & Stage, E. K. Early adolescents' structure of proportional reasoning. *Proceedings of the Fourth International Conference for the Psychology of Mathematics Education.* Berkeley, Calif.: Lawrence Hall of Science, 1980, 136–142.

Krutetskii, V. A. *The psychology of mathematical abilities in school children.* (Translated from the Russian by Joan Teller. Edited by Jeremy Kilpatrick & Izaak Wirszup.) Chicago: University of Chicago Press, 1976.

Kuhn, T. S. *The structure of scientific revolutions.* University of Chicago Press, 1970.

Kumar, D. Sample solutions. *Journal of Mathematical Behavior,* 1980, *3,* 183–199.

Larkin, J., McDermott, J., Simon D., & Simon H. A. Expert and novice performance in solving physics problems. *Science,* 1980, *208,* 1335–1342.

Lenat, D. B., & Harris, G. Designing a rule system that searches for scientific discoveries. In D. A. Waterman & Frederick Hayes-Roth (Eds.), *Pattern-directed inference systems.* New York: Academic Press, 1978, 25–51.

Lochheed, J. Faculty interpretations of simple algebraic statements: The professor's side of the equation. *Journal of Mathematical Behavior,* 1980, *3,* 29–37.

Matz, M. *Underlying mechanisms of bugs in algebraic solutions.* Paper presented at the Annual Meeting of the American Educational Research Association, San Francisco, Calif., 1979.

Matz, M. Towards a computational model of algebraic competence. *Journal of Mathematical Behavior,* 1980, *3,* 93–166.

Michener (Rissland), E. The structure of mathematical knowledge (Tech. Rep. 472). Cambridge, Mass.: Massachusetts Institute of Technology, Artificial Intelligence Laboratory, 1978. (a)

Michener (Rissland), E. Understanding understanding mathematics. *Cognitive Science,* 1978, *2,* 361–383. (b)

Miller, G. The magic number 7 ± 2. *Psychological Review,* 1956, *63,* 81–97.

Minsky, M. A framework for representing knowledge. In P. Winston (Ed.), *The psychology of computer vision.* New York: McGraw-Hill, 1975.

Minsky, M. K-Lines: A theory of memory. *Cognitive Science,* 1980, *4,* 117–133.

Minsky, M., & Papert, S. *Artificial Intelligence memo no. 252.* Cambridge, Mass.: Massachusetts Institute of Technology, Artificial Intelligence Laboratory, 1972.

Newell, A., & Simon, H. A. *Human problem solving.* Englewood Cliffs, N.J.: Prentice-Hall, 1972.

Norman, D. Memory, knowledge and the answering of questions. In R. L. Solso (Ed.), *Contemporary issues in cognitive psychology: The Loyola symposium.* Washington, D.C.: V. H. Winston and Sons, 1973.

Opper, S. Piaget's clinical method. *Journal of Children's Mathematical Behavior,* 1977, *4,* 90–107.

Parker, P. A general method for finding tangents. *Journal of Mathematical Behavior,* 1980, *3,* 208–209.

Pimm, D. Metaphor and analogy in mathematics. *Proceedings of the Fourth International Conference for the Psychology of Mathematics Education.* Berkeley, Calif.: Lawrence Hall of Science, 1980, 157–162.

Schank, R. C., & Colby, K. M. (Eds.) *Computer models of thought and language.* San Francisco: Freeman, 1973.

Skemp, R. R. *Intelligence, learning and action.* New York: Wiley, 1979.

Suzuki, K. Solutions to problems. *Journal of Children's Mathematical Behavior,* 1979, *2,* 159–163.

Winograd, T. *Procedures as a representation for data in a computer program for understanding natural languages* (Ph.D. thesis). Cambridge, Mass.: Massachusetts Institute of Technology, Artificial Intelligence Laboratory, 1971.

Winston, P. H. Learning structural description from examples. In P. H. Winston, (Ed.), *The psychology of computer vision.* New York: McGraw-Hill, 1975.

CHAPTER 7

GEOFFREY B. SAXE
JILL POSNER

The Development of
Numerical Cognition:
Cross-Cultural Perspectives[1]

This chapter is about the relation between culture and the development of numerical thought. It is the only chapter in this volume with a specific focus on culture, and for this reason, we believe that it serves an important function. It is a reminder that the development of numerical thinking is embedded in social life, and the way children form and apply numerical concepts is partly rooted in their social experience.

The chapter is divided into three sections. In the first section, we discuss some of the general properties of numeration systems and the way they are employed to represent number. In the second section, we review two major theoretical formulations of cognitive development, one advanced by Vygotsky, the other by Piaget. Each provides a basis for inquiry into universal and culture-specific processes in children's formation of numerical concepts. In the third section, we examine the strengths and weaknesses of the existing research on cross-cultural number development associated with each theoretical formulation. Finally, in our concluding remarks, we offer some suggestions concerning fruitful new research directions.

[1]This chapter was prepared while Geoffrey Saxe was supported by grants from the National Institute of Education (G-78-0076 and G-80-0119) and the City University of New York (PSE-CUNY #13473) and while Jill Posner was supported by a grant from the National Institute of Child Health and Human Development (5T 32 HDO 7196).

291

NUMERATION SYSTEMS

Virtually all cultural groups have developed or borrowed numeration systems for the purposes of communicating numerical information and mediating numerical problem solving. Despite the diversity of these systems and the purposes for which they are used, the organization of any numeration system can be understood in terms of a distinction between number symbols and number operations (see Gelman & Gallistel, 1978; Saxe, 1981[a]). This distinction is most easily understood by means of an example.

Consider a young shepherd who is entrusted with the task of comparing the relative size of two flocks of sheep. The shepherd first tries the most direct method of comparison by using a correspondence operation—he tries to establish a one-to-one matching between the flocks. He soon discovers, however, that such a pairing is difficult to accomplish because sheep do not remain stationary. Consequently, he chooses an alternative approach that is less direct but ultimately more powerful. The shepherd solves the task by employing an intermediate group of elements, a set of pebbles. With the pebbles he can represent the correspondence operation symbolically. He first establishes a one-to-one correspondence relation between the set of pebbles and the first flock. He then turns to the second flock. If he can establish an exact one-to-one correspondence relation between the pebbles and the second flock, he can infer that the flocks have the same number; if not, he can determine which flock has the greater number.

This example highlights several properties of the activity of number representation. The first is that individuals can and often do employ aspects of their environment (the pebbles) as symbolic vehicles in order to increase the power of their problem solving. In particular, the use of pebbles as an intermediary enables the shepherd to compare sets that are distant in space (or otherwise difficult to physically pair.) The second property is that individuals employ symbolic vehicles to represent logico-mathematical relations—relations that are not in the objects but are an inherent aspect of the subject's enumerative activities. In other words, a summation of the sheep is not in any way a feature of the sheep themselves, nor is it inherent in the shepherd's vehicle of representation. Rather, it is an abstract transformation imposed by the shepherd on the sheep. The third property illustrated by the example is that the intermediary, which initially has only a local function—to compare the two sets of sheep numerically—can become a symbolic object with which the individual himself interacts. For example, the shepherd can perform additions and subtractions directly on the pebbles in order to signify operations that could be performed on any set of discrete elements, real or imaginary. These numerical relations would be inconceivable without their embodiment in a representational system such as pebbles. Thus, the problem-solving intermediary and the numerical operations can interweave with one another in such a way that they become functionally indissociable.

Although these properties have been presented in the context of a shepherd's idiosyncratic form of number representation (the pebbles), the same properties are evident in an individual's use of culturally defined forms of numeration (such as the Western numeration system). However, in the latter case, the symbolic vehicle is a knowledge system that has evolved in social history to serve the collective needs of members of a particular cultural group.

DEVELOPMENTAL MODELS

Each of the three properties just discussed raises a set of fundamental questions about the development of numerical thought, questions that will provide a direction for the discussion to follow. Property 1 implies that during the course of development, children transform historical forms of number representation that are initially external to their cognitive repetoire into symbolic vehicles that become an inherent part of their problem-solving activities. This means that the study of number development must not only describe the child's acquisition of the numeration system of the culture, but more importantly, it must provide an analysis of the changing relations over the course of the child's development between the child's acquisition of the numeration system and the process of problem solving. Property 2 implies that children use number symbols to represent logico-mathematical operations such as addition, subtraction, multiplication, and division. Thus, the study of number development must also provide an account of the origins and development of these logical operations and the way they become manifest in the use and understanding of a numeration system. Lastly, Property 3 implies that numerical thought often entails operating with and upon a particular historical invention, the form of which may facilitate operations of certain types and limit others. For this reason, the study of number development should provide an account of the way that differences in the sociohistorical constructions of number will lead to variations in the ways individuals solve problems.

In this section, we will consider two theoretical formulations of cognitive development, one inspired by Vygotsky, the other by Piaget. Each, we believe, offer important insights about number development that address, with varying degrees of adequacy, these three issues.

Vygotsky's Approach

The Soviet approach, represented in the writings of L. S. Vygotsky (1962, 1978), is concerned with understanding cognition as a "mediated" activity. By "mediated" it is meant that individuals do not interact with the world directly, instead they interact with their personal representations of the world. These representations include such social constructions as linguistic signs and discourse, orthographies, and numeration systems. For Vygotsky, the formative processes that influence the development of representational activities and the organizing role that these activities have for the general development of intelligence provide a critical area for study. It is these formative processes that link the cognitive development of the individual with the collective practices of the social group. Though Vygotsky did not treat the development of numerical cognition directly in his work, he did offer a general treatment of language and its relation to thought. A brief outline of his developmental framework for language will provide an introduction to our extension of Vygotsky's approach to the development of number.

Vygotsky argued that an inherent property of the development of speech and thought is that the relations between them change. Early in development, speech and thought are rooted in different kinds of activities and develop independently of one another. Preverbal thought consists of practical goal directed action. At the same time, early vocalizations are prerational and serve a primarily expressive function. With development, the child gradually transforms the historical construction of language into a vehicle that mediates his or her own thinking, and at the same time, the child's thinking becomes a mediated activity linked to the social group. Vygotsky is careful to point out that the interpenetration of thought and speech is never complete; rather, there is a dialectical relation between the two throughout development such that the emergent properties of their union are continually changing.

If we extend Vygotsky's treatment to numerical cognition, his approach leads to an analysis of two of the three core issues presented in the example with the shepherd—in particular, how it is that a representational system for number, which has emerged in the social history of a cultural group, is transformed by the individual such that it becomes an intermediary deployed in problem-solving activities (Property 1) and a symbolic object with which an individual interacts (Property 3).

NUMERATION SYSTEMS AS PRODUCTS OF SOCIAL HISTORY

Since, in Vygotsky's formulation the acquisition of historical systems of representation should profoundly influence the cognitive development of the individual, it is important to consider the differences in the characteristics of numeration systems across cultural groups as well as some of the factors that contribute to the changes in numeration systems in the course of the social history of a group.

For our purposes, one way of imposing some order on the wide range of numeration systems is to categorize them on the basis of two main features—the vehicles used to signify numerical relations (physical or verbal representations) and the predominant organizational structure of the system (base or nonbase structure). We will briefly review examples of each of the four resulting categories.

Spatial Representation/No Base Structure.   Typical of numeration systems that employ physical entities for number terms and have no base structure are the body part counting systems used by various New Guinea highland groups (see Lancy, 1978). The Oksapmin of the West Sepik Province employ such a system (see Saxe, 1981b). Twenty-seven body parts are used to represent number, 13 on each side of the upper periphery of the body, and the nose. To count, the Oksapmin name each body part in a prescribed order beginning with the thumb on one hand and ending with the little finger on the other. Thus, *besa*, the word for forearm, denotes the number "7" when counted on one side of the body and the number "21" when counted on the other side. If an individual needs to enumerate beyond "27," the count continues back up the second side of the body. This system does not have a base structure, and it provides only a finite set of terms by which to represent numerical relations. Numeration systems organized in this way are rather uncommon today; Lancy (1978) estimates them to represent only 10% of the New Guinean counting systems.

Spatial Representation/Base Structure.   The second category of numeration system also employs physical–spatial terms to represent numerical relations, but differs from the first in that a base or generative structure is employed.

An excellent example of how changing economic conditions may lead to an organizational change in the form of counting systems comes again from the Oksapmin, who in recent years, with the introduction of Western currency to the area, have adapted their indigenous system to include a base structure. This adaptation has enabled individuals to represent much larger numbers, an essential requirement for accurately carrying out economic transactions with currency. In the adapted system, the count begins as it does traditionally, with the thumb; however, instead of proceeding to the little finger on the other side of the body (27), the individual stops at the inner elbow of the other side (20). The completion of the count at this point corresponds to the number of shillings in a pound. When there is a need to count further, one round or "one pound" is recorded and the count is begun again. The system, then, is a hybrid that retains its signifiers for numerical relations but makes use of the base structure inherent in the local system of currency (see Moylan, forthcoming; or Saxe, 1982, in press, for a more extensive discussion of this system).

Nonspatial Representation/No Base Structure.   Most examples that fall into this category are not true numeration systems; that is, they are not employed for representing cardinal number, although they do convey ordinal relations. The days of the week are a typical example. The days are ordered as a list of primary lexical terms and are used to designate the ordinal relation between cyclical 24-

hour periods. No base is involved, although the terms are used iteratively.

Similar ordinal systems are used for other relational purposes. Saxe (1981c) reported that the Ponam islanders name their children by using a system that marks birth order relations. For example, in all families, boys are given one of a set of eight names, depending upon the order of their birth with respect to other boys in the family; there is a different set of eight names that apply to girls. Like the days of the week, the birth order naming system does not encode cardinal relations but is used to designate order relations.

Nonspatial Representation/Base Structure. Verbal numeration systems, such as those employed in Western societies, use arbitrary number names to signify the units and employ a base principle to generate higher numbers. Unlike written place value numerals, which are rare inventions, verbal numeration systems of this type have been developed in many cultures and were in existence long before written numeration systems of a similar structure were invented.

The most common numeration system of this kind employs a base ten or twenty, often with vestiges of a subordinated base five. Zaslavsky (1973) reports that these two types predominate in Africa. The Dioula of the Ivory Coast have a prototypical base ten system in which the number words express the additive function directly: "Tan nin kele" (10 + 1), "Tan nin fla" (10 + 2), and so on (Posner, 1982). The Dioula rely on the regular structure of their numeration system in carrying out mental arithmetic. Adding and multiplying is done by regrouping numbers by tens. Thus, 56 + 49 would typically be added in the following manner: (50 + 40 = 90; 6 − 1 = 5; 1 + 9 = 10; 90 + 10 + 5 = 105). The formation of numerals in this exceedingly regular base ten system is suggestive of such a procedure.

An unusually complex base twenty system that employs both additive, multiplicative, and subtractive principles is that used by another West African group, the Yoruba of Nigeria. Consider the Yoruba verbalization of the number "315." It can be expressed as follows: (200 × 2) − (20 × 4) − 5. The expression for 525 is (200 × 3) − (20 × 4) + 5 (Armstrong, 1962; Zaslavsky, 1973). This system is not one reserved for elites, though it would seem to require considerable arithmetic dexterity to master. The Yoruba are urban people and accustomed to commercial interactions. Learning to count and to use the Yoruba number system for arithmetical problem solving apparently occurs in the context of everyday market activities.

Although base ten and twenty constructs dominate the field of world numeration systems (Smeltzer, 1958), a variety of unusual bases have been documented both in contemporary and ancient groups. A small sample of these include a base four system used by the Huku of Uganda (Zaslavsky, 1973), a base fifteen system used by the New Guinea Huli (Cheetam, 1978), and a base sixty system used by the ancient Babylonians (Menninger, 1969).

Our brief survey of different types of numeration systems highlights a major theme of Soviet psychology, namely, that children's cognitive development occurs within a social context that itself has evolved varying intellectual tools. The way that these tools become the child's own and mediate numerical problem solving are central issues for Vygotsky's formulation. Indeed, Vygotsky's analytic categories are geared toward illuminating this fundamental topic in cognitive development. We consider Vygotsky's formulation next.

THE ACQUISITION OF KNOWLEDGE SYSTEMS

A major theme of Vygotsky's developmental approach is that there are qualitative changes in children's use of culturally organized knowledge systems for problem-solving activities and that these changes are an inherent feature of the acquisition process itself. To explain this process from Vygotsky's perspective, it is important to distinguish between two types of learning experiences, those that occur from the "bottom up," giving rise to what Vygotsky has called *spontaneous concepts,* and those that occur from the "top down," producing what Vygotsky has called *scientific concepts.*

Bottom-up learning is described as resulting from the child's spontaneous attempt to understand aspects of social and physical reality without the direct aid of adult or peer tutoring. These types of experiences lead the child to acquire practical concepts; that is, the child achieves local solutions to particular problems. For example, in order to reproduce the number of beads in a set, a child might create a numerically equivalent set by establishing one-to-one correspondences. In contrast, top-down learning is described as resulting from interactions with adults or more capable peers. In these interactions, problems are posed for the child, and he or she is presented with concepts of general applicability that are valued in the culture. Top-down learning, such as that encountered in school, gives the child the opportunity to form general concepts that may be adapted to different problem types but are not necessarily grounded in immediate experience. For example, a child might learn how to use the Western counting system in order to solve numerical problems. According to Vygotsky's formulation, each kind of learning should inform the other, and these links can be an important mechanism of developmental change.

Recent efforts consistent with Vygotsky's approach have attempted to demonstrate that a fundamental aspect of adult–child interactions in problem-solving contexts consists of a "scaffolding" process (Wood, Bruner, & Ross, 1976), in which adults adjust their (top-down) dialogue in such a way that the child can relate his or her (bottom-up) experiences to the novel problem at hand (Gearhart & Newman, 1980; Wertsch, 1979; Wood et al., 1976; Wood & Middleton, 1975). Wertsch, for instance, has demonstrated four levels of scaffolding behavior apparent in mothers' interactions with their toddlers who are attempting to solve a puzzle. This scheme can be applied to the child's acquisition of numeration concepts in the following way. Early in the developmental process, adults engage children in number-related activities that children are not completely capable of doing on their own, but are within the grasp of their understanding—activities that are within what Vygotsky labeled *the zone of proximal development*. By guiding them through number tasks, adults introduce children to the number symbols and general numeration strategies that are specific to their cultural group. Although at first the strategies and symbols are regulated by the adult, over the course of development the child is able to integrate and connect these concepts with those derived from his or her own (bottom-up) experiences and thereby achieves a progressive understanding of the form and function of these symbols and strategies. With increasing experience in the form of interaction with adults and others, top-down concepts come to guide and organize bottom-up learning with the result that children's problem-solving behavior becomes increasingly independent of others and yet more closely mirrors the cognitive functioning of members of their cultural group.

This Vygotskiian-inspired account offers some important insights about the function and influence that mediational systems have on the child's developing number abilities. First, central to Vygotsky's account is a materialism. The subject organizes the material available in the environment in order to increase the power of problem solving. Some of this ''material'' is social and takes the form of culturally organized forms of representational systems. Thus, just as our shepherd transformed pebbles into a mediational system, the child transforms socially defined symbol systems for number into a vehicle for numerical representation (Property 1). Second, on the Soviet account, the nature of this material and the way it is used in cultural life in turn influences the character of numerical thought in a particular way (Property 3). Thus, the sociohistorical context of the child's development plays a powerful determining role in the character of the numerical thought of the child. Although the Soviet approach presents a treatment of the first and third properties of numeration, it does not offer an account of the second property—the logical foundations of numerical representation and numeration concepts. It is the Piagetian formulation that addresses these issues primarily, and by so doing, deemphasizes the role of specific cultural forms of representation in structuring the child's numerical concepts.

Piaget's Cognitive Developmental Theory

Piaget's theory is concerned with the origins of logical structures of thought and the characteristics of these structures. The major analytic categories of the theory as well as the empirical work that the theory has generated are focused on this core issue.

Central to Piaget's thinking is a rejection of both nativistic and empiricist formulations of the origins of logical structures. Instead, he argues that the origins of logical structures are elaborated in sensorimotor activities, which, through an epigenetic process, are transformed into mental operations in the course of development (see Langer, 1980). During the sensorimotor stage (infancy), these coordinations are practical and directed toward achieving immediate goals. As development proceeds, the logic of practical action is transformed into a logic of reversible concrete operations (during middle childhood). This stage is characterized by a set of classificatory and relational structures (formalized as logico-mathematical groupings) which, Piaget argues, are the basis of children's understanding of elementary numerical operations such as addition, subtraction, multiplication, and division. It is not until the stage of formal operations (adolescence and adulthood), however, that these two sets of classificatory and relational structures become integrated into a generalized reversible system. Piaget argues that this integration is the defining characteristic of hypothetico-deductive reasoning, the essence of mature cognitive functioning. It is Piaget's position that the progression from one stage to another is not the simple result of maturational, experiential, or social factors (such as the acquisition of a numeration or other symbol systems). Rather, progress is achieved by the child through an equilibration process whereby all the aforementioned factors exert an influence as the child strives to achieve a coherence between the dialectic of existing forms of understanding and new experience. Piaget's general definition of the problem of cognitive development, then, differs from Vygotsky's. Whereas Vygotsky's focus is on numeration as a mediated activity (Properties 1 and 3) that has its roots in social interaction, Piaget's focus is on the emergence of logico-mathematical structures that underlie the use of numeration (Property 2) and have their roots in sensorimotor activities.

As a part of his theory, Piaget has offered a developmental analysis of such logico-mathematical concepts as cardinal and ordinal number, the composition of numerical relations, and a wide range of measurement concepts (see Piaget, 1952; 1968). Piaget argues that developmental changes in each of these concepts is a manifestation of a general shift in the organization of logico-mathematical thought. We will review here, in some detail, Piaget's treatment of numerical equivalence (the conservation of one-to-one correspondence relations) that develops with the concrete operational stage, as it is this concept that has been examined most extensively in cross-cultural research.

To investigate developmental changes in children's understanding of the conservation of numerical equivalence, Piaget presented children with a model set of elements and asked them to construct a numerical copy with an available set. Piaget found that, during an early stage, children make judgments of equivalence on the basis of one-way functions, operations which have no inverse. On this basis, the child carries out a systematic evaluation of the quantities according to the spatial extent of one collection relative to the other ($x \rightarrow y$). For instance, a child might align the endpoints of two sets and disregard the discrete number of the elements in the two sets. Similarly, a child might reason that one collection contains more because it extends further than the other. At the next stage, children begin to coordinate evaluations based on two one-way functions. Thus, children begin to coordinate a numerical evaluation based on length, with one based on density. As a result, children begin to produce one-to-one correspondence relations between two sets to determine numerical equivalence. Nevertheless, at this stage, if one row is subsequently spread apart or crowded together, children generally assume that the equivalence relation is not lasting. Finally, as classificatory and relational operations assume their concrete operational form, the coordination of one-way functions is completed. At this stage, children understand the necessity of conservation.

It is important to point out that the operations entailed in the concept of conservation are also formally entailed in other fundamental numerical concepts such as associativity, commutativity, and transitivity. Thus, whether one spreads elements apart as in the classic conservation task, regroups them (assocativity), rearranges their order (commutativity), or uses an intermediate group of elements as a basis to compare two sets, at a formal level of analysis one is asking a subject about the same or related sets of operational structures. From the Piagetian perspective, understanding any one of these operations by definition implies understanding the others. When subjects exhibit differences in their ability to demonstrate an understanding of related concepts on assessment tasks, variation in performance is ascribed to differences in the way the object "resists" the child's understandings. That is, it is argued that with some materials and in some contexts, it is easier for a child to apply conceptual achievements than in others.

In his approach to cognitive development, Piaget subordinated experiential or cultural factors to the equilibration process. This has probably contributed to the lack of concern with treating experience, in particular cultural experience, as a differentiated theoretical construct. Genevans generally acknowledge the influence of culture as a factor that may influence the rate of progress through the stages or the achievement of the endpoint of the developmental process (i.e., formal operations). Yet they do not explicate the manner in which cultural factors contribute to the developmental process. Insofar as numeration systems are bodies of knowledge that are acquired from one's social group, they, like other collective practices, are understood to have relatively little influence on the way individuals construct numerical concepts. Thus, while offering a formulation about the way in which numerical operations develop, the Piagetian focus leaves unanalyzed the mechanisms by which social factors contribute to the formation of numerical thought.

CROSS-CULTURAL RESEARCH ON THE DEVELOPMENT OF MATHEMATICAL CONCEPTS IN CHILDREN

The Genevan and Soviet formulations each imply that there are universal and culture-specific processes in cognitive development, although the nature of these processes differ in each theoretical formulation. Cross-cultural research has often been informed by one or the other of these theoretical positions, either directly or indirectly, and research on numerical cognition is not an exception. We have organized our review of this literature accordingly and will consider a sample of studies motivated by Piaget's theory as well as selected studies cast in a Vygotskiian theoretical framework. In what follows, we attempt to characterize the researcher's view of the way in which culture interacts with the development of quantitative thought and how the research contributes to each theoretical formulation.

Empirical Work Consistent with the Cultural Practice Approach

The theory proposed by Vygotsky and his student Luria has been subsequently elaborated by a group of American psychologists (Cole, Gay, Glick, & Sharp, 1971; Cole & Scribner, 1974; Wertsch, 1979). It is important to point out that Cole's group has focused on certain of Vygotsky's theoretical constructs and not others. Cole's emphasis has been primarily on a contextual analysis—an analysis of the way the conceptual abilities of individuals are influenced by and adapted to particular problem-solving contexts (a version of Property 3), whereas Vygotsky, in addition, stressed a developmental analysis of the shifting functional relations between different aspects of problem solving over the course of development (Property 1). For the most part, the existing cross-cultural research follows the contextualist focus rather than the developmental one.

Two general types of methodological approaches can be distinguished among those who use a contextualist focus, both of which concern the influence of cultural variables such as schooling, economic specialization, and the form of numeration system on problem solving. However, the first type generally makes use of "culture" as an explanatory construct for differences in the way that individuals mediate their solution to numerical problems, whereas the second makes use of a more fine-grained analysis of the strengths and limits of mediational strategies that develop in particular cultural contexts.

Consider the first type of methodological approach. The logic of these studies follows from the position that societies develop different technologies or systems of knowledge in response to environmental factors. These bodies of knowledge accrue over the course of history and are transmitted to young members both explicitly (by means of specific childrearing practices) and implicitly (in the way the society organizes the activities that children experience). It is therefore hypothesized that groups with varying cultural practices should solve mathematical tasks differently from one another (see Laboratory of Comparative Human Cognition, 1979, for a comprehensive discussion).

A pioneering effort in the area of cross-cultural numerical cognition that assumes this approach was carried out by Gay and Cole (1967). They examined number and measurement concepts of the Kpelle, a group in Central Liberia, as well as their childrearing practices and traditional educational values in an attempt to determine if these were incompatible with formal arithmetic concepts taught in Western-style schools. They also conducted a series of psychological experiments that examined the Kpelle's skill at estimating quantities of different types. In general, they found that the requirements of everyday life dictated which numerical estimation skills were most fully developed. For instance, the Kpelle were better than American college students at estimating the number of stones in a pile, whereas American college students were better at measuring length by hand spans. Gay and Cole argue that these patterns of differences resulted from the fact that stones are commonly used by Kpelle for tallying the number of cups in a container of rice, as rice is an important staple for the Kpelle. Thus, Gay and Cole argue, the Kpelle excelled on the estimation tasks. In contrast, measuring length by hand spans and foot lengths, a system also in traditional use, was more difficult for the Kpelle than for the Americans. The reason for this, Gay and Cole claim, is that the Kpelle do not have a system for length that employs interchangeable units (e.g., 1 ft = 12 in.). The Americans, on the other hand, estimated the length more accurately, because they used the English system to translate hand spans into inches.

A series of studies conducted on the Ivory Coast also begins with the assumption that significant societal values and cognitive skills acquired by children in a culture are derived in part from the culturally organized activities and practices. The research systematically documents the development of addition problem solving and other mathematical skills, which occur both informally (Ginsburg, Posner, & Russell, 1981a; Posner, 1982) and as a result of schooling (Ginsburg, Posner, & Russell, 1981b). Subjects came from groups with different economies: merchants (Dioula) and agriculturalists (Baoule). It was hypothesized that unschooled children from the merchant culture would be more accurate and use more efficient strategies than the children from agricultural families, because mercantile society values mathematical skill and affords experiences that encourage its development. As expected, the results show that unschooled Dioula (merchant) children adopt more economical strategies than their unschooled Baoule (agricultural) counterparts. In particular, they use a greater number of memorized addition facts and regrouping by tens $(7 + 5 = 10 + 2)$, as compared to the Baoule. Schooling provides requisite experience with number for the Baoule who perform in most respects on a par with Dioula peers when they receive instruction.

The experiment that examined the assimilation of school mathematics by Dioula and Baoule children also provides a result that links the social context of development to the formation of numerical problem-solving strategies. When doing written arithmetic, children initially adopted informal counting methods and, in some cases, traditional regrouping procedures, but with increased school experience, they used the standard written algorithm almost exclusively. Together, these two sets of studies with Ivory Coast populations demonstrate how cognitive strategies for arithmetic problem solving are acquired both within and outside the school context. More generally, the research shows how individuals develop the symbolic skills that are most useful to them in their differing social contexts.

The difference between the studies just described and those comprising the second type is primarily methodological yet important to highlight, as it reflects a shift in the way in which cross cultural research questions are posed (see Glick, 1981, for a recent discussion). In the previous studies, "culture" was offered as the "explanation" of variable performance across population groups. In the studies we review next, the focus is on the experimental situation itself, and group comparisons, if employed at all, occur between closely related populations that differ on a well delimited dimension. A basic theme in these studies is that it is incumbent on the researcher to design an experiment that will draw out the cultural knowledge of subjects. As the experimental task is varied, the strengths, limitations, and characteristics of subjects' cognitive skills are progressively revealed. In this way, it becomes possible to evaluate more effectively the types of mediational strategies and problem-solving techniques specific to individuals in particular cultural contexts. Research of this type, carried out with several of the same ethnic groups as those in the Cole and Ginsburg studies, is described next in some detail.

Lave (1977) sought to determine whether problem-solving strategies learned in school would, as has frequently been suggested in the literature (Greenfield & Bruner, 1969; Scribner & Cole, 1973), transfer to unfamiliar problems that arise in informal learning environments, such as the tailor shop. In other words, Lave was testing the claim that schooling promotes generalized cognitive skills. The study was conducted with a group of tailors from the Vai and Gola tribes whose tailoring experience and years of formal schooling varied independently. Lave's tasks included arithmetic problems that resembled those practiced at school, those typically encountered by tailors in their work, and others which lay somewhere in between the two domains. She argued that her findings support what Cole and his colleagues call the cultural practice or functional learning approach to cognition. Specifically, Lave found that the extent of school experience was the best predictor of success for school-like problems, whereas successful resolution of tailoring problems was predicted best by years of tailoring. Thus, the findings point to the compartmentalization of skills or mediational strategies in particular contexts, an important finding and one that is at odds with previous interpretations of the influence of schooling on cognitive abilities (see Laboratory of Comparative Human Cognition, 1979).

Using both a descriptive and experimental approach, Lave in collaboration with Reed (Reed & Lave, n.d.), explored critical linguistic and experiential contributions to the Vai and Gola tailors' arithmetic knowledge. Based on a prior analysis of the numeration systems used by the tailors and the kinds of arithmetic problems they commonly encountered in different contexts, Reed and Lave were able to predict the type of errors committed by tailors with varying amounts of schooling. For example, unschooled tailors who use the Vai–Gola numeration system, which employs 5, 10, and 20 as a generative base (e.g., $75 = [3 \times 20]$ $+ [10 + 5]$, often adopt a regrouping strategy for addition that creates groupings around each of these relevant numbers (e.g., $17 + 3 = [10 + 5 + 2] + 3 = 10 +$ $5 + 5 = 10 + 10 = 20$). The propensity for "losing track" of the groups to be summed when complex numbers are involved is very great. By contrast, individuals with 5 or more years of schooling adopt school-like algorithms (e.g., those for column addition) and typically make "carrying" errors that are off by factors of a greater magnitude than their unschooled colleagues (e.g., a factor of 100 in a three-digit problem). Evidence that tailors with just a little schooling were switching to the school-like system was also apparent. Thus, the general finding is that schooling has an impact on the way that arithmetic gets done, although it does not necessarily lead to a more generalized understanding of arithmetic principles. Such generalized understanding, the authors contend, is more a function of practice than educational background, though, like other researchers, they do not detail the type of practice required.

Petitto's (1979) thesis work among the Dioula provides some data relevant to understanding the relation between the structure of practical activities and generalized arithmetic knowledge. She studied arithmetic problem solving in two professional groups, tailors and cloth merchants. Both groups typically manipulate cloth, money, and meter measures in their work, but in different ways. For example, the main arithmetic calculations of cloth merchants involve determining the price of fractions of a meter of cloth. This is a ratio problem which requires complex arithmetic. Tailors' use of the meter is more concrete and direct, mainly requiring a simple doubling or halving operation. Both the merchants and tailors were interviewed and administered simple tasks that supposedly tap the component skills required to solve proportion tasks. In addition, subjects were given problems that required using a proportion to determine the price of various lengths of cloth (familiar materials) and comparable quantities of oranges (unfamiliar materials). There were no differences observed in the component skills tasks between tailors and merchants. However, the cloth merchants outperformed the tailors in both types of proportion tasks, not only on the cloth problems familiar to the merchants. Thus, Petitto's data show that under certain conditions, skills do transfer from familiar to novel contexts.

Another dimension of Petitto's work (Petitto & Ginsburg, in press) concerns the strategies the Dioulas use to perform the four basic arithmetic operations and whether their calculational ability is tied to a tacit understanding of formal mathematical principles. To accomplish this, she devised eight pairs of problems such that for any given pair the solution of the second could bypass calculation by relying on mathematical principles of different types. For example, the law of commutativity could be invoked tacitly to provide an immediate answer to the problem $38 + 46$, after a subject had already calculated $46 + 38$. A similar relationship holds between the problems 6×100 and 100×6.

The results showed considerable variability in the understanding and use of formal principles. For instance, many subjects identified the commutative relationship in addition problems, but failed to do so with multiplication. This failure is attributed to the asymmetry in the Dioula's linguistic expression for multiplication (the gloss for 100×6 is given as 6 "added to itself" 100 times which, if one actually proceeds to add iteratively, is much more laborious than 100 added to itself 6 times). Given the linguistic construction for the terms of the system, it is not obvious to the Dioula that at a deeper conceptual level the multiplier and multiplicand can be exchanged without affecting the product.

The results with respect to calculational strategies corroborate and extend those of Ginsburg *et al.* (1981b). Petitto finds that most Dioula use a regrouping strategy for addition and that this basic approach is elaborated upon in carrying out the other three operations. Subtraction is conceptualized as addition in reverse, multiplication as repeated addition; and division, the most laborious, is carried out by estimating a number that when multiplied by the divisor yields the original quantity. The answer is then checked by addition and adjusted if inaccurate.

As we have seen, both Petitto and Lave have employed a methodological approach that differs from the standard comparison of groups across cultures. All experimental contrasts involved comparisons within the cultural groups themselves. Both researchers conducted a series of experiments, and Lave relied heavily on ethnographic analysis in framing experimental questions and designing tasks. There is a great deal of convergence in the two sets of results, which is not suprising, given the similarity of cultural setting (the urban tailor's shop), language (both are of the Mande family), and tasks. Though the question of how mathematical skills are acquired is ignored by the authors, in the case of the Dioula, prior work of Posner (in press) offers the explanation that strategies are acquired by means of direct experience manipulating (e.g., grouping) and counting objects. More observational investigations of the activities that account for children's adoption of particular strategies need to be done, however, if we are to have more than a suggestion of how problem-solving strategies develop.

In general, the research consistent with Vygotsky's approach has provided documentation concerning the way numerical skills are interwoven with particular numerational systems and culturally organized practices. This body of research, however, has provided little information about the process of development of these cultural-specific skills, a central concern of Vygotsky's original formulation. As we turn to the Piagetian inspired research, we will see somewhat the reverse trend in the way in which the theoretical approach is translated into empirical research. Here the major focus is on documenting the characteristics of qualitative changes in cognitive functioning that occur in the course of development and a relative lack of concern with the particular cultural context of development.

Piagetian Studies

The cross-cultural research on number development that is inspired by Piaget's theory can be categorized into three general types. First and most extensive is the body of research concerned with determining the validity of Piaget's claim that his proposed stage sequence for number development constitutes a universal developmental process. Second is the limited research that has been conducted to determine the cultural factors that contribute to different rates of progress through Piaget's stage sequence. Third are the rare studies concerned with determining whether and in what way Piaget's general stages of cognitive development are related to the acquisition of different forms of non-Western numeration systems.

A number of authors have provided reviews that cover the first general theme of cross-cultural Piagetian studies (e.g., Ashton, 1975; Carlson, 1976; Dasen, 1972, 1977a), and the consensus among them is that on the whole, in the range of cultural groups studied, there is support for the universality of development in children's understanding of concepts of number conservation through the concrete operational stage. However, time lags are widely cited, and there is considerable variability across cultures with respect to performance on various tasks within a stage. Sometimes the relative difficulty of tasks within a stage does not correspond to Piaget's findings with Western children.

A study by Etuk (1967) concerning the development of number concepts is a good example of this first class of studies. Etuk interviewed Yoruba schoolchildren from traditional and modern homes, and, guided by the Piagetian formulation, she assessed subjects' abilities to classify, seriate, and conserve number. She found that the rate of acquiring these concepts was somewhat slower among the Yoruba, particularly among those who come from traditional homes than among Western children, although Yorubas exhibited the same basic levels of understanding. Of more significance to the theory perhaps is the relative difficulty of the three tasks. Although conservation and seriation concepts were generally found to emerge at the same time across the sample, class inclusion was not well understood by any group. In a more recent study, Opper (1977) found a similar pattern for the acquisition of the same three concepts in rural Thai children.

As is the case with much of the Piagetian research conducted during this period, Etuk's study leaves us with very little information concerning the reason for the variation in performance between "traditional" and "modern" children. This is unfortunate since there are many aspects of Yoruba culture that make it particularly interesting from the perspective of number development. Etuk mentions in passing that the vast majority of the traditional mothers (87%) were petty traders, but she gives no indication of how this kind of environment might be expected to stimulate mathematical activities or why, given the potential advantage, the traditional child performs more poorly than children from more Western families. Moreover, no mention is made of the complicated numeration system used by the Yoruba described in a previous section of this chapter. This oversight may reflect the small role assigned to numeration systems in Piaget's formulation as well as the overriding concern of Etuk's study—documentation of the universal aspects of Piaget's stages, stages that are not expected to be a product of cultural factors.

A second general approach to cross-cultural Piagetian research was adopted as a response to the many studies like Etuk's, which find different rates of development but which offer little explanation, other than the well-packaged independent variable—culture—to account for them. This latter research is concerned with identifying factors that influence the rate of acquisition of logico-mathematical concepts. The notion is that it may be possible to specify the kinds of organism–environment interactions that influence cognitive development. Researchers who have taken this tack have attempted to identify environments that might encourage or discourage the development of specific concepts.

This general line of research was initiated by Price-Williams, Gordon, and Ramirez (1967), who studied the children of traditional potters in Mexico and found them to be more advanced in their ability to conserve quantities of clay than a control group of Mexican children. Other conservation concepts were not affected. Using this study as a model, subsequent research has delimited further the nature of experiences required to bring about cognitive structural change. Steinberg and Dunn (1976), who found no advantage for another group of potters' children in Mexico, point out that the actual experience that these particular children have with clay does not in fact leave weight or amount invariant. The loss of moisture when pots are fired in the traditional manner causes them to shrink in size and become lighter.

Another conservation study that included potters and their children also examined subjects from commercial and agricultural families. Adjei (1977) expected his Ghanaian subjects to excel on the conservation task most related to their occupational specialties. Although no specific hypotheses were given, the reader assumes that Adjei proposes merchants to be advanced on number tasks and potters on the weight and substance tasks. It is not obvious on which tasks the farming group would be expected to show superior performance. Adjei's results are not in perfect correlation with these predictions. Among the 7- to 9-year-olds, the only difference observed was for weight, on which the potters' children out-performed the other groups. Women potters excelled on the substance, weight, and volume tasks, all skills called upon in pottery making. The number task, however, did not differentiate among women with different occupations; all the groups demonstrated an understanding of number conservation. Although this finding does not support Adjei's initial prediction, it is consistent with the classical Piagetian formulation that number conservation is expected to be acquired before the other conservations. It may be, as Adjei indicates, that in the number conservation task there is a ceiling effect that masks potential group differences. If this were the case, one would expect to see some variation at an earlier age. Although Adjei administered an inequality and seriation task to a group of 4- to 5-year-olds, he unfortunately does not report the results. Thus, whether there is an effect of cultural orientation such as commerce on the development of numerical concepts in young Ghanaian children is left unanswered by this study.

Another set of studies conducted in Papua New Guinea investigated factors that enhance the acquisition of logical concepts. The Papua New Guinea Indigenous Mathematics Project (Lancy, 1978) was motivated by national concern about the poor acquisition of mathematical skills by schoolchildren. Several possibilities were advanced concerning the causes of poor performance. One was that native numeration systems might actually be subverting the acquisition of modern mathematics. The specific hypothesis put forth was that children from societies with "true numerical counting systems" (i.e., composed of a base structure and nonspatial representation), considered by Lancy to be more abstract than other systems in use in Papua New Guinea, would evidence more abstract thinking in general. The nature of the causal relationship between "abstract" thinking and the type of numeration system available in a culture was not specified. It is not clear, for instance, whether the use of a particular type of numeration system is taken to influence "abstract" thinking or whether groups using "advanced" systems are more likely to engage in other activities (like commerce) that may facilitate abstract thought.

In any case, to test the hypothesis, 11 cultural groups with numeration systems of various types were examined; they ranged from those similar to Western systems to others that use body parts. Both conservation and other cognitive tasks were given to schooled and unschooled children from the chosen groups. No clear relationship was found between the type of numeration system in use and performance on any of the general cognitive tasks administered. Although Lancy could not produce evidence supporting his hypothesis, the organization of numeration systems may in fact have more local effects on quantitative thought, as we have seen in Lave's and Petitto's studies concerning adult problem solving.

In another study reported by Posner and Baroody (1979), the economic orientation of one's cultural group was hypothesized to be a factor that influenced the rate of acquiring number concepts. Posner and Baroody (1979) used a design similar to Adjei's in order to investigate the relation between the acquisition of number conservation and the development of counting skill among children on the Ivory Coast. Preschool, schooled, and unschooled children from tribes with different economic orientations (merchant and agricultural) were selected for study. The results showed an interaction between schooling and cultural background. Unschooled merchant children (Bioula) performed on a par with school children who had received explicit instruction in counting and conservation. The unschooled children from the agricultural milieu (Baoule) developed these skills at a slower rate, although those attending school were not shown to be at a disadvantage. Thus, it appears, on the one hand, the activities and requirements of merchant culture play an important role in the development of numerical concepts, and on the other, that instruction provides experiences that are unavailable to children from less quantitative-oriented societies. Even when there is no direct instructional intervention, the activities and requirements of merchant culture can have an effect on the development of numerical concepts.

The role of counting in facilitating the development of Piagetian quantity concepts is unclear. In the study just reviewed, Posner and Baroody found that a variety of counting skills were related to children's ability to understand number conservation, and Saxe (1979a, 1979b, 1981e) has demonstrated that the ability to use counting to produce numerical comparisons and reproductions of sets precedes the development of number conservation, both in the United States and in three traditional Papua New Guinea societies. Nonetheless, it is difficult to imagine how the enumeration of sets before and after a transformation could be the sole basis for the development of number conservation. Young children miscount to varying degrees, and as a consequence, could not derive an understanding of the logical necessity of conservation with reference to their counting alone (see Saxe, 1979a, for a general discussion of this issue).

The third theme in Piagetian research deals specifically with numeration systems and the way that operational structures become interwoven with children's use and understanding of their numeration systems. In the few studies that bear on this problem, Saxe (1979b, 1981b, 1981c) has argued that wide variations in the organization of numeration systems may produce predictable kinds of conceptual difficulties for children who are in the process of acquiring them. In one study, Saxe (1981b) hypothesized that, due to the specific properties of the Oksapmin body part numeration system, Oksapmin children would experience conceptual confusions in understanding that each body part represents a distinct ordinal value in a progressive summation of values. Specifically, it was expected that they would evaluate the numerical relation between body parts, not on the basis of their ordinal position of occurrence, but on the basis of their physical similarity (i.e., equivalence of terms on right and left sides of the body). To test this hypothesis, Saxe required Oksapmin children to compare values of symmetrical body parts (e.g., the thumb on the right versus the left hand) and asymmetrical body parts (e.g., the right thumb versus the left elbow). The results support the hypothesis that, early in development, young children are seduced by the physical similarity of body parts in producing numerical evaluations on these tasks, but over the course of development, children come to use body parts to represent unique ordinal positions in a progressive summation of elements.

In another study conducted in Papua New Guinea (Ponam Island), Saxe (1981c) examined qualitative changes in children's understanding of an indigenous birth order naming system. In the Ponam birth order system (described earlier), names are assigned to children on the basis of both sex and birth order. The system has several interesting logical properties: For example, the relative age relation among individuals within a family can be determined on the basis of birth order names within a sex but not across sexes. On the basis of Piaget's theoretical formulation, Saxe reasoned that an understanding of the determinate age relations within sex should emerge with the advent of concrete operations since this understanding requires a subject to coordinate two series of asymmetrical relations (age and birth order name), a defining characteristic of concrete operational thought. It was also reasoned that an understanding of the indeterminate age relations across sexes would not emerge until the advent of formal operations since this understanding requires a subject to create a combinatorial system that, in principle, could generate all possible birth orders. Consistent with these hypotheses, the results showed that children acquire an understanding of the determinate relations before indeterminate relations and that an understanding of the indeterminate relations is not achieved until late adolescence. Both studies by Saxe represent attempts to incorporate an analysis of culture-specific knowledge systems within the framework of universal structural changes described by Piaget.

In summary, most studies inspired by the Piagetian approach succeed in documenting a shift from preoperational to concrete operational mathematical concepts, whether these concepts are assessed with conventional tasks or those which make use of an indigenous knowledge system as a task context. Problematic for the Piagetian formulation is the lack of empirical support for the construct of stage, specifically, that there is a psychological reality to a content-free description of structural competence. In general, the cross-cultural literature, like the Western literature, shows that individuals display developmentally distinct forms of reasoning as a function of the type of assessment task that is administered. In addition, some researchers have not found evidence of either concrete or formal operational reasoning with particular types of assessment tasks in various cultural settings (see Dasen, 1972; Lancy, Souviney, & Kada, in press). We will briefly consider some of the attempts to reconcile these findings with Piaget's theory before we offer our own concluding remarks.

In an attempt to reconcile the lack of support for the the psychological reality of the construct of stage, Dasen (1977a, 1977b) employs a distinction between competence and performance used by other cognitive developmental theorists (see Flavell & Wohlwill, 1969). He proposes that the competence for thought structures at the concrete operational stage is universal, though its manifestation in a given situation may be culturally determined. In this way, if a subject manifests a concrete operational understanding on some but not all tasks, it is assumed that a performance factor is blocking the expression of the subject's competence. Although the distinction between competence and performance is a crucial one, Dasen's solution creates a serious paradox in evaluating the significance of cross-cultural findings. If Dasen's position is invoked post hoc by researchers attempting to deal with findings that contradict Piaget's theory, what will constitute negative findings with respect to the claim for universal thought structures? Priswerk (1976) has voiced a similar concern. In order to support his view, Dasen and others (Dasen, Lavallee, & Retschitzki, 1979; Dasen, Ngini, & Lavallee, 1979; Lavallee & Dasen, 1980) have used the training study as a means to reveal latent competence. If subjects do possess competence in some areas of concrete operational thought but, due to a performance factor, do not exhibit the competence in response to some task, then the competence should be expressed given a minimum amount of training. These types of training studies have met with mixed success.

In a more recent attempt to address the lack of empirical support for the construct of "stage" in cross cultural research, Harris and Heelas (1979) have proposed a hybrid model incorporating both contextualist and structuralist themes. They argue that operational structures are constructed by subjects in local conceptual contexts that are related to particular experiential factors. Unlike the classical Piagetian claim, Harris and Helas argue that these constructions do not generalize across conceptual domains but rather are linked to particular areas of experience.

Piaget (1972) has also considered the problem of the relation between cross-cultural findings and this theoretical formulation and has adopted a position that embraces sociocultural context more closely than his earlier position. Concerning the acquisition of the formal operational stage during adolescence, he argues that individuals "reach this stage (formal operations) in different areas according to their aptitudes and their professional specializations (advanced studies or different types of apprenticeship for the various trades): The way in which these formal structures are used, however, is not necessarily the same in all cases [p. 10]." In fact, in this article, Piaget leaves open the issue as to whether the formal operational structures are an adequate characterization of the cognitive functioning of the adult and whether additional or alternative structures are yet to be uncovered.

Thus, it appears from this that Piaget himself has moved toward a greater relativism in considering endpoints of the cognitive developmental process. Certainly he has here, as he has elsewhere (1966), attempted to accommodate his theory to the results accumulated from cross-cultural research. Such attempts at accommodation notwithstanding, it should be reemphasized that although Piaget concedes that educational and cultural transmissions are contributing factors to cognitive development, the primary focus of his theory is on organism–environment equilibrations of a very general sort. These interactions result in a sequence of underlying structures that have universal features. The operational structures that are content-free inform intellectual activities that are enacted via culturally determined systems of knowledge. It is through this process that culturally defined systems of representation take on logical properties. The role that the cultural encodings or surface structure of particular concepts have on the development of cognitive structures has been largely considered unimportant and therefore ignored in the Piagetian tradition.

CONCLUDING REMARKS

In this chapter we have argued that from both Piaget's and Vygotsky's theoretical perspectives there are both universal and culture-specific processes that are entailed in the development of numerical cognition, although the nature of these processes vary across the two theories. Accordingly, the two bodies of research reviewed differ in the way in which researchers frame objectives, elaborate theoretical constructs, and interpret results. The research stemming from Piaget's perspective takes logical operations as the major focus of study. Researchers using this orientation have attempted to document universal changes in children's understanding of logical operations and to determine culture-specific factors that influence the rate of development as well as the representational contexts in which these operations are manifested. In contrast, those whose work is consistent with Vygotsky's approach have made the cultural context of problem solving the primary focus of study. Researchers in this tradition have attempted to gain a greater understanding of the general processes that foster culture-specific problem-solving strategies.

That each of these formulations has contributed to our understanding of the three core properties of numeration is undeniable. However, as we have pointed out, each approach lacks comprehensiveness, and where one theoretical approach tends to be strong, the other is weak. Piaget's account treats numeration systems as equivalent across contexts, focusing on the logical coordination of relations that must underlie the use of any system (Property 2), whereas Vygotsky's account focuses on the way in which the particular forms of number representation become interwoven with and influence the fabric of numerical thought (Properties 1 and 3).

We believe that a coordination of the objectives in the Vygotskiian and Piagetian formulations can be achieved that would give rise to more comprehensive research programs, programs that could lead to an understanding of the interrelatedness and interactions between the formation of Properties 1, 2, and 3. In a new series of studies among the Oksapmin, one of us has employed such a coordinated approach (see Saxe, 1981d, 1981e, 1982, in press). We will review only a single study here as an example (Saxe, in press).

Recall that the Oksapmin use a body part counting system. In traditional Oksapmin life, there is virtually no social context in which one needs to carry out arithmetic computations. However, in recent years, a money economy has been introduced in the region and this new economy has created a new social context within which individuals must create solutions to addition and subtraction problems involving currency. It was expected that this new social context would lead Oksapmin to construct novel ways of using their body system to mediate arithmetical computations, a constructive process involving each of the core properties of numeration. First, it was expected that traditional forms of number representation would be ill-suited to handle manipulations with currency; however, with the introduction of the money economy, it was expected that Oksapmin would transform the traditional "material" in such a way as to achieve their new problem-solving ends involving the addition and subtraction of currency (Property 1). Second, it was expected that this transformation would be associated with underlying structural changes in the logical properties of numerical operations (Property 2). Finally, the new forms of problem-solving strategies should represent general procedures for the conceptual manipulation of numerical relations (Property 3). The way in which this change in the organization of number representation was studied was to interview individuals with varying experience with the money economy; these included four groups ranging from tradestore owners to traditional adults who had little contact with the money economy. Each individual was asked to solve arithmetic problems involving currency.

An analysis of peoples' strategies showed differences in each of the core properties as a function of the amount of participation in the money economy. Traditional individuals typically considered only correspondence relations between a series of body parts and a set of real-world objects in their solution procedures. For example, to subtract 9 from 16, individuals with little experience would typically enumerate the corresponding series of body parts as an answer—for example, shoulder (10), neck (11), ear (12), eye (13), nose (14), eye on other side (15), and ear on other side (16). In other words, they used each body part to represent a coin and then simply iterated the body parts that would be left once the transaction was completed, without offering a single numeric representation for the answer. Similarly, to add 7 + 9 without coins present, an individual with little experience would typically enumerate the body parts—shoulder (10), neck (11), ear (12)—and simply estimate when the enumeration should be ended since they would not be able to tell when they had completed enumerating seven additional body parts. In contrast, people who had regular experience with economic exchange created new forms of correspondence operations that enabled them to keep track of the addition or subtraction of coins. These individuals would often establish correspondences between a series and a subseries of body parts so that they achieved an exact representation of the answer by keeping a running record of the progressive addition or subtraction of body parts. For instance, to subtract 9 from 16 without coins, a typical strategy would be to "count-on" from the shoulder (10) to the ear-on-the-other-side (16) by using a subseries of the terms: thumb (1), index finger (2), middle finger (3), . . . forearm (7). Thus, these people would achieve a precise numerical representation ("forearm" or 7) for the product of the subtraction. Similarly, to add 7 + 9, an individual would call the shoulder (10), the thumb (1), the neck (11), the index finger (2), the ear (12), the middle finger (3), and so forth until a correspondence was reached between the ear-on-the-other-side (16) and the forearm (7).

An inherent feature of the transformation in number representation among the Oksapmin is a change in each of the core properties. First, Oksapmin are reorganizing their traditional "material" of number representation (the body system) to solve more adequately numerical problems that arise with participation in the money economy (Property 1). For instance, body parts are shifting in their function from being signifiers of solely real-world objects to means of representing numerical values in the addition or subtraction of values. Moreover, the transformation is inherently linked to Oksapmins' elaboration of logico-mathematical structures of correspondence (Property 2), as discussed in the "counting-on" strategy. For instance, correspondence operations are now constructed between a series and subseries of body parts in order to achieve problem-solving ends. Finally, though the novel strategies are being elaborated in the context of particular exchanges with currency, the strategies are general ones that the Oksapmin use to mediate their thinking about the addition and subtraction of currency (Property 3).

The conclusion of the present review is that the formation of mathematical concepts is a developmental process simultaneously rooted in the constructive activities of the individual and in social life. If we are to devise comprehensive models of number development, we must build into these models an account of the way these factors are interwoven with one another in the developmental process. To date, our conceptual models and empirical research have not systematically addressed this fundamental concern. Either the cultural context of problem solving is the focus of study and research does not inform our understanding of structural developmental change, or culture is unanalyzed in cross-cultural research, and the focus is on the universal development of cognitive structures. It is our expectation that future work will present fresh approaches.

ACKNOWLEDGMENT

Appreciation is extended to Joseph Glick whose insights about some of the issues discussed in this chapter provided valuable direction.

REFERENCES

Adjei, K. Influence of specific maternal occupation and behavior on Piagetian cognitive development. In P. R. Dasen (Ed.), *Piagetian psychology: Cross-cultural contributions.* New York: Gardner, 1977.

Armstrong, R. G. *Yoruba numerals.* Ibadan: Oxford University Press, 1962.

Ashton, P. Cross-cultural Piagetian research: An experimental perspective. *Harvard Educational Review,* 1975, *45,* 475–506.

Carlson, J. Cross-cultural Piagetian research: What can it tell us? In K. Riegel & J. Meacham (Eds.), *The developing individual in a changing world.* The Hague: Mouton, 1976.

Cheetam, B. Counting and number in Huli. *Papua New Guinea Journal of Education,* 1978, *14,* 16–27.

Cole, M., Gay, J., Glick, J., & Sharp, D. W. *The cultural context of learning and thinking.* New York: Basic, 1971.

Cole, M., & Scribner, S. *Culture and thought.* New York: Wiley, 1974.

Dasen, P. R. Cross-cultural Piagetian research: A summary. *Journal of Cross-Cultural Psychology,* 1972, *3,* 23–40.

Dasen, P. R. (Ed.) Introduction. In P. R. Dasen (Ed.), *Piagetian psychology: Cross-cultural contributions.* New York: Gardner, 1977. (a)

Dasen, P. R. Cross-cultural cognitive development: The cultural aspects of Piaget's theory. In L. Adler (Ed.) *Issues in Cross-Cultural Research Annals of the New York Academy of Science,* 1977. (b)

Dasen, P. R., Lavallee, M., & Retschitzki, J. Training conservation of quantity (liquids) in West African (Baoule) children. *International Journal of Psychology,* 1979, *14,* 57–68.

Dasen, P. R., Ngini, L., & Lavallee, M. Cross-cultural training studies of concrete operations. In L.

Eckensberger, Y. Poortinga, & W. Lonner (Eds.), *Cross-cultural contributions to psychology.* Amsterdam: Swets & Zeitlinger, 1979.

Etuk, E. *The development of number concepts: An examination of Piaget's theory with Yoruba-speaking Nigerian children.* Unpublished doctoral dissertation, Columbia Teachers College, 1967.

Flavell, J. H., & Wohlwill, J. F. Formal and functional aspects of cognitive development. In D. Elkind & J. H. Flavell (Eds.), *Studies in cognitive development: Essays in honor of Jean Piaget.* New York: Oxford University Press, 1969.

Gay, J., & Cole, M. *The new mathematics and an old culture.* New York: Holt, 1967.

Gearhart, M., & Newman, D. Learning to draw a picture: The social context of an individual production. *Discourse Processes,* 1980, *3,* 169–184.

Gelman, R., & Gallistel, R. *The child's understanding of number.* Cambridge, Mass.: Harvard University Press, 1978.

Ginsburg, H. P., Posner, J. K., & Russell, R. L. The development of knowledge concerning written arithmetic. *International Journal of Psychology,* 1981, *16,* 13–34. (a)

Ginsburg, H. P., Posner, J. K., & Russell, R. The development of mental addition as a function of schooling and culture. *Journal of Cross-Cultural Psychology,* 1981, *12,* 163–178. (b)

Glick, J. *Piaget, Vygotsky, and Werner.* Paper presented at Heinz Werner Conference, Worcester, Massachusetts, 1981.

Greenfield, P., & Bruner, J. Culture and cognitive growth. In D. A. Goslin (Ed.), *Handbook of socialization theory and research.* New York: Rand McNally, 1969.

Harris, P., & Heelas, P. Cognitive processes and collective representations. *Archives of European Sociology,* 1979, *20,* 211–241.

Laboratory of Comparative Human Cognition. What's cultural about cross-cultural psychology? *Annual Review of Psychology,* 1979, *30,* 145–172.

Lancy, D. F. (Ed.) The indigenous mathematics project. *Papua New Guinea Journal of Education,* 1978, *14.* (special issue)

Lancy, D. F., Souviney, R. J., & Kada, V. Intra-cultural variation in cognitive development: Conservation of length among the Imbonggu. *International Journal of Behavioral Development,* in press.

Langer, J. *The origins of logic.* New York: Academic Press, 1980.

Lavallee, M., & Dasen, P. R. L'apprentissage de la notion d'inclusion de classes chez de jeunes enfants Baoules (Coted'Ivoire). *International Journal of Psychology,* 1980, *15*(1), 27–41.

Lave, J. Cognitive consequences of traditional apprenticeship training in West Africa. *Anthropology and Education Quarterly,* 1977, *8,* 177–180.

Menninger, K. *Number words and number symbols.* Cambridge, Mass.: MIT Press, 1969.

Moylan, T. *The social basis of language change in Oksapmin, Papua New Guinea.* Doctoral dissertation, City University of New York/Graduate Center. Forthcoming.

Opper, S. Concept development in Thai urban and rural children. In P. R. Dasen (Ed.), *Piagetian psychology: Cross-cultural contributions.* New York: Gardner, 1977.

Petitto, A. L. *Knowledge of arithmetic among schooled and unschooled African tailors and cloth-merchants.* Unpublished doctoral dissertation, Cornell University, 1979.

Petitto, A., & Ginsburg, H. Mental arithmetic in Africa and America: Strategies, principles and explanations. *International Journal of Psychology,* in press.

Piaget, J. *The child's conception of number.* New York: Norton, 1952.

Piaget, J. Nécessité et signification des recherches comparatives en psychologie genetique. *International Journal of Psychology,* 1966, *1,* 3–13.

Piaget, J. Quantification, conservation and nativism. *Science,* 1968, *162,* 976–979.

Piaget, J. Intellectual evolution from adolescence to adulthood. *Human Development,* 1972, *15,* 1–12.

Posner, J. The development of mathematical knowledge in two West African societies. *Child Development,* 1982, *53,* 200–208.

Posner, J. K., & Baroody, A. Number conservation in two West African Societies. *Journal of Cross-Cultural Psychology,* 1979, *10,* 479–496.

Price-Williams, D., Gordon, W., & Ramirez, M. Manipulation and conservation: A study of children from pottery-making families in Mexico. *Memorias del XI Congreso Interamericano de la Psicologia,* Mexico City, 1967, 106–121.

Priswerk, R. Jean Piaget et le relations inter-culturelles. *Revue Europeenne des Sciences Sociales,* 1976, *14,* 495–511.

Reed, H. J., & Lave, J. *Arithmetic as a tool for investigating relations between culture and cognition.* Unpublished manuscript, University of California, Irvine, n.d.

Saxe, G. B. Developmental relations between notational counting and number conservation. *Child Development,* 1979, *50,* 180–187. (a)

Saxe, G. B. A comparative analysis of the acquisition of numeration: Studies from Papua New Guinea. *The Quarterly Newsletter of the Laboratory of Comparative Human Cognition,* 1979, 37–43. (b)

Saxe, G. B. Number symbols and number operations. Their developments and interrelations. *Topics in Language Disorders,* 1981 (Dec.), 67–76. (a)

Saxe, G. B. Body parts as numerals: A developmental analysis of numeration among remote Oksapmin village populations in Papua New Guinea. *Child Development,* 1981, *52,* 306–316. (b)

Saxe, G. B. When fourth can precede second: A developmental analysis of an indigenous numeration system among Ponam Islanders in Papua New Guinea. *Journal of Cross-Cultural Psychology,* 1981, *12,* 37–50. (c)

Saxe, G. B. *Social change and cognitive growth: The invention of body part algorithms among Oksapmin children in Papua New Guinea.* Paper presented at the 1981 Bienniel Meeting of the Society for Child Development, Boston, Mass., 1981. (d)

Saxe, G. B. *Culture, counting, and number conservation.* The Graduate Center of The City University of New York, 1981. (e)

Saxe, G. B. Culture and the development of numerical cognition: Studies among the Oksapmin. In C. J. Brainerd (Ed.), *Children's logical and mathematical cognition.* New York: Springer-Verlag, 1982.

Saxe, G. B. Developing forms of arithmetic thought among the Oksapmin of Papua New Guinea. *Developmental Psychology,* in press.

Scribner, S., & Cole, M. Cognitive consequences of formal and informal education. *Science,* 1973, *182,* 553–559.

Smeltzer, D. *Man and number.* New York: Collier, 1958.

Steinberg, B. M., & Dunn, L. A. Conservation competence and performance in Chiapas. *Human Development,* 1976, *19,* 14–25.

Vygotsky, L. S. *Thought and language.* Cambridge, Mass.: MIT Press, 1962.

Vygotsky, L. S. *Mind in society: The development of higher psychological processes* (M. Cole, V. John-Steiner, S. Scribner, & E. Souberman, Eds.).Cambridge, Mass.: Harvard University Press, 1978.

Wertsch, J. V. From social interaction to higher psychological processes. A clarification and application of Vygotsky's theory. *Human Development,* 1979, *22*(1), 1–22.

Wood, D., Bruner, J. S., & Ross, G. The role of tutoring in problem solving. *Journal of Child Psychology and Psychiatry,* 1976, *17,* 89–100.

Wood, D., & Middleton, D. A study of assisted problem solving. *British Journal of Psychology,* 1975, *66,* 181–191.

Zaslavsky, C. *Africa counts.* Boston: Prindle, Weber & Schmidt, 1973.

BARBARA S. ALLARDICE
HERBERT P. GINSBURG

CHAPTER **8**

Children's Psychological Difficulties in Mathematics[1]

The aim of this chapter is to examine children's learning difficulties in elementary mathematics. After discussing the traditional concept of *learning disabilities,* we review research on neurological and environmental antecedents of learning problems. Next we describe investigations adopting a cognitive approach and conclude with some findings derived from our own case studies.

THE CONCEPT OF LEARNING DISABILITIES

Developmental psychologists have generally devoted little research attention to children's learning difficulties, although they are often encountered and treated in schools and clinics. The topic has been addressed mainly by workers in special education who, employing the concept of learning disabilities, have conducted much research on problems of reading and less on mathematics. Of course not all researchers in special education accept the concept of learning disability; in recent years it has come under serious criticism (e.g., Coles, 1978). Yet the notion of learning disability remains a popular one and therefore requires examination. The learning disability paradigm usually begins with a "defini-

[1]Preparation of this chapter was supported by a grant to Herbert P. Ginsburg from the National Institute of Education (NIE-G-78-0163).

319

tion.'' This involves a child (*a*) who is of normal intelligence, as defined by an IQ score (say 90 and above); (*b*) who is performing poorly in school as defined by scores on standard achievement tests (say two grade levels below his age group); (*c*) who is characterized by the absence of severe emotional problems; and (*d*) who has no sensory impairment such as blindness or deafness. The definition is intended to eliminate from consideration retarded and psychotic children, so that one can focus on children of normal intelligence who nevertheless fail to learn.

Next, the concept of learning disability assumes that children defined in this way suffer from a kind of disease entity, a unitary syndrome, which has a large neurological component, sometimes referred to as *minimal brain damage*. This learning disability syndrome then ''explains'' the child's poor performance in school. Why can't Johnny add? Because he ''has'' a learning disability, just as he might have a cancer.

Several points should be made concerning the concept of learning disabilities (LD). The definition of LD in terms of normal IQ and low academic achievements explains nothing. The definition is only a rough screening device—it eliminates from consideration some children, but does not explain the behavior of those remaining. In particular, the child's possession of normal IQ and low achievement test scores does not necessarily imply that he or she suffers from neurological deficit or minimal brain damage, nor that there is any kind of unitary disease syndrome causing the difficulties. The existence of brain damage and its presumed causative role must be independently assessed. We shall see next that, in fact, there is little evidence supporting a neurological view. In brief, the definition serves no explanatory function; it merely attempts to locate a group of children whose behavior needs explaining.

The group of children defined by the standard criteria as *learning disabled* have been shown to be extremely heterogeneous; the definition is ambiguous and in practice excludes few children of any type. LD may include emotionally disturbed children. Keogh, Major, Omori, Gandara, and Reid (1980), examining research reports, found that:

> LD subjects in the sample articles were frequently described as evidencing symptoms of frustration, aggression, inappropriate affect, poor social and/or school adjustment, and behavior disorders. They were also described as immature and irritable, and as having sleep disorders and emotional problems. . . . The high frequency of social–behavioral symptoms in these LD samples raises the question of the distinctness or independence of the various emotional, behavioral, and learning disability conditions [p. 24].

Owen, Adams, Forrest, Stoltz, and Fisher (1971) also find a high degree of emotional disturbance in their LD subjects. In these studies, it is not clear whether the observed emotional disturbance is cause, effect, or correlate.

LD may include normal children with temporary difficulties. Belmont and Belmont (1980), in a longitudinal study, found that children's achievement test scores fluctuate considerably over time. Many children may therefore be defined as learning disabled on one occasion but not another. Few children show a *consistent* LD pattern. "If children had been classified as learning disabled on the basis of failure at a given time, then the vast majority of sometimes poor performers (25% of the class) would have been mislabeled since they were quite adequate at other times [p. 497]."

Diagnoses of LD are often made on the basis of poor or incomplete evidence. Adelman (1978) has shown that "of 208 school-labeled learning disabled children, 37% did not meet a criteria [sic] of normal intellectual ability [p. 718]-." Furthermore, in another sample, "only one out of fifteen diagnoses resulted from a consistent pattern of data from tests. . . . Nine of fifteen diagnoses were not based on any test findings. . . . Five of fifteen diagnoses were arrived at despite contradictory evidence [pp. 720–721]."

Children defined as LD by the standard criteria may nevertheless exhibit different underlying cognitive processes. There are numerous combinations of skills and weaknesses that could combine to give a child an IQ test score of a "normal" child and the achievement test scores of an underachiever. For example, a third-grade student taking a math achievement test may receive a below-grade score because he or she hasn't been introduced to multiplication and therefore leaves multiplication questions blank. A second child can do multiplication by finger counting. However, the child will not have a chance to complete the test items because finger counting is slow. The child knows more about math than he or she is permitted to demonstrate. A third child misses many problems because he or she ignores the operator signs and mistakenly does addition instead of multiplication. Although these children receive the same scores, their problems are different.

In view of the heterogeneity of the LD population and the fact that diagnoses are often arrived at in a capricious fashion, one can only conclude that the LD definition is mainly a *political* device, not a scientific one. As Farnham-Diggory (1980) put it:

> In the public sector, learning disability is not a clinical entity; it is a budget category used to channel funds into local educational programs. The category is intended to supplement budget categories for children who are retarded, socially or emotionally disturbed. . . . What did not exist for a long time was a category for children who were none of the above, but who still had difficulties . . . severe enough to warrant special educational attention. The category *learning disabled* was invented to meet that need [pp. 571–572].

So LD explains nothing, includes a heterogeneous group of children, and is often determined in an inconsistent manner: Its main purpose is to channel into the schools money for children who might need special help. Yet, as we shall see next, the effort is often counter-productive.

The concept of learning disabilities may serve as yet another means of "blaming the victim." Labeling the child as learning disabled in effect places the blame within the child; it asserts that he or she possesses some defect that is responsible for the observed problem; the fault is within the person (Coles, 1978). Such an approach ignores the possibility that the child's difficulties may often result in good measure from deficiencies in the system that is supposed to train him or her, and thus may be counter-productive.

The label learning disability implies some kind of defect in the *learning* process itself. Yet to our knowledge, this has never been demonstrated. The typical research strategy involves the examination of *performance* failures, not the direct investigation of *potential for learning*. The label learning disability is fundamentally misleading as to the nature of the deficit involved.

In view of all these considerations, we take the following approach. We believe that standard tests (IQ and achievement) can be used as a rough device to screen out retarded and severely emotionally disturbed children and to identify a broad class of "normal" children performing poorly in school. To identify these children as learning disabled is misleading. It is not clear that there is anything wrong with their learning. Also the term may bring with it a lot of unnecessary and indeed dangerous theoretical baggage including neurological deficit, unitary disease syndromes, and blaming the victim. Hence, we refer to these children— to normal children performing poorly in school—as experiencing psychological difficulties. We intend no implication to the effect that the "fault" is the child's; the psychological difficulties may stem in large measure from poor instruction. At the outset we know only that the children with whom we are concerned are not retarded or severely emotionally disturbed and that they are not performing at a normal level in school. The research task is to discover the reasons for school failure in a group that is most likely extremely heterogeneous. It is conceivable that, within this group, some children fail because they are lazy, others because they suffer from neurological problems, others because they cannot learn, others because they suffer from thought disorders, and still others because they have been taught badly.

Consider now research that focuses on various of these possibilities. We begin with studies examining neurological and environmental "causes."

NEUROLOGICAL AND ENVIRONMENTAL CAUSES

Some studies assume that learning disabilities, as defined in the traditional manner, are "caused" by neurological defects. Other studies, taking a radically different point of view, are based on the notion that poor performance in mathematics, as in other academic subjects, is caused by a "deprived environment." Consider first the neurologically oriented research.

Neurological Studies

We have identified three studies that attempt to relate neurological deficits to mathematics learning problems. Cohn's (1971) view is that mathematics learning disabilities represent a special case of language dysfunction, which itself results from "incoordination at various complexities of neurological behavior [p. 388]." Cohn conducted case studies of some 31 children referred to a clinic because of severe learning problems and other abnormalities. Cohn obtained neurological data consisting of EEGs, measures of motor behavior, measures of orientation, etc., on the subjects and from these data compiled a neurological index for each subject. The combination of EEGs and the various "soft signs" yielded an estimate of neurological damage in the children under study. Cohn also obtained samples of reading, writing, written arithmetic, and speech functioning. The subjects were seen several times over a period of years.

Cohn reported that improvement over time in the neurological index was directly related to improvement in the various language activities that he defines broadly to include arithmetic. Thus, he concluded that disturbances of organization characterized the learning disabled child and accounted for the various disruptions in the developmental course of language functioning, including arithmetic behavior.

Unfortunately, Cohn does not present his data in a systematic fashion, so that it is difficult to evaluate his argument. For example, Cohn does not make clear how the neurological index is computed, nor how improvements in the index are determined. In a similar fashion, he does not make clear how the level of mathematics functioning is determined, nor how improvements in it are assessed. Hence, it is obviously difficult for this research to demonstrate a relationship between neurological functioning and children's learning of mathematics. Second, Cohn's descriptions suggest that many children were characterized by low intellectual functioning and/or severe emotional disturbance. Thus, the children may not be typical of those usually termed learning disabled. Third, the study fails because Cohn lacks a coherent theory of the learning and understanding of mathematics. Consequently, Cohn fails to offer adequate interpretations of children's mathematical behavior. For example, he places great emphasis on the child's ability to obtain correct answers to written addition, subtraction, and multiplication problems. We learn nothing from his data about the reasoning that led to the responses. Cohn suggests at one point that learning disabled children ought to gain reasonable facility with arithmetic because, unlike the language activities, drill will suffice for learning to occur. Such a view ignores evidence that even children who appear to be performing in a rote, mechanical fashion are attempting to impose some sense on their mathematical activities [Erlwanger, 1975].

Kosc (1974) suggests that mathematics learning problems, which he terms *developmental dyscalculia,* result from "a genetic or congenital disorder of those parts of the brain that are the direct anatomico-physiological substrate of the maturation of the mathematical abilities [p. 165]." Kosc's study, done in Czechoslovakia, began with the screening of 374 children in the public schools. They were given a variety of mathematics tasks, such as adding columns of numbers arranged in a triangle, copying a complex figure, successively subtracting 7 from 100, and performing arithmetical reasoning (as measured by a test from the Terman-Merrill intelligence scale). On the basis of the screening procedure, 68 children were identified as possibly suffering from developmental dyscalculia. The children were then given various neurological tests, involving left–right and spatial orientation, finger agnosia, and laterality. Of the 68 children performing poorly on the mathematics tests, 24 also displayed the neurological "soft signs" (e.g., difficulty with left–right orientation, finger agnosia, etc.). Those subjects who performed most poorly on the screening tasks were more likely to display the neurological soft signs than were the children who did somewhat better. Kosc concludes from all this that neurological deficiencies are at the basis of developmental dyscalculia.

Unfortunately, Kosc's study suffers from several major weaknesses. First, Kosc generally presented his data and procedures in a sketchy fashion, so that they are hard to evaluate. He fails to compare his experimental subjects with a normal control group, with respect to the prevalence of the soft signs. Hence, we do not know whether the experimental subjects exhibit a frequency of soft signs significantly different from that expected in the normal population. Furthermore, from the soft signs it is difficult to make clear inferences concerning the existence of neurological deficits, and it is virtually impossible to make statements concerning abnormalities in specific neurological substrates. Third, like Cohn, Kosc fails to employ a sophisticated theory of children's mathematical thinking, so that we cannot be sure that low performance on Kosc's mathematical tasks indicates anything fundamental concerning difficulties in mathematics learning or understanding.

A third study, conducted by Weinstein (1978), attempts to relate mathematics learning problems to a developmental lag in certain left hemisphere processes. Weinstein proposes that the left hemisphere is more suited to handling the sequential, operational processing of mathematics tasks than is the holistic processing mode of the right hemisphere. Children whose left hemisphere functioning develops more slowly are thus likely to encounter lags in the development of mature mathematics behavior.

Weinstein's subjects were fifth- and seventh-graders of normal intelligence and reading ability but below average mathematical achievement. Control subjects were matched for IQ and reading ability, but were achieving normally in mathematics. On written calculations, place value tasks, counting and ranking of large numbers, story problems, and Piagetian tasks, the mathematically disabled subjects performed somewhat more poorly than the control subjects. The behavior of the disabled group resembled that of younger, normal children. Two of six soft sign neurological tasks—finger localization and left–right discrimination—also differentiated the groups. In a follow-up study, Weinstein specifically tested the hypothesis that a delay in the development of left hemisphere analytic processing is related to the poorer performance of the mathematically disabled subjects. She recorded subjects' eye movements as they responded to verbal (e.g., ''Name four famous scientists'') and spatial (e.g., (''Which way does the head of President Lincoln face on a penny?'') questions. This procedure purports to reveal which hemisphere is being used for processing the answer, with an eye movement to the right signifying left hemisphere processing and a movement to the left signifying right hemisphere processing. Both groups (math disabled and control) displayed more leftward eye movements, but the tendency was significantly greater in the math disabled children than in the controls. Weinstein concluded that the results supported the hypothesis that mathematics learning problems are caused by a lag in left hemisphere functioning, although she admitted that additional research is needed for developing a theory of the neurological basis for developmental dyscalculia.

We find Weinstein's hypothesis to be unconvincing because of the indirectness of the soft sign neurological measures and their undemonstrated relationship with actual central nervous system dysfunctioning. Furthermore, since there was a predominance of left hemisphere processing even in normal subjects, it is not in itself a problem in mathematics performance; perhaps some, unspecified, critical ratio of left to right hemispheric processing is important. What this ratio might be and how left hemisphere functioning relates to *specific* mathematics tasks are questions left unanswered in Weinstein's study.

The three studies reported fail on a number of grounds to support a direct relationship between neurological deficits and mathematics learning problems. First, the Cohn and Kosc studies are not based on a sound theory of children's mathematical behavior. Hence, the behavior elicited by their mathematical tasks may not measure anything significant concerning children's mathematical understanding. If measurements are not guided by sound theory, the resulting observations may be tangential to central aspects of mathematical understanding, and the results, even if statistically significant, may be of little conceptual significance. Second, all of the studies described attribute to neurological soft signs greater validity than they merit. There is little evidence suggesting that the soft signs are accurate indices of neurological function. In a recent review of the literature, Coles (1978) concludes that there is little research demonstrating a relationship between neurological soft signs (whatever their validity) and learning disabilities. Similarly, Owen *et al.* (1971) found that: ''Definitive damage to the central nervous system was not an important antecedent for learning disorders [p. i]." Also: "Of the 264 subjects examined, only 4 children were found to have definitive signs of neurological abnormality [p. 63]." Even in the occasional study where a positive relationship is found, the differences are too small to make useful clinical inferences. And, of course, even if a correlation between soft signs and learning disabilities is demonstrated, one cannot make conclusive inferences concerning causality. Obviously, severe damage to the brain might result in cognitive impairment, including disorders in mathematical thinking as Luriya (1969) has elegantly shown. Yet this in no way implies that difficulties in mathematics achievement stem from brain damage. In particular there is now little evidence to support the view that ordinary or even severe mathematics difficulties necessarily stem from minimal brain damage.

Environmental Causes

If the literature concerning neurological causes of learning disabilities is inconclusive, what is known concerning the role of environmental causes? Consider first studies concerning the "deprived environment" and social class. Work in this area does not stem from the learning disabilities tradition, but from certain positions in developmental psychology. J. McV. Hunt proposes that certain groups of children, especially the poor and the black, suffer from *cognitive deficits* that prevent adequate understanding of academic subject matter. According to Hunt's environmentalist theory, poor children are said to develop in a deprived environment, which stunts their intellectual growth and hence prevents them from attaining adequate levels of achievement in school. According to Hunt (1964), "Cultural deprivation may be seen as a failure to provide an opportunity for infants and young children to have experiences required for adequate development of those semiautonomous central control processes demanded for acquiring skill in the use of linguistic and mathematical symbols [p. 236]." In brief, a deprived environment is said to lead to generally deficient intellectual functioning, which in turn leads to the poor learning of academic knowledge, including mathematics.

An alternative approach (Cole & Bruner, 1971; Ginsburg, 1972) proposes that poor children possess fundamentally sound cognitive skills and exhibit a few cognitive differences; their school failure must be explained on grounds other than cognitive deficit. According to the *cognitive difference* view, the evidence suggesting a deficit is faulty. Typically, it derives from standard tests, which may be culturally biased. Adequate testing procedures would reveal cognitive strengths or differences, not deficits. Why then do these children do so badly in school? Ogbu (1978) suggests that many blacks perceive themselves as members of a caste who can realistically anticipate little in the way of economic opportunity. Hence, poor blacks and other lower class individuals have little incentive to work hard at academic studies. The problem is primarily economic and motivational, not cognitive.

In brief, cognitive deficit theory proposes that school failure—at least some subset of learning disabilities—can be explained on the basis of cogntive deficit stemming from a deprived environment. Cognitive difference theory denies that the deficit exists and proposes that school failure must be explained on entirely different grounds. Although these views have aroused considerable controversy, little research has been conducted on the fundamental issue of whether the cognitive deficit in fact exists. The question is whether poor children, who the evidence shows are extremely likely to fail in school, in fact suffer from a general cognitive deficit or from a deficit with respect to mathematical reasoning. Unfortunately, few relevant studies exist in the area of mathematical thinking. Kirk, Hunt, and Volkmar (1975) used tasks involving counting and enumeration to compare the numerical abilities of lower class and middle class children. In this study, lower and middle class 4-year-olds were asked to perform the following number tasks: (a) to construct arrays of blocks that were "the same" (numerically) as the examiner's, both with a model present and absent; (b) to produce an array, given a numerical command (e.g., "Put out three blocks."); and (c) to identify verbally the numerical size of presented arrays, which varied from one to six blocks. In general, the performance of lower class children was significantly poorer on all tasks. These results, then, seem to support the cognitive deficit position. However, we argue that the study is not conclusive. It deals with a small set of mathematical skills in a theoretically limited way. Most of the tasks required the child to be familiar with number names to six, and no investigation was made of number concepts that might have developed regardless of whether the child had been coached on number names. It is also unclear whether the children understood that a numerical response was required on the first two tasks. Finally the authors admit that the tasks were unpleasant and that the children tried to avoid them. M & M's were necessary to motivate the children. Thus, it is doubtful that the procedures were maximally effective in eliciting children's competence. It may well be that the observed social class differences were not so much a result of cognitive processes but of a willingness to comply with adult demands in an unpleasant situation or to "catch on" as to what is desired in the experimental situation.

Ginsburg and Russell (1981) have also conducted an investigation in this area. The aim of their study was to assess the "informal mathematical thinking" of black and white preschool children, both lower and middle class. According to the cultural deprivation view, lower class black children, because they have been raised in a deprived environment, should display serious weaknesses in informal mathematical thinking. Yet, according to Ginsburg and Russell, recent cross-cultural research suggests that many aspects of mathematical thinking are extremely widespread across cultures, if not universal. Hence, one might not expect lower class American blacks to display serious deficiencies in informal mathematical thinking.

In examining these issues, the investigators made use of a conceptual scheme developed by Ginsburg (1977). This view describes the development of two kinds of "informal mathematical knowledge." The notion of *informal knowledge* (or *spontaneous concepts*) has been stressed by several cognitive theorists, notably Piaget and Vygotsky. These writers point out that in their natural surroundings, preschool children spontaneously develop informal concepts and strategies, many of which are related to the academic knowledge taught in school. Thus, the preschool child develops elementary notions of mathematics, physics, causality, and the like. In Piaget's view, the child assimilates academic material into the already existing spontaneous concepts. The child's informal knowledge serves as a kind of cognitive underpinning or scaffolding for school learning.

Children's informal knowledge of mathematics may be described in terms of two systems operating concurrently in individual children. In the case of *system 1*—before entering school, or outside the context of formal education—children develop "intuitive" mathematical concepts and techniques for solving quantitative problems, especially those that do not demand a numerical response. For example, consider the *perception of more*, originally described by Binet (1969). Given two randomly arranged collections of blocks, a child of 3 or 4 years of age can easily determine which of the two collections has more than the other, at least when relatively small numbers of elements are involved. The child's speed of judgment (only a second or two is needed) indicates that counting is not necessary to solve the task. (The child seems to solve the task by employing such aspects of stimulus information as area covered and density of elements.) Since System 1 develops outside the formal school setting, it may be termed *informal*. Furthermore, since System 1 does not appear to involve counting or other specific information or techniques transmitted by culture, it may be termed *natural*.

In the case of *System 2*, the child during the preschool years also learns counting, a culturally derived tool, and uses it in the service of problem solving. Thus, Hebbeler (1977) has shown that preschoolers can use counting, sometimes on the fingers, to solve elementary problems of addition, usually when objects are involved. In this volume, Fuson and Hall (Chapter 2) describe the development of counting in the preschool years, and Resnick (Chapter 3) describes its use in school arithmetic. System 2 is *informal* insofar as it develops outside the context of schooling; but it is *cultural* since it depends on social transmission (by adults, television, books, etc.).

Finally, in our culture the child goes to school and begins the formal study of mathematics. *System 3*, mathematical thinking, then results from this formal tuition as it is assimilated into the child's already existing mathematics knowledge (Systems 1 and 2). Academic knowledge is the combined result of teaching and already existing concepts (as also stressed by Resnick in Chapter 3). System 3 may be described as both *formal* and *cultural* in nature.

Given this theoretical framework, Ginsburg and Russell asked the following questions: Do children who are likely to experience difficulty in school—especially lower class and black children—possess the informal knowledge (defined in terms of System 1 and System 2 skills) that could serve as an adequate basis for learning school arithmetic? Or, conversely, can school failure be explained by weak informal knowledge of mathematics or by spontaneous concepts that are not sufficiently powerful to assimilate what is taught in school? The investigators hypothesized that System 1 and System 2 skills are extremely widespread and robust and are exhibited by the vast majority of young children, regardless of social class, race, or culture. If this is so, then virtually all children may be said to possess a naturally occurring—even biologically based—set of cognitive skills that could serve as the basis for later mathematical learning. Evidence to this effect would contradict cognitive deficit theories.

Ginsburg and Russell's subjects were lower and middle class, black and white, preschool and kindergarten children from the Washington, D.C. and Baltimore metropolitan area. Most of the lower class black children resided in inner city areas of Washington D.C. Each child, seen individually, was given a large number of mathematical tasks (17 in all), measuring various aspects of Systems 1 and 2 knowledge. The tasks were administered in several sessions by interviewers who were familiar with the children. The subjects seemed to enjoy the tasks and made no efforts to leave the testing situation.

System 1 tasks included Binet's *perception of more* task, as just described, and Piaget's conservation of number. System 2 tasks included addition story problems, involving small quantities. Some problems were given with objects present (e.g., pennies in the case of, "Teddy Bear had two pennies, and he found another one. How many does he have now?"), whereas other stories did not provide the relevant objects.

The results were extensive, and only a few main features can be summarized here. In general, in the vast majority of cases, Ginsburg and Russell found no social class differences and at most statistically insignificant trends favoring middle class over lower class children. In the vast majority of cases, children of both social classes demonstrated competence on the various tasks and similar strategies for solving them. If these competencies and strategies were not evident at the preschool level, they emerged by kindergarten age. Furthermore, the research shows fewer racial than social class differences (which themselves were few). In general, the study demonstrated that basic mathematical thought develops in a robust manner among lower and middle class children, black and white. School failure cannot be explained by an initial deficit in basic cognitive skills, specifically System 1 and 2 mathematical skills. Instead, poor children display important cognitive strengths, developed "spontaneously" before the onset of schooling. These strengths should provide a sufficient foundation for later understanding of school mathematics. The results also stressed the need for employing flexible techniques designed to put children at ease and elicit their genuine competence.

We may conclude that a deprived environment does not necessarily produce children deficient in certain basic intellectual skills. Of course, we do not argue that a deprived environment—specifically, lower class poverty—is beneficial for those growing up in it, nor that it exerts no effects on psychological functioning. We maintain only that a simple causal relation between a deprived environment and deficiencies in informal mathematical thinking has not been established.

Educational Factors

Curiously, workers in the area of learning disabilities have failed to stress a very plausible kind of environmental influence that may exert deleterious effects on children's mathematical functioning. We refer to schooling. Can there be any doubt that poor educational practices are responsible for some children's learning difficulties? Ordinary observation in the classroom quickly reveals a variety of poor educational practices: teaching in the same manner to children with varied needs; the failure to connect new material to what children already understand; teaching some children too quickly and others too slowly; and using textbooks that are confusing and mathematically inaccurate. Furthermore, it is well known that many elementary school teachers are themselves uncomfortable with mathematics and feel that they can do a better job at teaching reading than arithmetic. The failure of mathematics education in this country is obvious; no research is needed to demonstrate that it is a major cause of children's learning difficulties.

Conclusion

Studies attempting to demonstrate a neurological basis for mathematics learning difficulties are inconclusive for several reasons. First, they employ poor measures of neurological dysfunction. Second, they generally fail to employ sophisticated theories or measures of mathematical functioning. If the measurement of the "effect" is inadequate, how can one conduct a legitimate study of the "cause?" Studies of environmental causes fail to show that a deprived environment results in serious deficiencies in informal mathematical knowledge (although it may have other effects). Lower class, black children appear to enter school with the basic informal knowledge necessary for success in learning school mathematics. The most obvious environmental cause of poor mathematics achievement is schooling that is especially inadequate in the case of mathematics.

COGNITIVE ANALYSES

In this section we examine studies employing a cognitive approach to the analysis of low mathematics achievement. Consider first the nature of such analyses.

The Nature of Cognitive Analyses

We begin with a population of children, screened by IQ tests and standard achievement measures, whose failure in school mathematics needs to be explained. A cognitive approach, as opposed to a learning disabilities approach, proceeds in the following manner. It attempts to analyze the cognitive processes responsible for the child's immediate behavior, specifically his or her poor performance on mathematical tasks. The cognitive analysis focuses not on product—that is, the child's test scores or answers to problems—but on the cognitive processes underlying the behavior. The product, the behavior itself, is of interest primarily insofar as it sheds light on underlying cognitive processes. The cognitive analyses can be done at several different levels: general cognition, mathematical processes, and metacognition. Consider how these various levels of analysis might work in a particular case, namely to explain Johnny's failure to add, as when, given 2 + 2, he responds "5." In this case, we might propose that:

1. Johnny cannot reason about or remember anything in general (a general cognitive explanation).
2. He has no intuitive concept of addition (System 1).
3. He lacks counting skills necessary for addition (System 2).
4. He does not remember the addition facts as taught in school (System 3).
5. He lacks skills for organizing things to be remembered (the metacognitive explanation).

These are all conceivably legitimate cognitive explanations, operating on different levels. Cognitive factors are *immediate* causes. They refer to generative events immediately preceding the behavior in question. Such cognitive causes may themselves have a varied etiology. For example, the child may lack an intuitive concept of addition (System 1) because of (*a*) brain damage, (*b*) a deprived environment, (*c*) poor teaching, or (*d*) emotional conflict. The cognitive perspective is neutral with respect to etiology. All of the causes described, alone or in combination, may exert some influence. In the cognitive perspective, it is premature to begin research with a consideration of long-range causes, like a deprived environment; instead, it is first necessary to achieve an understanding of current cognitive processes, so that one can determine what the putative long-range causes are causes of.

A focus on cognitive processes is necessary because they mediate the workings of the various causative factors described. Neurological factors exert their influence through cognitive processes; a deprived environment may lead to dysfunction in cognitive processes; inadequate instruction may produce disordered knowledge systems; emotional conflict may result in bizarre modes of thought. Hence, cognitive analysis is central. Furthermore, disordered cognition may *result in* emotional conflict. A child who knows that his or her learning is inadequate may experience anxiety, which in turn may further disrupt cognitive processes. In brief, while different types of explanation and causal analysis are legitimate, cognitive factors are always involved one way or another, either as mediating links for other causes or as fundamental factors in themselves.

We now examine cognitive analyses on several levels, beginning with Piagetian structures.

Piagetian Structures

To what extent does the child's failure in school mathematics result from cognitive deficiencies, that is, disorders in the basic structures of thinking described by Piaget? According to this possible explanation, the learning of school mathematics may be guided by underlying cognitive structures like the concrete operations or formal operations. These structures are not task-specific, but rather regulate behavior across a range of tasks. The child lacking or slow in certain underlying structures—for example, the concrete operations—may find it impossible or difficult to learn certain branches of mathematics.

We will consider two types of evidence concerning the way in which Piagetian structures may be implicated in mathematics difficulties. First, we consider evidence relating to the issue of whether children limited to preoperational reasoning (that is, who have not attained the stage of concrete operations) are capable of knowing and using certain mathematical ideas. Second, we consider evidence concerning the relationship between conservation skills and mathematics learning.

The first group of studies asks whether preoperational children are capable of certain types of mathematical thinking. For example, in the Ginsburg and Russell (1981) study described earlier, involving black and white children from various social classes, it was found that preschool children generally failed conservation of number tasks, but nevertheless were capable of various mathematical activities, like solving elementary addition and subtraction problems.

Many other studies of preschool children, although not directly concerned with Piagetian stages, also show that preschool children, sometimes as young as 2 and 3 years of age, display unsuspected strengths with respect to early mathematical thinking. Thus, Gelman and Gallistel (1978) show that very young children—who are most unlikely to be in the stage of concrete operations—employ certain general mathematical principles to guide their counting activity. It is abundantly clear that young preoperational children do indeed possess mathematical skills: The concrete operations are not *necessary* for at least simple forms of mathematical thinking.

Can the same point be made concerning somewhat older children at the kindergarten level? Some insight into this issue is provided by Mpiangu and Gentile (1975) who investigated preoperational kindergarten children's ability to learn mathematics. Mpiangu and Gentile conducted a training study, reasoning that if number conservation skills were necessary for mathematics understanding, then attempts to teach new number skills to nonconservers would be unsuccessful, whereas conservers would benefit from the instruction. Their subjects were 116 kindergarten children who were given eight conservation of number tasks and also an arithmetic screening task to assure that they were not already skilled on the concepts to be taught. Assignment to training or control conditions was random. Training group subjects received ten 20-min sessions in small groups where they were taught counting, including counting by two's; number recognition; relations (before, just before, after, just after, between); number synthesis, and analysis (finding the correct answer and justifying small number addition and subtraction items). The control group spent ten 15-min sessions playing a game with picture cards. The authors reported that all experimental group children, regardless of status on conservation tasks, benefitted from the instruction and outperformed the control group. There was no experimental-control group interaction effect between conservation status and post-training arithmetic scores. This suggests that subjects low in conservation skills benefitted from instruction as much as those high on conservation skills. The authors conclude that conservation of number is not crucial for early mathematics training. Of course, the generality of such studies is limited by the age and nature of subjects, as well as the type of instruction attempted. The Mpiangu and Gentile results do not speak to the question of whether older children or children with mathematics difficulties, both at the preoperational stage, can profit from instruction in more difficult types of mathematics than that taught in the Mpiangu and Gentile study.

In brief, we can conclude from the evidence considered to this point that concrete operational thinking is not *necessary* for at least simple types of mathematics learning in young children.

A second group of studies deals with the relationship between conservation skills and mathematics learning. Several studies (e.g., Hood, 1962; Wheatley,

1970) have reported positive relationships between attainment of conservation skills and mathematics achievement in the early school grades. For example, in the Mpiangu and Gentile study just described, there was a positive correlation between conservation score and arithmetic ability on both the pre- and post-training arithmetic tests. Subjects doing better at conservation likewise scored higher on arithmetic screening tests. Similarly, in the Weinstein study (1978) discussed earlier, there was a positive correlation between the attainment of advanced conservation skills (e.g., volume and area) and mathematics achievement in older children. Weinstein's learning disabled subjects performed more poorly than the normal controls on area and volume conservation (although all of the subjects could conserve number). Thus, slower development of mathematics concepts was accompanied by slower development of complex conservation skills. Perhaps the learning disabled subjects were experiencing difficulty with elementary aspects of formal operations, and therefore with school mathematics.

Generally, the studies purporting to demonstrate a relationship between the development of conservation and mathematics skills have used correlational designs. This research procedure leaves open the possibility that some underlying factor is responsible for the development of both conservation and mathematics skills and that the skills are unrelated but simply develop concurrently. To settle issues of causality, one requires more direct forms of evidence, such as that provided by training studies like Mpiangu and Gentile's.

In this section we have seen that preoperational children can develop mathematical concepts and that certain concepts can be taught in the absence of concrete operational thinking. Conservation is not *necessary* for at least some aspects of mathematics learning. Furthermore, there appears to be a correlation between success on conservation tasks and mathematics achievement, but the mechanisms underlying this relationship—the role of Piagetian structures in mathematics learning—have not been specified.

Mathematical Processes

A second form of cognitive analysis looks at the development of informal and formal systems of mathematical knowledge. Consider the possibility that elementary school mathematics failure might result from retardations in informal knowledge (Systems 1 and 2), those skills that provide a cognitive underpinning for school arithmetic. This may be termed a *developmental lag*. Alternately, these informal skills may develop adequately and difficulties in mathematics performance might be related only to the cognitive processes directly involved in formal mathematics.

These possibilities were investigated by Russell and Ginsburg (1981). The subjects were 27 children in each of three groups: fourth-grade children of normal intelligence who were achieving at least 1 year below grade level in mathematics, as measured by a standardized achievement test; fourth-graders matched in IQ but achieving at grade level in mathematics; and randomly selected third-graders (since test data were not available for them). Each child was seen individually for three sessions and was given a large number of tasks. Since the results are extensive, we will describe only a few representative findings here.

First, consider data bearing on the extent of informal knowledge, specifically *mental addition*. Each child was asked to solve eight addition problems without using paper and pencil. The sums ranged from under 20 to over 100. The third-grade (TG) and math difficulty (MD) groups were moderately accurate with mean scores of about 5 out of a possible 8. The fourth-grade control (C) group, with a mean of over 6, was significantly more accurate than the MD group. As one might expect, children had more difficulty with larger problems. Note the similarity between the TG and MD groups, and the fact that the C group was not much different from the others.

Considering the strategies used to solve these addition problems, we find first that the proportional use of the various strategies (mental algorithm, counting on, regrouping, etc.) is remarkably similar among all three groups. Second, among all three groups there is an adaptive deployment of strategies for different sized sums. Counting on is used for about half or more of the small sums and declines in use until it is never in use for sums over 100. The mental algorithm (that is, picturing in one's head the usual written algorithm) is used for a minority of the sums under 50 but increases in use to nearly half or more of the sums over 50. We also find that whereas the MD group selected adaptive strategies on most problems, these children executed the strategies with less accuracy than the C group, but with accuracy similar to the TG group.

Consider next data bearing on the issue of whether MD children display cognitive disorders—that is, unusual patterns of thought. Subjects were examined on *the use of principles,* an "insight" task. Each child was presented with addition and subtraction problems in which commutativity of addition or the reciprocity of addition and subtraction could be used as a shortcut to finding the answer. There were two problems involving commutativity and two involving reciprocity. Most of the children in all three groups used commutativity to solve at least one problem. Reciprocity was much harder and the spread among the groups was greater; yet even here, a majority of the MD group used reciprocity at least once. Thus, we see that a majority of the MD group had knowledge of simple mathematical principles—commutativity and reciprocity—and were able to use it to shortcut the process of problem solving. This knowledge and behavior may be considered examples of insightful problem solving (Wertheimer, 1954). Use of principles is highly "intelligent" and demonstrates that, at least in this case, MD children do not exhibit bizarre modes of thought.

In brief, with respect to mental addition and use of principles, MD children behave in sensible ways, show no sign of cognitive disorders or bizarre thinking, use strategies quite similar to those employed by normal children, but do so somewhat less efficiently.

Although displaying some relatively impressive skills, MD children exhibited weaknesses in other areas. One weakness involved *written calculation*. The MD children made more errors than C subjects and about the same number as TG. The strategies underlying the errors—for example, faulty counting and algorithms—were quite similar for the MD and TG groups. Another difficulty involved elementary *addition facts*. Each child was asked to give quick answers to 10 addition fact problems with sums under 20. The MD group knew somewhat fewer facts than the TG group and substantially fewer than the C group. Over 50% of the MD group knew only 4 or fewer such facts, whereas the percentages for the TG and C groups were 44 and 15, respectively. Clearly then the MD children exhibit a serious problem with respect to memory of elementary number facts. Does this signify some kind of memory disorder? The matter requires further investigation.

Consider now the question of whether MD children suffer from a developmental lag in informal mathematical knowledge.[2] The data show that, in general, MD children's *performance* levels on many tasks—for example, the number of correct responses on mental addition—were more similar to those of younger children than same age peers. Of course, this developmental lag in *performance* should come as no surprise since by definition MD children score below the norm on standard tests. The MD children also exhibited a developmental lag in the *strategies* underlying written calculation: MD and TG children made similar numbers of mistakes and used similar strategies to produce the mistakes. Thus, the typical MD child not only gets wrong answers on written computational problems, as one might expect, but he or she also arrives at these wrong answers through common error strategies—*bugs* exhibited by younger children.

Do MD children exhibit a similar lag with respect to *informal* mathematical concepts and strategies? Not necessarily. On at least some tasks—for example, mental addition—MD children use the same strategies as normal peers, but do so with less accuracy. On other tasks—for example, use of principles—MD children use the same strategies (shortcuts via principles) as same-age peers, but do so less frequently. There is little evidence of a lag in informal concepts. Even though their *performance* is retarded, MD children's mathematical cognition—particularly *informal* concepts and strategies—appears to be essentially "normal."

---

[2]Note that the issue under investigation is not whether the developmental lag involves a retardation in underlying cognitive processes of the type described by Piaget. The Russell and Ginsburg data do not bear on this issue since their study did not include measures of Piagetian operations or other basic cognitive processes.

We have seen then that MD children lag behind in performance, but not necessarily in mathematical cognition, particularly informal skills and concepts. Does this mean that there is nothing distinctive about MD children? Russell and Ginsburg show that there is at least one area—number facts—where MD children seem to experience unusual difficulty. The reason for this is as yet unknown. It is conceivable that some kind of memory disorder is responsible for the low level of performance.

METACOGNITION

Cognitive analyses of mathematics learning may also proceed by asking about the role of *executive processes* in mathematics difficulties. School learning requires an active learner who uses executive strategies to control the processes of thinking, remembering, and understanding. Lacking or failing to use these executive strategies, children may fail at various academic tasks, including mathematics. Their failure may lead the observer to the erroneous conclusion that they lack skills and conceptual knowledge. In reality, however, the problem may be one of performance failure stemming from lack or poor use of executive processes.

Some memory research with mentally retarded groups illustrates this point. Generally, we observe that the mentally retarded do not recall information with the proficiency of normal children or adults. In a review of training studies, Robinson and Robinson (1976) report that when explicit metacognitive training for a particular memory task is given to mentally retarded subjects, their performance equals that of normal controls. Yet this training appears to be highly task-specific with little transfer to related tasks. The mentally retarded subjects do not appear to learn how to employ the metacognitive strategies in new situations. Furthermore, there are even some conditions under which the mentally retarded and normal control groups do not differ in memory. These include situations where no active metacognitive strategies are required and long-term memory situations where the original knowledge of mentally retarded individuals equals that of normal controls.

Torgesen suggests that similar failures to use metacognitive strategies may account for difficulties encountered by children who are *reading disabled*. For example, second-grade reading disabled children were observed to rehearse less than normal controls in a task requiring recall, over a 15-sec interval, of a sequence of pictures (Torgesen & Goldman, 1977). The learning disabled children also recalled significantly less accurately than did the control group. Yet, when training was given in the use of rehearsal strategies, the differences in recall between the two groups disappeared. Similar findings emphasizing the importance of metacognitive strategies have been reported for older, learning disabled children. For example, when instructed to use category organization to assist in memorization, learning disabled children showed memory performance identical to that of normal children (Torgesen, Murphy, & Ivey, 1979). Learning disabled children do not seem able to apply metacognitive strategies to tasks when it would be appropriate to do so.

We are not aware of studies that deal specifically with the role of metacognition in children's mathematics learning. However, the Russell and Ginsburg findings that MD fourth-graders have relatively poor knowledge of number facts may suggest a failure in metacognitive processes. Thus, at least some of the MD children may have been unaware that there are things one can do to remember number facts.

Overview

Very little is known about cognition and MD children. It has *not* been demonstrated that they are seriously disabled in Piagetian thought structures, nor that these structures play a major role in the learning of school arithmetic. Indeed, research suggests that these children possess reasonably adequate informal concepts of mathematics; there is no developmental lag in this area. Perhaps MD children suffer from particular difficulties in knowing the number facts; the matter requires further study. Apparently there have been almost no studies of metacognition in MD children. The limited amount of available data may be interpreted as showing that the majority of MD children are essentially "normal" from the cognitive point of view. Presumably, factors other than cognitive ones must be invoked to explain school failure in these children. This of course is not to deny that some small subgroup of MD children may be characterized by genuine cognitive and/or neurological problems.

COGNITIVE ANALYSES: CASE STUDIES

Thus far, we have considered group studies offering cognitive analyses of mathematics difficulties. We turn now to case studies employing clinical interviews. Clinical interviews have certain advantages that make them a desirable tool for investigating cognitive aspects of mathematics difficulties. First, they permit a flexibility that generally is not present in the standard procedures used in traditional experimental design. This flexibility may be highly useful for identifying the cognitive processes guiding the performance of a particular child and for assessing *competence* (see Chapter 1 by Ginsburg, Kossan, Schwartz, and Swanson). Second, in a case study, the interviewer observes the child over time, typically ranging from several weeks to several months. This provides a depth of knowledge usually unavailable in experimental studies. In the following section, we offer case study analyses of mathematics difficulties.

Piagetian Structures

Group studies demonstrate a correlation between concrete operations and mathematics achievement, but do not shed light on the cognitive mechanisms underlying this relation. Case studies conducted by Lesh (1980) suggest however that there might be a relationship between particular aspects of concrete operations and certain specific mathematics skills. For example, a second-grader who exhibited difficulty in seriation—the Piagetian task dealing with the understanding of order relations—also had difficulty in distinguishing between numerals in different sequences (e.g., 537 from 573).

The age of the child exhibiting preoperational thought must also be considered in any attempt to establish a relationship between Piagetian structures and mathematics achievement. One of Lesh's case studies concerned a sixth-grade girl who was a good student in every subject except mathematics. She tested at the second-grade level in mathematics achievement and was described as having no intuitive feelings for the relative size of numbers. She failed several Piagetian conservation tasks. For example, when her teacher tested conservation of length by comparing two identical strips of paper, one of which was then divided into centimeters, the girl believed that the two were no longer of identical length. A sixth-grader of normal intelligence who lacks conservation skills may well be considered *deficient* in terms of the attainment of concrete operations; and this deficiency may prevent mastery of the sixth-grade mathematics curriculum. On the other hand, failure to master conservation skills may not be considered a *deficit* in kindergarten or first-grade students. Thus, Mpiangu and Gentile's findings that 5-year-old nonconservers can master grade level curriculum tasks may not be true of older nonconservers—who may be considered deficient in concrete operational thought—working with a more complex curriculum.

In summary, Lesh's case studies suggest some important considerations for understanding possible relationships between Piagetian cognitive structures and mathematics performance. First, statements about these relationships must be specific rather than general. The nature of failure on Piagetian tasks must be specified and related to specific mathematics task performance, rather than to mathematics achievement in general. Second, the relationship between the presence or absence of concrete operational skills and mathematics performance may be an age-related, rather than an *absolute* phenomenon. The lack of concrete operations may not hinder the development of many mathematics skills at an early age, but may indeed interfere with more complex forms of mathematics learning in older children.

Cognitive Factors in Mathematical Learning

We have discussed considerable evidence that informal mathematics skills develop routinely in children, regardless of social class or race and that they develop even in groups of children who are experiencing learning difficulties in mathematics. Clinical cases also support these findings. Even among children who are referred for special assessment and remediation, informal skills are well developed and may be used as the basic structures to which formal knowledge about place value and algorithms is assimilated. If informal skills are reasonably well developed, then what kinds of cognitive difficulties do we encounter in children who comprise a clinic population? In this section we look closely at attention and memory, often cited by teachers as causes of learning difficulties.

ATTENTIONAL FACTORS IN MATH LEARNING

Among clinic referrals, a sizeable group of elementary school children and even some junior high students are considered by their teachers to have "short attention spans," which are basic to their learning difficulties. Descriptive statements made by the teachers include: "never finishes work," "distractable," and simply, "short attention span." Our observations suggest that "attentional" difficulties in fact encompass several different types of phenomena, which we now describe.

Getting Started.   Beginning work on an assignment should be easy. The teacher has just demonstrated or explained what is wanted and now the child has to carry out the assignment. This routine is well practiced, occurring on a daily basis. Some children manage when the problems are already arranged on a ditto sheet but encounter great difficulty when they must copy problems from the book to the paper. Consider Kerry. She sat, doing nothing until the interviewer urged her to write the problem. Then her hand wandered over the page, the pencil touching down here and there, but never long enough to write the problem. Kerry simply did not know where to begin. Kerry also had other "getting-started" difficulties. She was not consistent in her starting place for finger or block counting and therefore frequently lost track of what had been counted. Similarly, Grace could never remember whether her name was to be written on the left or right side of the page and often sat wondering. She reported that the teacher's admonishment to get something down on her paper didn't help; she would be glad to, if only the teacher would say where.

It appears that getting-started difficulties are not uncommon. They explain why many assignments are not completed and why some children appear to be daydreaming or otherwise inattentive. Some attention difficulties may best be considered organizational problems: In a relatively unstructured situation the child is not able to start up and organize problem-solving efforts. Since the behavior does not take place, the child appears to be inattentive. Sometimes, getting-started difficulties may be overcome by explicit instruction. Thus, Kerry was helped to place an X on the paper where the first problem should go and to ask herself, "Where do I begin?"

Burdensome Procedures.   David was given the problem $8 + 4$. He said, "8 plus 8 is 16 . . . 15 . . . 14 . . . 13 . . . 12. The answer is 12." This indirect procedure placed great demands on David's attentional capacity. If someone called to him or there was a commotion, he lost his place and had to begin over again. It seemed to us that he was not generally more distractable than many children, but if he were distracted, the indirect procedure was thoroughly disrupted and he had to begin all over. We suggest that some attention difficulties are specifically tied to use of unnecessarily cumbersome procedures and are not general characteristics of the child.

Work That is Too Difficult.   It seems that nothing is more likely to produce apparently inattentive behavior than work that the child perceives as being beyond his comprehension. Sometimes the work *is* too difficult. It is meaningless to the child because the basic schemes necessary for comprehension are not present; such a child needs work at an appropriate level of difficulty. Other children may incorrectly estimate the difficulty of the task and need some help with the first problem to see that they do indeed know how to respond.

We have suggested that what are referred to as "attention" problems may be understood in terms of organization, burdensome procedures, and inappropriate difficulty levels. Attentional problems should not be thought of as an immutable characteristic of the child. Indeed, many of these problems can be remedied by simple educational interventions.

MEMORY

Memory defects are frequently named by teachers as the reason for math failure, and the frequency with which this problem is named seems to increase in the intermediate grades and junior high. Paul was typical of these students who are described as "knowing it one day and not the next." The interviewer spent some time reviewing fraction computations with Paul, at his request. He worked well with finding least common denominators, reducing, changing improper fractions, and with computations. However, the next week he could not recall many of the things he had previously done correctly. For example, he was uncertain whether common denominators were necessary in addition and whether the denominators were added. The interviewer then asked Paul about fractions in daily situations: Suppose you had two quarters of an apple; how else could you say it? Suppose you had eight candies; how many would there be in a quarter of the set? Paul demonstrated little knowledge concerning fractions. Thus, he had no intuitive basis for deciding whether denominators should be added in written computation. When we asked Paul, and others like him, about what is involved in learning mathematics they frequently cite the memorizing of algorithms as the central aspect. If you can remember the algorithm, you are "set." The notion that one can apply informal knowledge in an intelligent manner to assist in obtaining answers is an unfamiliar idea. Hence, algorithms are learned in rote, meaningless ways, and are easily forgotten. Paul can do a procedure one day but not the next because the procedure makes no sense.

In brief, we suggest that forgetfulness may frequently result from trying to do mathematics in a rote, meaningless manner. Were the conceptual framework made available, then forgetfulness would be reduced.

Metacognition

One aspect of metacognition is the deliberate use of strategies to facilitate learning. Such active learning is particularly helpful in the acquisition of the common number facts. Russell and Ginsburg (1981) showed that math difficulty students have particular problems in knowing the elementary number facts. In our clinic work, we also find that many of these students possess limited knowledge of the number facts and that this results in the necessity for extensive calculations. Without easily available number facts, students must engage in time-consuming and tiring calculations. Often these students do not complete their assignments and are exhausted by the effort. Three questions are of interest here: What do MD students understand about remembering number facts? Can number facts learning skills be taught? Do such skills help children in classroom work?

Consider the cases of Arnold and Ron. Arnold, a third-grader, was acknowledged to have general difficulties with mathematics. Ron, a second-grader, concerned his teacher because he had difficulties in knowing the number facts, but otherwise was an adequate mathematics student. When Ron was questioned about how one remembers number facts, he said that some people just seem to know them, some don't, and he was one who didn't. He seemed unaware that something could be done to learn the facts. When Arnold was asked the same question, he responded, "My teacher says just keep going over it." Arnold seemed to acknowledge the need for active drill. Although Arnold practiced with flash cards both at school and at home, "going over it" did not result in increased number fact knowledge. However, Arnold was stoic: Someday going over it would pay off. Thus, Ron did not seem aware of any strategy that could assist learning; Arnold opted for a strategy that did not seem to work.

Can an effective strategy be taught to such children? We attempted to instruct both Ron and Arnold, separately, in a principled method for producing the number facts. The method exploits the boys' memory of the "doubles" (e.g., $3 + 3 = 6$, $4 + 4 = 8$, etc.). Essentially, the child is shown that if he knows $4 + 4 = 8$, he can get the answer to $4 + 5$ by *reasoning* that it will be the answer to $4 + 4$ and one more. Note that the method does not involve *calculating* $4 + 5$ from scratch, but producing the answer, in a quick and efficient manner by reasoning from what is already known (the doubles). During this instruction, Ron was excited about the idea that he could do something to remember. We know that the teaching was successful for Ron because he could soon explain the method to his teacher and use it to remember various number facts. Indeed he devised similar methods for memorizing the facts that did not fall into the doubles plus or minus one category. Within 2 weeks, he knew his addition facts to sums of 16. By contrast, Arnold, more phlegmatic in nature, was less active during the instruction and had difficulty in using the reasoning procedure to produce number facts. A week after the initial instruction, Arnold could not remember the procedure for reasoning about number facts; indeed he could not recall any such lesson. He repeated his teacher's admonition to "keep going over it." When he was shown the work papers used in the previous lesson, he claimed that they did not look familiar. In the following weeks, attempts to help Arnold become an active memorizor proceeded very slowly.

Ron and Arnold represent extreme responses to the number facts instruction. Most children to whom we have demonstrated the method show some understanding in the first lesson and require several weeks before the method becomes truely functional for them. Why the extreme differences between Ron and Arnold? We can only speculate. Ron came to the task with apparently successful mathematics experiences and may well have assimilated the technique into a general expectation that mathematics is reasonable and systematic. Hence, for Ron, *reasoning* about mathematics was a sensible activity. Arnold, for whom math was more difficult, perhaps had none of these expectations. For him, mathematics was not a sensible activity, and therefore the method may have seemed quite foreign and unconnected to anything he may have experienced. For Arnold, "going over and over it" was the only acceptable approach. If so, the teaching of a reasoning method may have interfered with a procedure that Arnold was trying to use.

In summary, we see that some children like Ron lack an effective procedure for learning the number facts but can be taught a reasoning method that is extremely effective. Other MD children, like Arnold, attempt to learn the number facts by rote, drill procedures, which are unsuccessful. Yet teaching them a more effective reasoning procedure may not be easy. There are important individual differences in the approach to learning number facts.

EXTRICATION FROM DIFFICULTIES

Another aspect of metacognition is the ability to reflect upon one's work, particularly when one is in difficulty so as to extricate oneself from an unsuccessful problem-solving attempt. What do children do when they "get stuck?" When this occurs, we routinely ask: "What do you do to get the answer?" The following are some typical strategies.

Some children do not seem to analyze their problem-solving procedures at all: They simply say that they would ask the teacher. Of course, such passive appeals to authority are often encouraged by the schools.

Sometimes, children suggest that the problem could be solved by counting of some type, for example, on the fingers or with blocks. These children are in effect proposing an informal substitute for written calculations. Implicit in this suggestion is a recognition that the informal mathematics of counting is equivalent to the written algorithm of the school. Unfortunately, few mathematics difficulty children make this fruitful connection. Too often, they do not trust their own informal procedures, or do not see a relationship between them and school techniques, or are discouraged by the schools from using unorthodox techniques.

Very seldom do we observe math difficulty children engaging in a successful analysis of their own unsuccessful attempts at written calculation. Seldom can these children identify where their procedures have gone wrong, and seldom can they suggest ways of getting out of their current difficulties. The lack of self-reflection and analysis in these cases is striking. So far, the only effective technique we have developed to help such children is the demonstration of a successful solution *in progress*. Watching a successful computation being carried out or a word problem being transformed into numerical terms may enable the child to extricate himself or herself from difficulty and successfully complete a series of similar problems. Indeed, such demonstrations enable the child to identify his or her own difficulty. One procedure that does not seem to help is presenting the child with a completed problem, given as a static representation. The completed solution—for example, a successfully completed calculation - does not convey the steps to be undertaken and hence is of little value to children in difficulty.

In brief, when experiencing difficulties in problem solving, many children appeal to authority; some attempt informal substitutes, and few can identify reasons for their own difficulties. These children may sometimes be helped by a demonstration of successful techniques in progress.

Learning Potential

Vygotsky (1978) recognized that not only does the learner possess well-developed and functional knowledge, but also certain understandings that are immature or partially developed and therefore not fully functional. He referred to this "embryonic" knowledge as belonging to the zone of proximal development. Recently, Brown and French (1979) have suggested that assessing potential development may be an important diagnostic approach, and Feuerstein (1979) has developed a test battery designed to measure learning potential in culturally disadvantaged and learning disabled groups. Except for these researchers, there has been surprisingly little interest in investigating *learning* ability in children exhibiting poor performance in school. With respect to such children, a key question is whether under favorable environmental conditions—that is, conditions that they do not usually experience in school—they can learn specific academic material. In investigating learning potential, the notion of a zone of potential development may be quite useful. In many cases, children with mathematics difficulty possess unsuspected cognitive resources in the form of informal concepts and skills. This embryonic knowledge can be exploited as a scaffolding on which to build formal understanding. In the clinic, the sensitive clinician often attempts to identify a zone of potential development and to build instruction around it.

For example, Lanny, 10 years old, did not seem to understand the standard place value notation. He could not answer questions dealing, for example, with the number of tens represented in the number 39. The clinician[3] found that Lanny was skilled at counting money. When asked, for example, to get $349, he correctly took three $100 bills, four $10 bills, and nine $1 bills from the cash box. Lanny seemed to have a form of place value knowledge operating in the context of money. But his knowledge did not extend to situations involving written and spoken numbers. The clinician's next task was to determine whether, by providing cues that would link the two tasks, Lanny could extend and generalize his understanding of place value to a new realm, namely written and spoken numerals. At first, Lanny did not spontaneously make the connection between the money system and the numerical problem. The clinician provided the link by pointing out the similarities between money and numbers. The clinician required the child to translate between one system and the other. Soon Lanny was able to use the money model for interpreting place value notation. Thus, he drew upon available cognitive resources to "learn" a new concept.

This situation is not unusual. In our experience, most mathematics difficulty children have areas of learning potential that may be quite easily developed. Often, these children possess informal concepts or skills that, with minimal effort, may be extended to the formal learning of mathematics. For most mathematics difficulty children, *learning* is not a significant problem. It is a mistake to refer to these children as *learning* disabled.

On the other hand, some children who appear to have the necessary underlying knowledge sometimes do not easily extend their understanding even as a result of direct tuition. There are what Vygotsky referred to as individual dif-

---

[3]The clinician was Debby Petrivelli. Other interviews described in this chapter were generally carried out by Barbara Allardice.

ferences in the width of the zone of potential development. Children with wide zones, like Lanny, can profit from exposure to even minimal cues, whereas children with narrow zones may profit little even from directed instruction.

We have seen that many MD children possess a zone of potential development. They often can learn by exploiting their already available informal knowledge. Many, if not most, MD children do not have difficulty in learning. Yet there are individual differences among these children. Some children learn only with great difficulty; perhaps deep-seated psychological characteristics of the child are responsible. Obviously, a good deal of systematic research is necessary to investigate the nature and development of learning potential in children with mathematics difficulties.

Personality Factors

Finally we consider several personality characteristics that may contribute to mathematics difficulties. Sometimes children avoid struggling with mathematics so as to protect their self-esteem, and sometimes children appear to be "unmotivated."

PROTECTING SELF-ESTEEM

Stacy, a third-grader was initially asked by the interviewer to describe what she was doing in school mathematics. This question often serves to relax the children, possibly because it allows them to initiate conversation about some concrete subject matter. Stacy, however, cast her eyes downward and hunched her shoulders, seeming to protect herself from a whipping. The major portions of the initial three interviews with Stacy were spent in extinguishing this response so that mathematics could become the focus of attention. In the classroom, such withdrawal seemed to protect Stacy from having to confront her lack of mathematics understanding. The teacher would approach Stacy to offer some individual attention but, by her own admission, could not tolerate Stacy's agony and moved on to a more receptive student. We can only speculate that at an earlier time, Stacy experienced failure with mathematics and discovered this protective mechanism. The avoidance may have served to protect her from situations that revealed her lack of understanding, but it also made learning about mathematics nearly impossible.

Julie attempted to protect herself from mathematics failure by frequently telling the interviewer, "we haven't had that." In school, Julie had to avoid dealing with difficult problems by telling the teacher that the problems were impossible to do, that she was stupid. The result was that the teacher assigned Julie easier work. This in turn contributed to Julie's sense of incompetence: She knew that the rest of the class was doing multiplication and division, whereas she was still working on addition and subtraction. At the same time, Julie was in fact capable of reasonable work. At one point, the interviewer introduced a game that essentially dealt with notions of "fractions," but without using the standard terminology. Under these conditions, Julie rather quickly mastered several basic fraction concepts. Thus, she seemed able to learn, but used several techniques to avoid expected failure.

These children may be somewhat unusual in the degree to which they avoid mathematics, but many children demonstrate a milder version of avoidance and their general unwillingness to explore the new because of concern about being wrong. Many children have a fear of failure, and the schools reinforce it in many children. Yet a prerequisite for learning in general is a willingness to experiment and make mistakes.

MOTIVATION

Teachers frequently believe that students requiring special help in mathematics suffer from motivational problems. The student *could* do it, but doesn't want to. He or she prefers to fool around. The student may be lazy. He or she hasn't got the proper motivation.

Yet, in our clinic work with these children, we rarely see what we think of as unmotivated or lazy behavior. In the clinic setting, the children frequently work hard, often for a solid hour, and do not seem to want to leave. Indeed, they are willing, some even eager, to return. Thus, these children are capable of hard work and intensive motivation under the proper conditions. Why then do they often show a lack of motivation in school, and sometimes in the clinic?

Jackie usually worked hard in the clinic session, but occasionally showed such signs of low motivation as asking for permission to go get a drink of water or discussing her mother's tennis lessons. Her teacher said that in school Jackie often preferred to "fool around" or disturb other children in lieu of doing her mathematics assignment, and that she seemed unconcerned about her failing grades. Is the low motivation some kind of character trait? When we looked closely at what preceded Jackie's request for water or her irrelevant conversations in the clinic, it became apparent that these behaviors came at a point when Jackie realized that she did not know how to do the problem at hand. Thus, the behavior interpreted as indicating low motivation seemed to be attempts to avoid failure.

Mickie often hurried through her work, doing it incorrectly, according to her teacher. We found that Mickie worked carefully when she knew exactly what to do, but hurried when she did not understand. Again, the behavior that might be taken as an index of poor motivation was an avoidance mechanism.

Of course, there may be some children who have true motivation problems.

However, our observations suggest that what frequently appears to be lack of motivation may more usefully be seen as an unwillingness to stay with a task that is seen as impossible. The "lack of motivation" may well have developed as a *response* to initial failures in mathematics. It may then contribute to continued failure by making it difficult for the child to deal directly and openly with mathematics. And, like attention problems, motivation difficulties are viewed by teachers as being immutable characteristics of the child and not subject to remediation. Yet in fact, changes in classroom procedures may quickly produce significant alterations in the child's behavior.

FUTURE DIRECTIONS

In addition to continuing research focused on the cognitive activities of children with mathematics difficulties, we should be concerned with three relatively new—or at least unexplored—areas of investigation.

The first is *learning potential*. So far as we have been able to determine, there have been virtually no studies of whether children experiencing cognitive difficulties suffer from a deep-seated and genuine problem in learning. We do not know whether these children can learn abstract mathematical material and, if so, how they go about learning it. Investigations of learning style, technique, and process will add significantly to our understanding in the future.

The second area for investigation involves *personality factors*. Cognition operates within the context of personality. We have shown through case study examples how such factors as the need to protect self-esteem can be as much as part of learning mathematics as schemes of reasoning. Similarly, mathematics difficulties may involve anxiety, or poor self-concept, or defensive cognitive style. Success in mathematics may involve falling in love with it, emotional commitment, and a flexible cognitive style. In either event, we need to take a broad psychological approach, rather than a narrow cognitive one, to understanding children's work in this area.

Finally, we need to consider the social context of learning. Learning occurs in a variety of contexts; the classroom is only one environment in which cognitive skills may be acquired. Many individuals who perform poorly in the classroom or do not attend school at all nevertheless acquire basic mathematical concepts and skills. Even a learning disabled child may function quite effectively in out-of-school situations requiring complex cognitive skills (Cole & Traupman, 1980). What is the nature of such out-of-school learning and why does it sometimes proceed more effectively than classroom instruction? These are among the interesting questions for future research on the social context of learning. Taking these questions seriously would be an important step in moving away from blaming the victim and toward acknowledging the complex interactions involved in learning.

REFERENCES

Adelman, H. S. Diagnostic classification of learning problems: Some data. *American Journal of Orthopsychiatry*, 1978, *48*, 717–726.

Belmont, I., & Belmont, L. Is the slow learner in the classroom learning disabled? *Journal of Learning Disabilities*, 1980, *13*, 496–499.

Binet, A. The perception of lengths and numbers. In R. H. Pollack & M. W. Brenner (Eds.), *The experimental psychology of Alfred Binet*. New York: Springer, 1969.

Brown, A. L., & French, L. A. The zone of potential development: Implications for intelligence testing in the year 2000. In R. J. Sternberg & D. K. Detterman (Eds.), *Human intelligence: Perspectives on theory and measurement*. Norwood, N.J.: Ablex, 1979.

Cohn, R. Arithmetic and learning disabilities. In H. R. Myklebust (Ed.), *Progress in learning disabilities* (Vol. 2). New York: Grune & Stratton, 1971.

Cole, M., & Bruner, J. S. Cultural differences and inferences about psychological processes. *American Psychologist*, 1971, *26*, 866–876.

Cole, M., & Traupman, K. *Comparative cognitive research: Learning from a learning disabled child*. Unpublished manuscript, 1980.

Coles, Gerald S. The learning disabilities test battery: Empirical and social issues. *Harvard Educational Review*, 1978, *48*, 313–340.

Erlwanger, S. Case studies of children's conception of mathematics, Part I. *Journal of Children's Mathematical Behavior*, 1975, *1*, 157–283.

Farnham-Diggory, S. Learning disabilities: A view from cognitive science. *Journal of the American Academy of Child Psychiatry*, 1980, *19*, 570–578.

Feuerstein, R. *The dynamic assessment of retarded performers*. Baltimore, Md.: University Park Press, 1979.

Gelman, R., & Gallistel, C. R. *The child's understanding of numbers*. Cambridge, Mass.: Harvard University Press, 1978.

Ginsburg, H. P. *The myth of the deprived child*. Englewood Cliffs, N.J.: Prentice-Hall, 1972.

Ginsburg, H. P. *Children's arithmetic: The learning process*. New York: D. Van Nostrand, 1977.

Ginsburg, H. P., & Russell, R. L. Social class and racial influences on early mathematical thinking. *Monographs of the Society for Research in Child Development*, 1981, *46*, no. 6, serial no. 193.

Hebbeler, K. Young children's addition. *Journal of Children's Mathematical Behavior*, 1977, *1*, 108–121.

Hood, H. B. An experimental study of Piaget's theory of the development of number in children. *British Journal of Psychology*, 1962, *53*, 273–286.

Hunt, J. McV. The psychological basis for using pre-school enrichment as an antidote for cultural deprivation. *Merrill-Palmer Quarterly*, 1964, *10*, 209–248.

Keough, B. K., Major, S. M., Omori, H., Gandara, P., & Reid, H. P. Proposed markers in learning disabilities research. *Journal of Abnormal Child Psychology*, 1980, *8*, 21–31.

Kirk, G. E., Hunt, J. McV., and Volkmar, F. Social class and preschool language skill: V. Cognitive and semantic mastery of number. *Genetic Psychology Monographs*, 1975, *92*, 131–153.

Kosc, L. Developmental dyscalculia. *Journal of Learning Disabilities*, 1974, *7*, 164–177.

Lesh, R. A. Mathematical learning disabilities: Considerations for identification, diagnosis, remediation. In R. Lesh, M. Kantowskii, & D. Mierkiewicz (Eds.), *Applied mathematical problem solving*. Columbus, Ohio: ERIC Center, The Ohio State University, 1980.

Luriya, A. R. On the pathology of computational observation. In J. Kilpatrick & I. Wirszup (Eds.), *Soviet studies in the psychology of learning and teaching mathematics* (Vol. I). Chicago: University of Chicago, 1969.

Mpiangu, B. D., & Gentile, J. R. Is conservation of number a necessary condition for mathematical understanding? *Journal for Research in Mathematics Education,* 1975, *14,* 179–192.

Ogbu, J. U. *Minority education and caste.* New York: Academic Press, 1978.

Owen, F. W., Adams, P. A., Forrest, T., Stolz, L. M., & Fisher, S. Learning disorders in children: Sibling studies. *Monographs of the Society for Research in Child Development,* 1971, *36*(4).

Robinson, N. M., & Robinson, H. B. *The mentally retarded child* (2nd ed.). New York: McGraw-Hill, 1976.

Russell, R. L., & Ginsburg, H. P. *Cognitive analysis of children's mathematics difficulties.* Rochester, N.Y.: University of Rochester, 1981.

Torgesen, J., & Goldman, T. Rehearsal and short-term memory in reading disabled children. *Child Development,* 1977, *48,* 56–60.

Torgesen, J., Murphy, H., & Ivey, C. The influence of an orienting task on the memory performance of children with reading problems. *Journal of Learning Disabilities,* 1979, *12,* 396–401.

Vygotsky, L. S. *Mind in society: The development of higher psychological processes* (M. Cole, V. John-Steiner, S. Scribner, E. Souberman, Eds.). Cambridge, Mass.: Harvard University Press, 1978.

Weinstein, M. L. Dyscalculia: A psychological and neurological approach to learning disabilities in mathematics in school children (Doctoral dissertation, University of Pennsylvania, 1978).

Wertheimer, M. *Productive thinking.* New York: Harper, 1954.

Wheatley, G. Conservation, coordination, and counting as factors in mathematics achievement. In I. J. Athey & D. Rubadeau (Eds.), *Educational implications of Piaget's theory.* Waltham, Mass.: Ginn-Blaisdell, 1970, 294–301.

GUY GROEN
CAROLYN KIERAN

CHAPTER **9**

In Search of Piagetian Mathematics[1]

Unlike most of the preceding chapters, this does not deal with our own research. Rather, it is intended to be a general treatment, providing a kind of postscript to the rest of the book. Piaget seems to deserve a chapter in a book such as this. Mathematics lies at the core of his work. Indeed, to discuss his contribution to children's mathematical thinking is really equivalent to discussing his entire theory. Conversely, virtually all contemporary research in children's mathematical thinking has been influenced by, or is a reaction to, Piaget's work.

However, the reader will hardly fail to notice a curious fact. A few years ago, research on childrens' mathematics was dominated by Piaget. To many in the field, the task was to extend Piaget's theory or reinterpret it. Even where there was disagreement (as, for example, in the extended controversey over the development of number concepts), there was a consensus that Piaget's questions were reasonable ones to ask, even though the answers might be wrong. The priority seemed to frequently lie in "disproving" Piaget rather than in developing a radically different approach. In contrast, the research described in this volume relegates Piaget to the background. Information-processing theory, broadly conceived, has replaced the Piagetian framework as a broad explanatory

[1]The authors would like to thank Seymour Papert for much helpful advice, especially with respect to the direction this chapter should take.

351

model. This is not an artifact of editorial selection. As Beilin (1981) points out, the significance of information-processing theory in cognitive development has grown concomitantly with a retreat from the Piagetian framework.

It is the purpose of this chapter to consider why such a drastic change occurred. One possibility is that it is simply the outcome of a process of natural evolution. However, it can be argued that Piaget was not an embryonic information processor. There appear to be major differences in the questions that are asked, as well as in the way they are answered (Beilin, 1981; Cellerier, 1972). For example, Beilin asserts that Piaget was primarily interested in questions about structure, whereas information processing is primarily concerned with questions about function. Although we will return occasionally to this issue, we will devote the bulk of our attention to an alternative possibility. This is that Piaget's theory is so subtle, ambiguous, and open to misinterpretation (Groen, 1978) that it has simply been abandoned for a more manageable framework.

One reason for this may be the fact that the information-processing approach is highly amenable to bottom-up analysis, in which the theories or models are generated by specific tasks. Thus, it is possible (in principle) to take any piece of mathematics and analyze it from an information-processing point of view, albeit in a highly task-specific fashion. In contrast, Piaget's approach is top-down. It approaches mathematics from more general notions about intellectual structures. It is possible that these notions impose constraints upon the kind of mathematics that can be accommodated by the Piagetian framework. As a result, there may exist a special kind of Piagetian mathematics. Moreover, it may be quite different from much of that to which Piaget's theory has been applied in the past.

In order to come to grips with this possibility, we have decided to adopt a somewhat unconventional approach. We will consider a set of questions that are quite simple, but peculiar in their degree of interrelatedness. We will not attempt to provide a set of answers. Rather, we will indicate the difficulties and show how the process of formulating an answer (or thinking about what it might contain) leads to more questions, which have to be answered before we are able make some reasonable inferences regarding the nature of Piagetian mathematics.

IS PIAGET RELEVANT TO SCHOOL MATHEMATICS?

There are many kinds of mathematics. There is the kind that professional mathematicians do. There is the kind that preschool children do. Somewhere in between, there is the kind of mathematics that children encounter at school. However, the list does not end there. The kind of mathematics that children are taught may be different from what they do (e.g., Groen & Resnick, 1977). There is also Ginsburg's distinction between intuitive, informal, and formal mathematics (see Allardice & Ginsburg, Chapter 8). Where does Piagetian mathematics fit into all this? An immediate difficulty is that there is no neat way of distinguishing between many of these different kinds of mathematics. In particular, there is a mixture of behavioral criteria and ones based on mental processes. An exception is school mathematics, which we define as consisting of a set of tasks (what children are taught or required to do) and a set of processes (what children actually do in performing those tasks). One could, at least in principle, simply look at a curriculum and make a list of the tasks that are taught. The processes might vary drastically, but one would at least have a well-defined set of tasks. Of course, such tasks would vary, depending upon the particular curriculum. However, there is a body of mathematics that forms a common core of most school mathematics curricula. In this connection, it is convenient to distingush between conventional school mathematics (such as arithmetic or solving equations), which forms this common core, and unconventional school mathematics (such as computer programming), which is currently not a part of this core. In this section, we will only be concerned with conventional school mathematics.

On the surface, the answer to our question concerning the relevance of Piaget might appear to be obvious. After all, Piaget and his collaborators have produced a vast number of empirical studies of conservation, seriation, classification, and the like that have appeared in books dealing with children's logic, number concepts, and geometry (Inhelder & Piaget, 1958, 1964; Piaget & Inhelder, 1958; Piaget, Inhelder, & Szeminska, 1960; Piaget & Szeminska, 1952). These books also contain some developmental notions about how the ability to perform on such tasks progresses through stages. Finally, there is a theory that appears to be about mathematical entities such as groups, relational structures, and topological spaces (e.g., Piaget 1949).

The problem with an answer along these lines is that there are so many gaps in what Piaget provides that it is impossible to proceed without creating a theory of one's own. The most serious is the gap between Piaget's tasks and the tasks of school mathematics. Although there may be a connection, it is not an explicit one. They lack the face validity or direct correspondence possessed by a task such as addition or solving equations.

It might be tempting to claim that the correspondence exists not at the surface, but at a deeper level. Piaget's tasks might be essential components of tasks in school mathematics, or they might somehow define necessary conditions for success. Approaches along these lines have been extremely common, especially in North America, where they have frequently been attributed to Piaget. However, Piaget himself has never been explicit on this point. His theory provides no apparatus for bridging the gap between his empirical work and school mathematics.

The confusion arises because there is a subset of Piaget's theory that appears to provide this apparatus. This is stage theory—the familiar notion that intelligence develops through a sequence of stages and substages. The bridge is provided by the following assumptions:

1. Piaget's "intelligence" is the same as that required by school mathematics.
2. Intelligence develops according to Piaget's "main sequence" of stages (sensorimotor, preoperational, concrete, and formal operations).
3. The stages of this main sequence, together with their substages, define a partially ordered set of slots into which the tasks of school mathematics can be inserted. This defines the level of intelligence necessary for successful performance on any given task.
4. A given level is attainable only by going through all prior stages and substages.
5. Performance in school mathematics can be improved by explicitly teaching appropriate Piagetian tasks, as generated by Assumption 3 (e.g., improving addition by teaching conservation of number).

The numbering of these assumptions is not accidental. They are listed in order of increasing departure from Piaget's theory. The first is more or less consistent with it. The last is in direct contradiction. Because of this latter fact, we will call such approaches quasi-Piagetian.

We do not intend to deal with these quasi-Piagetian approaches in this chapter. Since this involves ignoring the vast majority of "Piagetian" research on children's mathematical thinking, a few words of explanation are necessary. We do not view this research as useless. If interpreted with sufficient care, it can yield valuable information even about Piaget. There are plausible historical reasons for its existence (Groen, 1978). It might even be viewed as a necessary stage in the development of our understanding of Piaget. The assumptions listed previously were perfectly reasonable inferences from the literature available in English (e.g., Flavell, 1963; Piaget, 1950b, 1953) at the time of Piaget's rediscovery in the late 1950s and early 1960s. There were few ways, short of an extended stay in Geneva, of discovering that they were erroneous. Even Piaget and his collaborators only began to make their position clear on these issues 10 years later (e.g., Piaget, 1967).

Despite all this, there are a number of good reasons for terminating their discussion at this point. First, the problems with those assumptions have been treated at length elsewhere (e.g., Groen, 1978). Second, it seems more appropriate to clarify the issues involved in Piaget's own theory rather than to cloud them by discussing theories that are fundamentally different, despite all the confusing resemblances. Paradoxically, the existence of these resemblances may imply that an accurate assessment of these variant theories will only be possible when this clarification occurs. Finally, recent research on children's informal arithmetic

(Gelman & Gallistel, 1978; Ginsburg, 1977) yields empirical evidence that there is something seriously wrong with Assumptions 3, 4, and 5, where school mathematics is concerned. These results (some of which are discussed in Chapter 2 by Fuson and Hall and in Chapter 3 by Resnick) indicate that children initially develop many number concepts and a considerable ability in simple arithmetic through the use of counting processes, such as those postulated by Groen and Parkman (1972) and Groen and Resnick (1977). These processes do not develop according to Piaget's main sequence and cannot be fitted into the slots of Assumption 3. In particular, a child appears to be able to perform these operations efficiently while still unable to solve conservation problems. Such results may not be restricted to arithmetic. Papert's informal observation of children working in the LOGO environment (Papert, 1980) suggest that they may possess an informal geometry that is also procedurally based, and which also may not follow Piaget's main sequence in its development. By logical extension, we can entertain the possibility that much of school mathematics fails to follow the main sequence.

At first sight, such results may seem embarrassing to Piaget's own theory. However, this is not so. To see this, it is necessary to make a slight digression into the philosophy of science. The classical positivistic view of a scientific theory is that it can be verified or refuted by some crucial empirical evidence. This view has changed in recent years. The notion of verification has been cast into doubt. The notion of falsification remains, but even this has problems. In his well-known book, Kuhn (1962) has pointed out that theories remain despite the accumulation of large numbers of anomalies. Lakatos (1972) has suggested some specific properties that a theory or research program might posess that enable this kind of situation to occur. To Lakatos, a theory consists of an invariant hard core surrounded by a protective covering. Negative evidence only affects the latter. The hard core and the protective covering are related through a set of heuristics. Some of these are so-called negative heuristics, which serve to deflect anomalies. Groen (1978) has shown that Piaget's theory fits neatly into a Lakatosian framework of this kind. One of its most important features is a very powerful negative heuristic: the notion that experiments involving tasks that are epistemologically trivial (in the sense that they do not reflect important knowledge) are irrelevant to the research program.

It is possible to argue that much of school mathematics is epistemologically trivial. Facts and rules can be learned without understanding or thinking about what one is doing, and this would seem to indicate that they are not coordinated with the intellectual structures that Piaget views as important. However, this argument raises a difficult problem. We suddenly seem to be leaving psychology: The whole notion of what knowledge is trivial or nontrivial would seem to be more philosophical than psychological in nature. It is this that motivates our next question.

TO WHAT EXTENT IS PIAGET A PSYCHOLOGIST?

It is important to note that the issue is not whether Piaget conducted empirical psychological studies. Much of the effort of Piaget and his collaborators was devoted to the application of a well-defined, though subtle method (the *revised clinical method*) with children. This would appear to satisfy any reasonable atheoretical definition of empirical research in psychology. The issue is the extent to which these studies, the underlying theory and, and more generally, the overall research program are relevant to psychology.

There are two ways in which such relevance might occur: (*a*) deliberately or directly, because Piaget's research program was designed, in whole or in part, to answer psychological questions; (*b*) accidentally or indirectly, because some aspect of Piaget's research influenced an investigator engaged in a different program of psychological research. The virtue of this distinction is that it divides the main question into a difficult problem and an easy one. In the first, we are asking, whether Piaget was really a psychologist. In the second, we only ask whether some psychologists have been able to adapt some aspects of Piaget's work for their own clearly defined purposes in a productive fashion. The history of psychology provides a number of instances that show that this could be the case even if Piaget were not a psychologist.

The second is an easy question because it can be unambiguously answered in the affirmative. It is simple to point to a number of psychologists who have used Piaget's ideas in this way. For example, Ginsburg's clinical method evolved out of Piaget's. It uses the same kinds of probes to examine very different kinds of tasks. A concern with the nature of stages and stage transitions motivated the work of Case (1978) and Pascual-Leone (1970). A number of investigators have developed specific information-processing models for performance on some of the better known Piagetian tasks (e.g., Klahr & Wallace, 1976; Young, 1973). It would also be possible to point to the quasi-Piagetian approaches discussed in the preceding section. However, these form a peculiar category since they cannot be understood or evaluated without an extensive analysis of Piaget's theory. In contrast, the research we have cited stands on its own. Where Piaget's ideas are used, they are explicitly redefined in very different contexts. As a result, such research is self-contained and can be understood without reference to Piaget.

The first is a difficult question because of the all-embracing nature of Piaget's research program. It encompasses all of scientific knowledge, and psychology is not the central focus. Piaget calls his field genetic epistemology and has emphasized on a number of occasions that it is not to be confused with psychology. For example, Piaget (1972) quotes approvingly from his citation for the Distinguished Scientist Award of the American Psychological Association, which refers to his psychological research as a by-product of his more general philosophical preoccupations.

The common approach among psychologists has been to view such statements as charming modesty on Piaget's part. Genetic epistemology has been viewed as a subset of psychology or as an attempt to transform important philosophical questions into important psychological ones. Indeed, the term itself was coined by the early-twentieth-century psychologist, Baldwin, for a theory of cognitive development that greatly influenced Piaget (in particular, it introduced the mechanisms of accommodation and assimilation). However, there are good reasons for taking Piaget's statements considerably more seriously. We mentioned earlier the existence of serious gaps in Piaget's work, when viewed from the standpoint of psychology. A deeper analysis of how philosophy and psychology are interwoven into genetic epistemology may yield some insight as to why these gaps exist. In particular, if Piaget is correct, then the empirical tasks are simply a by-product of a set of philosophical questions. There is therefore a possibility that the empirical techniques may have been designed to answer questions very different from those normally asked in developing a psychological theory. On the one hand, this could, of course, be stimulating for psychology. On the other hand, if the difference is too great, then Piaget's psychology may only be capable of answering Piaget's philosophical questions. If this were the case, then Piaget would truly be an accidental psychologist—good for ideas, perhaps, but not with a theory that is directly applicable in a psychological context.

Of course, the fact that a philosopher undertakes a program of psychological research cannot be regarded as automatically dooming his work to indirect relevance. The same question can be asked in both a psychological form and a philosophical form. Alternatively, the psychology may be separable from the philosophy in the sense that one can be understood without reference to the other. This is true, for example, in the work of Patrick Suppes who, though a professional philosopher, has conducted major research programs in both areas. One may have motivated the other, but the relevance of his work in psychology can be evaluated quite separately. The problem with Piaget is that he is neither a psychologist nor a philosopher, but a genetic epistemologist. Are all of Piaget's philosophical questions psychological ones in disguise? Given Piaget's own differentiation between philosophy and psychology, this seems unlikely. Hence, the basic issue is whether genetic epistemology is a monolithic research program or whether it is separable into components. If such components exist, then it may be that some actually do consist of attempts to answer genuinely psychological questions. If not, then, whereas it might not be possible to flatly state that Piaget had absolutely no direct relevance to psychology, it would certainly be most difficult to find out of what it consisted.

WHAT IS GENETIC EPISTEMOLOGY?

It is best to allow Piaget, as much as possible, to define the area in his own words. The clearest treatment available in English[2] is a series of lectures delivered in the United States (Piaget, 1971a), which has the additional virtue of having been translated by a psychologist (Eleanor Duckworth) familiar with the work done in Geneva. The following quotations provide a reasonably precise idea of what Piaget means by genetic epistemology:

> Genetic epistemology attempts to explain knowledge and, in particular, scientific knowledge, on the basis of its history, its sociogenesis, and especially the psychological origins of the notions and operations upon which it is based [p. 1].

> All epistemologists refer to psychological factors in their analyses, but for the most part their references to psychology are speculative and are not based on psychological research. I am convinced that all epistemology brings up factual problems as well as formal ones, and once factual problems are encountered, psychological findings become relevant and should be taken into account [p. 7].

> The first principle of genetic epistemology, then, is this—to take psychology seriously. Taking psychology seriously means that, when a question of psychological fact arises, psychological research should be consulted instead of trying to invent a solution through private speculation [p. 9].

> So, in sum, genetic epistemology deals with both the formation and the meaning of knowledge. We can formulate our problem in the following terms, by what means does the human mind go from a state of less sufficient knowledge to a state of higher knowledge? [p. 12].

> Our problem, from the point of view of psychology and from the point of view of genetic epistemology, is to explain how the transition is made. . . . The nature of these transitions is a factual question. The transitions are historical or psychological or sometimes even biological. [p. 13]

All this is quite straightforward. The issue is how scientific knowledge (which is taken to include mathematics) changes to a higher state. Whether or not one state of knowledge is higher than some other is to be determined by "logicians or specialists in a given realm of science." The purpose of psychology is to explain the transitions.

Two tasks remain to be accomplished. The first is to develop a way of describing knowledge. This is accomplished by viewing knowledge as consisting of structures. Piaget's notion of *structure* has always been more or less identical to the mathematician's notion in the definition, for example, of a relational structure, or a set whose elements satisfy certain axions. Such structures are viewed as providing models or representations of whatever it is that the knowledge is about. In particular, the structures of physics model physical phenomena.

---

[2]The standard introduction (Piaget, 1950) is available only in French.

The structures of mathematics are models of entities that are themselves structures.

The notion of structure is so important to Piaget that it is advisable to define it more precisely. This is to be found only in his later work (Piaget, 1971d). In his earlier work, he refers to it either by analogy or example, as in his extensive discussions of groups and "groupements." In Groen's (1978) terms, the following is a reasonably accurate paraphrase of various definitions to be found in Piaget's later work: A structure is a system consisting of a set of states (which a mathematician would call the state space or the underlying set), a set of transformations (or mappings) between states, and a set of global laws governing the application of the transformations. It has the following properties:

1. The system is a *totality,* to use Piaget's term. The global laws apply to all elements of the system.
2. The system is *self-regulating,* again using Piaget's term. The state space contains every state that a transformation can act upon.
3. Structures are *modular* (our own term, not Piaget's). A given structure can contain substructures, each with its own state space, transformations, and global laws.

It is important to note that this definition contains absolutely no psychology. Indeed, it is very close to being standard mathematics or mathematical logic, except for the idiosyncratic terminology used by Piaget. Piaget is chiefly concerned with the elementary structures of logic, algebra, and topology. However, the notion is far more general. It is the basic conceptual tool in his analyses of physics and biology. It also applies in many areas not considered by Piaget. The second task is to define what makes one structure better than another. In considering this issue, Piaget is considerably more idiosyncratic and begins to leave the realm of standard logic and mathematics. Despite the claims that it is a matter to be decided by subject-matter specialists, he has his own unique criteria. The best are those in which the transformations are what he calls *operations.* An operation is a transformation that is reversible. This is an extremely difficult notion to define. Essentially, it means that an operation can be put into reverse and so reconstruct the states that it originally transformed, without undoing the effects of the original operation. Associated with this is the notion that, for a transformation to be reversible, it is necessary that certain properties of the state space be conserved when the transformation occurs. Examples of this latter property occur widely in Piaget's writing—they are the basis of the familiar conservation tasks. Unfortunately, it is impossible to give a precise example of reversibility or its connection with conservation without developing an elaborate notation.[3] Howev-

---

[3]The currently popular Rubik's Cube provides an interesting example of the importance of reversible operations. Its well-known difficulty is largely due to the need to perform complex backtracking while retaining information about preceding states.

er, what has been stated is sufficient to make the point that conservation and reversibility originate in epistemological rather than psychological notions. They grew out of formal considerations of the structure of knowledge rather than some theory of childrens' thinking.

Where, then, does psychology enter? Piaget's (1971a) answer is surprising, given the preceding discussion. "The fundamental hypothesis of genetic epistemology is that there is a parallelism between the progress made in the logical and rational organization of knowledge and the corresponding formative psychological processes. . . . It is with children that we have the best chance of studying the development of logical knowledge, mathematical knowledge, physical knowledge and so forth [p. 13]." Why study the organization of knowledge in such an indirect fashion? Surely the epistemology we have sketched out provides the bases for a direct attack on the psychology of scientific knowledge based on empirical studies examining the thinking of actual mathematicians and scientists. Instead, we find what would appear to be a theory of child development.

There are two reasons. The first is that Piaget is interested in major changes in the structure of knowledge. Most scientists are not so successful as to cause a major breakthrough. If one follows Kuhn (1962), one might say that they are doing normal science, following an established paradigm, rather than generating a scientific revolution. The history of science is relevant, but provides only limited information. In particular, many of the changes in which Piaget appears to be most interested may have occurred before there was any written record. On the other hand, children can be presumed to have theories about the world around them and also theories about their own thinking. To Piaget, these correspond to the structures of physical knowledge and logico-mathematical knowledge. It seems reasonable to expect such theories to undergo major changes as a result of the child's interactions, not only with its environment, but also with its own thinking.

The second reason is subtle, but extremely important. One of the chief reasons why Piaget may call himself a philosopher is his extensive use of the dialectical method of reasoning. In this, one progresses by creating a tension between opposites and striving for a resolution. Although little discussed, it is pervasive throughout Piaget's work and colors not only the questions he asks, but the concepts he formulates to answer them. Gruber and Voneche (1977) suggest that genetic epistemology is essentially a dialectical working out of the tensions between two opposing notions: *static structuralism* and *dynamic geneticism*. The hypothesis of parallelism (resonance may be a better word) between structural change and cognitive development is a way of resolving the tension. The dialectic does not stop there. The search for evidence creates new tensions. To resolve them, Piaget develops a logic and a psychology. These, in their turn, create their own tensions.

There are a number of places in which Piaget discusses this logic (Piaget, 1949, 1957) and this psychology (Piaget, 1952; Piaget & Inhelder, 1969) in a self-contained and comprehensive fashion. In these, the dialectical method does not specifically appear. Indeed, there is little mention of epistemology at all. He appears to be doing his best to restrict himself to the normal tools of more conventional psychologists and mathematicians and to stress that the same criteria should be applied to this work as are used to assess the worth of psychological theories and formal axiomatic systems in general. The outcome is essentially two subdisciplines, in which the canons of scientific method are strictly adhered to, and embedded in the larger discipline of genetic epistemology, in which the principal method is dialectical. It is almost as if Piaget is developing another dialectical tension, this time between science and the dialectic itself.

Now that it seems reasonable to speak about a separate psychological component, is it possible to concentrate on this and ignore the dialectic? It may be necessary to refer to the broader context if one asks *why* Piaget did something. It seems possible to avoid it by focusing on *what* he did. However, there are two problems with such a strategy. The first is that the dialectic molds the psychology in a way that causes unexpected twists and turns. These may be totally baffling without some insight into the process that created them. The second is that there is a danger of losing sight of the reasons why the psychology is there in the first place. It may be that it is really an attempt to answer questions about the dialectical process. Its relevance may depend on the extent to which opportunities for such a process actually occur in the world of the child.

IS THERE MORE THAN ONE PIAGET?

Dialectical reasoning has a somewhat disconcerting property. Because it operates by resolving tensions between opposites and expands its scope by going through successive stages of this kind (and hence creating new opposites), it is very tolerant of inconsistency and contradiction. These may arise simply because opposites are being juxtaposed with the intention of achieving a resolution. For example, we have already encountered Piaget, the philosopher and Piaget, the psychologist pursuing very different methodologies. In this case, we find the resolution in Piaget, the genetic epistemologist.

However, incongruities may also arise because the resolution achieved at one point in time is inconsistent with the resolution achieved later on (one might use a Piagetian parallel and think of the difference between preoperational and formal operational explanations). The long-term result might be a sequence of very different theories. Of course, any scientific research program tends to generate different theories over the long run. However, they are consistent in the sense that any contradictions are usually due to the need to explain empirical data or to overcome logical inadequacies. Here, on the other hand, the contradictions appear to be quite arbitrary unless the dialectical process is taken into account.

We do not wish to go too deeply into the Piagetian dialectic. We simply make the point that even though Piaget was attempting to follow scientific method in his psychology, the nature of the dialectical process makes us expect not one such psychology, but several.

It is conventional to divide Piaget's work into three phases: the early work up to about 1930, the period from 1930 to 1957 that saw the development of the major aspects of what is usually viewed to be his nature theory, and the period since 1957 of his major international impact. Do the first two phases constitute two different psychologies? There are some similarities. The early work does deal with children's thinking and begins to develop the notion of development through assimilation and accommodation. However, there are far more differences. The best known is that there was a major change to a less verbal empirical methodology, but there are others that are equally important. As Gruber and Voneche point out, the main theme of the early work is egocentric thought. There is no insistence on a clear progression of stages and no appeal to explicit logical models that are assumed to underlie the child's thought. In contrast, the main theme of the middle period is the role of these models and the child's progression through a sequence of stages defined by them. All this makes it impossible to treat the work of the two phases in the same context, the conventional view being that there are indeed two Piagets—a "young" Piaget and a "mature" one.

Do the last two phases also constitute two different psychologies? It is usually assumed that they do not. The period since 1957 is usually viewed as one in which Piaget reacted more directly to stimulation and criticism from the outside world and, as a result, produced a major clarification of his theory. There is much truth to this point of view. It is certainly much easier to understand Piaget after reading, for example, *Biology and Knowledge* (Piaget, 1971c) or the books on genetic epistemology referred to in the preceding section. However, is this enhanced understanding really due to a reworking of old themes, or does it reflect the development of a new, but clearer, theory? Did Piaget merely assimilate the reactions of the outside world or did some major accommodations occur?

To examine this question, it is necessary to be more precise in our chronology. The picture of three phases is somewhat oversimplified. There was actually a transitional period in the 1930s consisting of the three books on the sensorimotor stage (e.g., Piaget, 1952). These can be viewed as either an investigation of the infant's development out of egocentricity or as an outline of the child's progression through the first of the main sequence of stages. Although consistent with the earlier work, there is a distinct change of theme. There is also a change of theme in a series of books on perception, imagery, and memory that appeared (in their original French editions) in the 1960s (Piaget & Inhelder, 1971, 1972). These are concerned with much simpler kinds of thinking. In Piaget's terms, the work is concerned with the figurative aspects of thought, rather than the operative aspects[4] that are the main focus of the middle period. Also, the stages through which they develop have no relationship to the main sequence that the operative aspects progress through.

All this is quite consistent with the middle period. A different stage sequence simply reflects the fact that a different aspect of thought is being studied. The attention to issues closer to the domain of traditional experimental psychology might simply reflect a desire on Piaget's part to tie up the loose ends of his psychology. Indeed, there appeared at this time *The Psychology of the Child* (Piaget & Inhelder, 1969) which was, to quote from the preface, an attempt "to present, as briefly and clearly as possible, a synthesis, or summing up, of our work in child psychology." This book, which is tightly organized around the notion of progression through the main sequence of stages (the figurative aspects of thought are inserted at the preoperational stage), is perhaps as intriguing for what it omits as for what it includes. Genetic epistemology might as well not exist. An examination of the index reveals a complete absence of reference to such terms as *equilibration, regulation,* and *reflective abstraction,* which recur time and time again, not only in Piaget's epistemological work, but also in his later writings on psychology. It is therefore surprising to find, in the last paragraph of the book that "equilibration by self-regulation" constitutes the "formative process" by which progression through stages occurs. Something new is being pointed to that does not fit in with the rest of the book.

The Psychology of the Child is a watershed between the early Piaget and the late Piaget. On the one hand, the synthesis it presents is psychology of the middle period, giving us a concise picture of its main characteristics. First, there is a strong stage theory based on a parallel theory of logical competence. Second, there is an emphasis on tasks that are essentially concrete embodiments of mathematical and physical concepts. Third, there is a lack of explicit connection with the top level of genetic epistemology, in the sense that the psychology is developed as an independent entity, rather than as something that exists to answer epistemological questions.[5] On the other hand, the newer work is also summarized, and this is clearly creating problems for the synthesis.

Two tensions are particularly apparent. One is due to an increased emphasis on the preoperational period in the empirical work, much of it concerned with what Piaget and Inhelder call *the semiotic or symbolic function*—the figurative aspects of thought just referred to, together with language. The problem is that, previously, the semiotic function was defined negatively, in terms of an absence of concrete operations, whereas it is now possible to define it positively, in terms of structures built upon the object concepts of the sensorimotor stage. The other

---

[4]These are low-level structures that the operative structures ultimately coordinate.

[5]This is a property of the entire middle period. The first clear exposition of genetic epistemology dates from the early 1950s, yet none of the books devoted to psychology refer to it explicitly.

is a tension between the theory of the middle period, which is essentially a characterization of what a child can or cannot do in terms of various logical criteria and a very different kind of theory concerned with mechanisms of development, which is outlined in the final chapter of the book. One is essentially static, whereas the other is dynamic. The problem is whether they are really compatible.

These tensions are worked out during the final period. This was an extremely productive period in Piaget's life. It was not a period of nature reflection and clarification, as one might expect in someone of Piaget's age. There were definite changes occurring—one involved in the genetic epistemology. It now had three components rather than two: a logic, a psychology, and a biology. He began to write explicitly about evolution and biological adaptation, and this interacted with the psychology much as the logic did during the middle period. The result was one of the major themes that we have already noted: a preoccupation with the mechanisms of development and the processes underlying the acquisition of knowledge. As Furth (1981) puts it, there was a shift from an emphasis on structure to an emphasis on function.

This took a number of forms. One was an investigation of reflective abstraction, which Piaget felt to be a basic process underlying the acquisition of formal operations (Piaget, 1976). Another was a theoretical analysis of the adaptive processes underlying equilibration (Piaget, 1977). Finally, in his last years, Piaget appears to have turned the dialectical process explicitly upon itself and to have embarked on an investigation of its function in the child's acquisition of knowledge (Piaget & Voyat, 1979).[6]

This work shows evidence of other changes too. There appears to have been a breakdown of the compartmentalization between the psychology and the other components of genetic epistemology that was present in the middle period. In fact, much of Piaget's psychological theory now begins to appear in books, such as *Biology and Knowledge,* that are concerned with more general aspects of genetic epistemology. There is also a change in the kinds of tasks used in Piaget's empirical work. Rather than tasks that are embodiments of physical and mathematical concepts, one tends to find tasks, such as the Tower of Hanoi, that are essentially abstract puzzles. Although there are exceptions, such as the work on causality, such of the empirical work gives the appearance of being designed to study the process of change per se, rather than the development of specific concepts.

Such changes are sufficient to indicate that new questions were being asked and a new theory was being created. However, this may only have indicated a shift of emphasis on Piaget's part, or a desire to explore new domains. Did any

---

[6]Piaget structured his research around a series of symposia held every summer. Three of the last were entitled La Contradiction, La Nécessité, and La Dialectique.

actual inconsistencies or contradictions result from this shift? If such exist, then they are most likely to arise from the two tensions that were being worked out during this final period. Up to now, we have only considered one of these, and that in a positive fashion. We have concentrated on the dynamic theory and avoided considering what that did to the static theory of the middle period. There is a reasonably strong possibility that there was a corresponding change in Piaget's logical models from notions derived from classical logic to notions dealing more with mappings and transformations. The issue of whether these new notions are consistent with the previous ones is intriguing, but too technical to be pursued here. It is the second tension that concerns us more. There are grounds for believing that some changes occurred in Piaget's whole conception of stages and their necessity. It is to these that we now turn.

ARE STAGES NECESSARY?

We begin by indicating that there is a problem. In 1975, Piaget put forward a new theory of equilibration. It is clear that he intended this to be taken extremely seriously. Indeed, he proposed, not entirely facetiously, that it be viewed as his doctoral thesis in psychology. As Furth (1981) puts it, it is the keystone that holds together, both logically and psychologically, the edifice of his theory. Some commentators have found considerable ambiguity in this theory (e.g., Boden, 1979). One of the major problems appears to be a difference between the notion of stages used in this theory and that used in the theory of the middle period.

The point has been made most succinctly by Brown in his foreword to a collection of experiments entitled, somewhat appropriately, *Experiments in Contradiction* (Piaget, 1981), which appeared at the same time as the theory of equilibration and was highly influenced by it. To Brown, one of the major difficulties in interpreting this book is an ambiguity in what is meant by the passage from irreversibility to reversibility. On the one hand, there is progression through the main sequence of Piaget's middle period, where concrete operations are defined in terms of partial reversibility, and the stage of formal operations is defined as that where everything is reversible. On the other hand, Brown feels that, in the new theory, Piaget indicates that the passage from irreversibility to reversibility occurs within each main stage.

Brown attempts to resolve this by suggesting that Piaget is offering a choice between biological development (which accounts for the main sequence) and development by constructing new structures through the equilibration process. However, this is inconsistent with Piaget's position in *Biology and Knowledge,* where he states that the development of operational (and hence fully reversible) structures is determined by epistemological rather than biological factors. An alternative explanation is that Piaget had, by then, abandoned the whole notion of progression through a fixed main sequence of stages. In particular, concrete operations were no longer a prerequisite for formal operations. Formal operations and sensorimotor schemes remained, but more as definitions of maximum and minimal competence. The theory is attempting to explain the transition from minimal to maximum competence, rather than a transition between stages. Explaining transitions to and from the stage of concrete operations has ceased to be an issue because there is no longer anything to prevent formal operations from occurring prior to concrete operations.

This is different from the position of the middle period. Thus, in *The Psychology of the Child* we find that "nascent formal thought restructures the concrete operations by subordinating them to new structures." The idea here is not only that formal operations somehow evolve out of concrete operations, but that the concrete operations are substructures of the formal operations. It is this latter notion that disappears in the later period. Thus, at an American conference on mental testing (Green, Ford, & Flamer, 1971) Piaget explicitly denies that concrete operations play any role at all in formal operations. To achieve formal operations, the child must tear down his or her existing partially reversible structures and build new reversible ones. From this, the inference is unavoidable that concrete operations are things that get in the way. They have a retarding, rather than a facilitating effect upon the acquisition of formal operations.

Did such a change actually occur? Apart from the fact that the new equilibration theory fits more satisfactorily into this looser framework, there are other arguments that could be put forward. It is consistent with the claim of some Piagetians (e.g., Kamii & DeVries, 1977) that the goal of preschool education should be to enhance the acquisition of structures that lead to formal rather than concrete operations. At another level, it is a logical consequence of the training study by Inhelder, Sinclair, and Bovet (1974) who found that training in concrete operations did not result in the development of the kind of hierarchical structure with relatively noninteracting components that one might anticipate if reversible structures were built up from nonreversible ones. Finally, it evolves naturally, given Piaget's penchant for dialectical reasoning, from the increased emphasis on the preoperational stage. In particular, it was found that quite coherent structures, which Piaget calls *semi-logics,* were being formed during this period. Their connection with formal operations was clearer than their connection with partially reversible concrete operations. This resulted in a tension between the preoperational stage and the stage of concrete operations that could only be resolved by dropping the assumption that concrete operations were logically necessary for formal operations.

Does all this indicate that stages were not really necessary in Piaget's final theory? To discuss this issue, it is important to distinguish between various ways in which stages can be used. At one extreme, it can be simply a technique for describing data. It is clear that Piaget used stages in this fashion until the end of his life. At the other extreme, there is what Gruber and Voneche term the hypothesis of a single, orderly, and universal succession of stages of cognitive development. This is the part of stage theory which we argue he dropped during his last years. It might be unwise, however, to infer that the stage concept had been reduced to mere data analysis. The dialectical process lends itself naturally to an expression in terms of stages, with each new synthesis corresponding to a new stage. It is possible that Piaget's main use of stages may have always been as a means of developing and testing a theory of the dialectical process, rather than a specific hypothesis about the stages of cognitive development.

WHAT IS PIAGETIAN MATHEMATICS?

Having explored the broader aspects of Piaget's theory, we are now in a position to return to the issue that was raised in the introduction to this chapter. Do Piaget's general notions about intellectual structures result in only a special kind of mathematics being compatible with the Piagetian framework? If so, can we define it in a reasonably precise fashion and give some nontrivial examples? In how broad a context is it likely to be useful? In light of our preceding discussion, we restrict ourselves to the framework provided by Piaget's final period.[7]

We begin by considering school mathematics. At the beginning of the first section, this was defined as consisting of a set of tasks (what children are taught or required to do) and a set of processes (what children actually do in performing those tasks). For the purposes of the present discussion, it will be useful to extend this definition to include a third component: a set of instructional techniques (how children are taught or are enabled to teach themselves). This is because we will be discussing unconventional as well as conventional school mathematics, and the differences between the two are based on teaching techniques as well as curriculum content.

The discussion in the first section was limited to school mathematics of the conventional kind. Our main concern was to show that what might be called a curriculum driven, bottom-up approach to applying Piaget leads to serious difficulties, especially when combined with a narrow interpretation of his theory. There is, however, a rather direct link between Piaget's theory (in the broad interpretation we developed in subsequent sections) and certain kinds of *uncon-*

---

[7]Much of Piaget's previous work is, of course, compatible with this framework or can be made compatible with relatively minor reinterpretation. This is especially true of the material included in Gruber and Voneche (1977).

ventional school mathematics. It stems as much from Piaget's epistemological and dialectical concerns as from his work in psychology. It is developed in his writings on education. These consist of a series of essays dealing, at a very general level, with the relationship of his work to practical educational matters that might be encountered in schools. Although they all touch upon mathematics in one way or another, a discussion of the more general essays (Piaget, 1971b, 1974) would lead us too far afield. They have been extensively treated by one of us elsewhere (Groen, 1978). Here, we restrict our attention to one that deals specifically with the topic of mathematics education (Piaget, 1973).

Although Piaget makes some proposals in this essay regarding both teaching methods and what might be taught, the most explicit discussion is really concerned with the processes used by the child in learning mathematics. What Piaget proposes is that the basic processes by means of which the structures underlying formal operations develop are the same as those that underlie the ability to think mathematically (in the sense that professional mathematicians and scientists do). The latter are the so-called logico-mathematical structures, which are essentially structures that coordinate or contain knowledge about structures. These originate (even in young children) in what Piaget calls logico-mathematical experience, which consists of the observation of actions and coordinations between actions. They develop by a process of what Piaget calls reflective abstraction, which is essentially the self-referential use of existing structures to construct new ones by observing one's thoughts and abstracting from them or, as Papert (1980) phrases it, thinking about thinking. To Piaget, this "corresponds precisely to logical and mathematical abstraction [Piaget, 1973]."

Piaget goes on to make some general comments about teaching. He seems to feel that there are two general problems that the teacher is faced with. The first is to facilitate the transition from observing one's own actions to thinking about them and thence to thinking about thinking. The second is to achieve some kind of satisfactory coordination between the logico-mathematical structures of the pupil and those of the teacher. Three "general psycho-pedagogical principles" underlie the solution of these problems.

1. Real comprehension of a notion or a theory implies its reinvention by the pupil. As Piaget (1974) puts it elsewhere, "to understand is to invent."
2. Many of the processes a child uses in solving a problem are unconscious. Thus, the pupil will be far more capable of "doing" and "understanding in actions" than of expressing himself or herself verbally. Mathematical thinking can develop only if the pupil can become aware of these unconscious processes.
3. Formal mathematics (i.e., that which uses abstract notation and other aspects of mathematical formalism) utilizes structures that may be quite different from those utilized in the informal "natural" mathematics of

the child. What must be developed is a new structure reflecting a satisfactory coordination between the formal and the informal. Hence, formalization should be kept for a later moment as a systematization of the
informal notions already acquired.

Piaget is much vaguer concerning the specific content that should be taught.
However, he does make one general comment that is highly significant: This is
that the logico-mathematical structures are far closer to those being used in
modern mathematics than to those used in traditional mathematics. Superficially
interpreted, this could be misleading. Piagetian mathematics might be viewed as
a combination of discovery learning with the *new mathematics*. This is incorrect
for two reasons. First, it can be argued that, when Piaget speaks of "understanding a theory by reinventing it," he had something considerably more dialectical
in mind than *discovery learning,* as the term is commonly used in mathematics
education. The latter is based on heuristic techniques primarily developed by
Polya (1948). Although not incompatible with the Piagetian framework, it is
problem solving rather than theory building. Piaget's idea seems more akin to the
evolutionary cycle of conjecture, proof, and refutation by counter-example proposed by Lakatos (1976). Second, the modern mathematics that Piaget is referring to consists of the universal "mother structures" of Bourbaki (1962) and the
category theory of MacLane (1972). This is quite different from the set theory
that forms the basis of the new mathematics. It might be possible to identify the
"easy" aspects of Bourbaki and category theory and bring them down to an
elementary level, as was done with set theory. However, this ignores and may
well compound a very non-Piagetian aspect of the new mathematics (and one
reason why it is frequently viewed nowadays as having failed). This is that it
imposes its structures from above. It does not build upon the informal theories
that the child already possesses. One might say that it attempts to teach the child
to think about somebody else's thinking rather than his or her own.

The point of Piaget's comment is not that the curriculum should include his
brand of modern mathematics. Rather, it is that what is taught should, in principle, lead naturally to these notions, and others like them, even if many of them
were not encountered explicitly before in, say, graduate school. What this implies is the existence of less abstract mathematical domains that are also "close"
to Piaget's logico-mathematical structures. To identify them, it is necessary to be
more explicit about what such closeness consists of. Piaget was clearly interested
in category theory and the mathematics of Bourbaki because they are primarily
concerned with transformations of simple mathematical structures into more
complex ones. However, they share a property with Piagetian structures that is
far more important for our purposes. They are built up from axions in which
transformations appear as primitive terms. In other words, transformations can
be defined independently of the states of the mathematical objects being trans-

formed. If the definitions we provided in the third section are viewed as being the "axions" for Piagetian structures then here, too, transformations appear as primitive terms. Indeed, some way of defining transformations independently from states seems to be required in order to talk about logico-mathematical structures that are based on *direct* experience of actions.

This property may seem subtle, but it is what gives Piagetian mathematics its uniqueness. Much of modern mathematics (most importantly, the kind that was used as a basis for the new mathematics curricula) is built up from conventional set theory, where notions such as *function* or *transformation* or *morphism* do not appear as primitive terms in the axioms. Hence, it cannot be readily accommodated in the Piagetian framework. On the other hand, there exist domains that clearly do have this property. At quite a high level of abstraction, there is, for example, the theory of transformation groups. At a somewhat less abstract level, there is lattice theory. Less abstract yet are graph theory and other aspects of combinatorial mathematics. Such examples are sufficient to demonstrate that Piagetian mathematics is nontrivial and may also convey something of its unique flavor. However, they are still a long way from school mathematics as Piaget views it. To re-invent such domains in their entirety requires formal operational thinking of a high level. They are research areas for professional mathematicians rather than children. What is still needed is some kind of Piagetian mathematics that provides a medium for the construction of formal operations out of the building blocks of logico-mathematical experience.

Papert (1980) has suggested a specific way in which this can be accomplished. This is through the use of what he calls *microworlds,* which are essentially mini-domains of Piagetian mathematics. They can be formulated as mathematical systems with axioms and theorems (e.g., Abelson & diSessa, 1981), but these lie beneath the surface as far as the child's direct experience is concerned. The transformations manifest themselves as commands that result in changes in the states of concrete objects. The effects of these commands are governed by the axioms (which are frequently simply descriptions of the outcomes of commands). The theorems manifest themselves as general properties of combinations of commands.

Papert's Turtle geometry is probably the most highly developed example of a microworld. The states are the positions and orientations of an arrow (called the Turtle) on the screen of a computer terminal. The primitive commands are embedded in the programming language LOGO. The principal ones are FORWARD X, which causes the Turtle to move X units forward in the direction it is pointing, and RIGHT X, which causes the Turtle to turn X degrees to the right. BACK X and LEFT X have the obvious corresponding effects. Finally, PENUP and PENDOWN cause the Turtle to begin and to cease drawing a line, respectively, as it moves. These are the axioms. Combinations of commands have the effect of drawing pictures on the screen. Thus, the theorems are statements about what kind of computer programs will generate a given class of pictures.

This example may serve to clarify what we mean by a transformation that is defined independently of the states. The basic action of FORWARD X is always the same. It moves the Turtle forward X units, regardless of its current position, and keeps its orientation invariant. Similarly, RIGHT X always turns the Turtle through X degrees and keeps its position invariant. Thus, the action of these transformations is defined completely independently of the absolute position and absolute orientation of the Turtle. The notion of a basic action may seem somewhat fuzzy, but it has an extremely well-defined concrete representation where FORWARD is concerned. It is the straight line drawn by the Turtle. Similarly, though slightly more indirectly, the basic action of RIGHT is the angle between two arbitrary lines. This kind of concreteness can also help indicate why the notion of state independence is important, despite its rather technical nature. It is what enables transformations to join together to form an independent structure that is invariant with respect to the states being transformed. This notion of an independent structure becomes, once again, concrete in Turtle geometry. The sequence FORWARD 100 RIGHT 120 FORWARD 100 RIGHT 120 FORWARD 100 will always create a triangle regardless of the initial state of the Turtle. This sequence also provides an example of how transformations can, in a sense, "become" objects. The structure "is" an object called a triangle. It even has properties: It is an equilateral triangle with sides of length 100. This example also illustrates what the coordination of structures consists of. What we have been talking about is a structure consisting of two coordinated substructures: the sequence of transformations and the lines and angles on the screen. The coordination consists of the fact that, when the transformations are executed, lines are drawn and angles are turned through by the Turtle.

Microworlds can exist outside the computer. It could be said, for example, that informal arithmetic based on counting is a microworld. In the basic model proposed by Groen and Parkman (1972), the state space can be viewed as numbers in a counter. The operations of setting to a number and incrementing by one are then state independent in the sense that their action is independent of the state the counter is in. The counting algorithms essentially involve the coordination of two such structures. However, if we wish to take Piaget's "general psycho-pedagogical principles" seriously, then the computer is extremely important. First, a programming language is a system of notation in which transformations and transformational structures have natural representations as operations and procedures. Thus, it can provide an introduction to mathematical formalism that is better coordinated with the natural structures of the child. Second, the process of writing a computer program encourages thinking about how one would perform the actions that are being embodied in the program.

Third, and most importantly, the pupil may invent a grossly incorrect or "buggy" theory about the microworld. In a noncomputer situation, it may be difficult for the student to become aware that something is wrong, due to a lack of adequate feedback regarding counter-examples and anomalies (Sleeman & Brown, 1982). In contrast, a computer-based microworld is naturally self-correcting. If a program does not execute as anticipated, it is clear immediately that something is wrong. The nature of the errors may yield additional information. If the cause is nontrivial, the task of debugging or discovering the cause of the error may lead to major modifications in the theory.

At this point, we can make some claim to have found Piagetian mathematics. What it is depends upon the level at which one wishes to proceed. We have considered two such levels: a mathematical level and a psychological level. At the mathematical level, it is simply a set of mathematical domains in which transformations are treated directly rather than subordinated to the notion of state. We have treated this topic at considerable length, but two things should perhaps be emphasized. First, the distinction is nontrivial (the new mathematics, for example, does not have this property). Second, this kind of mathematics is Piagetian because Piaget's logico-mathematical structures are state-independent in exactly the same way. Thus, we have some guarantee that it fits into the Piagetian framework.

At the psychological level, Piagetian mathematics is a set of microworlds. Once again, this notion has already been extensively discussed, but a few of the main points need to be emphasized. The first is that the microworld is essentially a concrete embodiment of the kind of mathematical domain just described. The second is the importance of an interactive computer environment in the practical implimentation of a microworld. Such an environment renders tangible two crucial notions that Piaget himself could only express in a very abstract fashion. It makes it possible to observe actions and coordinations of actions. Hence, a suitably constructed computer-based microworld provides a medium for direct logico-mathematical exerience. Also, writing and debugging a computer program involves observing one's thoughts and abstracting from them. Hence, the microworld can also be viewed as a medium for inducing the kind of spontaneous reflective abstraction that leads to the construction of new logico-mathematical structures.[8]

One thing remains to be considered. This is the issue of how our characterization of Piagetian mathematics relates to the information-processing approach. In the introduction to this chapter, we observed that the latter appeared to be providing a more manageable framework for research in childrens' mathematical thinking. On the other hand, it seems quite reasonable to claim that the Piagetian

---

[8]In Papert's (1980) terminology, thinking about thinking about programs and their outcomes leads to powerful ideas.

framework is quite manageable for the kind of mathematics we have discussed in this section. However, much of our effort in this chapter has been devoted to showing that there are several possible Piagetian frameworks. It is important to note that the one we are using here is different from the one found unmanageable in the past. Rather than relying so heavily on stage theory, it is built up from the epistemology and the more procedural notions of Piaget's final period. This has both a disadvantage and an advantage. The disadvantage is that this late theory exists in a rather fragmentary form. Much of it has not even been published; it has certainly not been assimilated, and may be incomplete. The advantage is that it is potentially more compatible with the information-processing approach. Indeed, one of the themes of this late period is a the working out of a dialectical tension between process and structure. Because of this, the two frameworks might best be viewed as complementing each other rather than competing against each other. The information-processing approach is based on processes that have structure, but has problems describing global structural properties. The Piagetian approach is based on structures that function, but has problems in describing what this functioning consists of. The study of Piagetian mathematics is unlikely to lead to a reestablishment of the Piagetian framework and the abandonment of the information-processing approach. It might, however, lead to a synthesis that provides a more adequate framework than the two in isolation.

REFERENCES

Abelson, H., & diSessa, A. *Turtle geometry.* Cambridge, Mass.: M.I.T. Press, 1981.

Beilin, H. Piaget and the new functionalism. *Eleventh Symposium of the Jean Piaget Society.* Philadelphia, 1981.

Boden, M. *Piaget.* London: Harvester Press, 1979.

Bourbaki, N. L'architecture des Mathematiques. In F. Le Lionnais (Ed.), *Les grands courants de la pensee Mathematique.* Paris: Blanchard, 1962.

Case, R. Intellectual development from birth to adulthood: A neo-Piagetian approach. In R. S. Siegler (Ed.), *Children's thinking: What develops?* Hillsdale, N.J.: Lawrence Erlbaum Associates, 1978.

Cellerier, G. Information-processing tendencies in recent experiments in cognitive learning. In S. Farnham-Diggory (Ed.), *Information processing in children.* New York: Academic Press, 1972.

Flavell, J. H. *The developmental psychology of Jean Piaget.* New York: D. Van Nostrand, 1963.

Furth, H. *Piaget and knowledge* (2nd ed.). Chicago: University of Chicago Press, 1981.

Gelman, R., & Gallistel, C. R. *The child's understanding of number.* Cambridge, Mass.: Harvard University Press, 1978.

Ginsburg, H. *Children's arithmetic: The learning process.* New York: D. Van Nostrand, 1977.

Green, D. R., Ford, M., & Flaser, G. (Eds). *Measurement and Piaget.* New York: McGraw-Hill, 1971.

Groen, G. J. The theoretical ideas of Piaget and educational practice. In P. Suppes (Ed.), *Impact of research on education: Some case studies.* Washington, D.C. National Academy of Education, 1978.

Groen, G. J., & Parkman, J. M. A chronometric analysis of simple addition. *Psychological Review,* 1972, *79,* 329–343.

Groen, G. J., & Resnick, L. Can pre-school children invent addition algorithms? *Journal of Educational Psychology,* 1977, *67,* 17–21.

Gruber, H. E., & Voneche, J. J. *The essential Piaget.* New York: Basic Books, 1977.

Inhelder, B., & Piaget, J. *The growth of logical thinking from childhood to adolescence: An essay on the construction of formal operational structures* (A. Parson & S. Milgram, trans.). New York: Basic Books, 1958. (Originally published, 1955.)

Inhelder, B., & Piaget, J. *The early growth of logic in the child: Classification and seriation* (E. Lunzer & D. Papert, trans.). New York: Harper & Row, 1964. (Originally published, 1959.)

Inhelder, B., Sinclair, H., & Bovet, M. *Learning and the development of cognition* (S. Wedgewood, trans.). Cambridge, Mass.: Harvard University Press, 1974.

Kamii, C., & DeVries, R. Piaget for early education. In M. C. Day & R. K. Parker (Eds.), *Preschool in action* (2nd ed.). Boston: Allyn & Bacon, 1977.

Klahr, D., & Wallace, J. G. *Cognitive development: An information-processing view.* Hillsdale, N.J., Lawrence Erlbaum Associates, 1976.

Kuhn, T. *The structure of scientific revolutions.* Chicago: University of Chicago Press, 1962.

Lakatos, I. Falsification and the methodology of scientific research programmes. In I. Lakatos & A. Musgrave (Eds.), *Criticism and the growth of knowledge.* Cambridge, England: Cambridge University Press, 1972.

Lakatos, I. *Proofs and refutations: The logic of mathematical discovery.* Cambridge, England: Cambridge University Press, 1976.

MacLane, S. *Categories for the working mathematician.* New York: Springer, 1972.

Papert, S. *Mindstorms.* New York: Basic Books, 1980.

Pascual-Leone, J. A mathematical model for transition in Piaget's developmental stages. *Acta Psychologica,* 1970, *32,* 301–345.

Piaget, J. *Traite de logigue.* Paris: Colin, 1949.

Piaget, J. *Introduction a l'epistemologie genetique* (3 vols.). Paris: Presses Universitaires de France, 1950. (a)

Piaget, J. *The psychology of intelligence* (M. Piercy & D. E. Berlyne, trans.). London: Routledge & Kegan Paul, 1950. (b) (Originally published, 1947.)

Piaget, J. *The origins of intelligence in children* (M. Cook, trans.). New York: International Universities Press, 1952. (Originally published, 1936.)

Piaget, J. How children form mathematical concepts. *Scientific American,* 1953, *189*(5), 74–79.

Piaget, J. *Logic and psychology.* New York: Basic Books, 1957.

Piaget, J. Review of *Studies in cognitive growth* (by J. S. Bruner, R. Olver, & P. Greenfield). *Contemporary Psychology,* 1967, *12,* 532–533

Piaget, J. *Genetic epistemology.* New York: Norton, 1971. (a)

Piaget, J. *Science of education and the psychology of the child.* New York: Viking, 1971. (b)

Piaget, J. *Biology and knowledge.* Chicago: University of Chicago Press, 1971. (c) (Originally published, 1967.)

Piaget, J. *Structuralism* (C. Maschler, trans.). New York: Basic Books, 1971. (d) (Originally published, 1968.)

Piaget, J. *The principles of genetic epistemology* (W. Mays, trans.). London: Routledge & Kegan Paul, 1972. (Originally published, 1970.)

Piaget, J. Comments on mathematical education. In A. G. Howson (Ed)., *Developments in mathematical education: Proceedings of the Second International Congress on Mathematics Education.* Cambridge, Cambridge University Press, 1973.

Piaget, J. *To understand is to invent: The future of education.* New York: Viking, 1974. (A translation of two works written for UNESCO in 1948 and 1971.)

Piaget, J. *The grasp of consciousness.* Boston: Harvard University Press, 1976. (Originally published, 1974, as *La prise de conscience.*)

Piaget, J. *The development of thought: Equilibration of cognitive structures* (A. Rosin, trans.). New York: Viking, 1977. (Originally published, 1975, as *L'equilibration des structures cognitives.*)

Piaget, J. *Experiments in contradiction* (D. Coltman, trans.). Chicago: University of Chicago Press, 1981. (Originally published, 1974, as *Recherches sur la contradiction.*)

Piaget, J., & Inhelder, B. *The child's conception of space* (F. J. Langdon & J. L. Lunzer, trans.). London: Routledge & Kegan Paul, 1956. (Originally published, 1948.)

Piaget, J., & Inhelder, B. *Psychology of the child* (H. Weaver, trans.). London: Routledge & Kegan Paul, 1969. (Originally published, 1967.)

Piaget, J., & Inhelder, B. *Mental imagery in the child.* London: Routledge & Kegan Paul, 1971. (Originally published, 1966.)

Piaget, J., & Inhelder, B. *Memory and intelligence* (A. J. Pomerans, trans.). New York: Basic Books, 1972. (Originally published, 1968.)

Piaget, J., Inhelder, B., & Szeminska, A. *The child's conception of geometry* (E. A. Lunzer, trans.). New York: Basic Books, 1960. (Originally published, 1948.)

Piaget, J., & Szeminska, A. *The child's conception of number* (C. Gattegno & F. M. Hodgson, trans.). New York: Humanities Press, 1952. (Originally published, 1941.)

Piaget, J., & Voyat, G. The possible, the impossible and the necessary. In F. Murray (Ed.). *The impact of Piagetian theory.* Baltimore, Md.: University Park Press, 1979.

Polya, G. *How to solve it.* Princeton, N.J.: Princeton University Press, 1948.

Sleeman, D., & Brown, J. S. Intelligent tutoring systems. New York: Academic Press, 1982.

Young, R. *Children's seriation behavior: A production system analysis.* Doctoral dissertation, Carnegie-Mellon University, 1973.

Author Index

Subject Index

DEVELOPMENTAL PSYCHOLOGY SERIES

Continued from page ii

ROBERT L. SELMAN. *The Growth of Interpersonal Understanding: Developmental and Clinical Analyses*

BARRY GHOLSON. *The Cognitive-Developmental Basis of Human Learning: Studies in Hypothesis Testing*

TIFFANY MARTINI FIELD, SUSAN GOLDBERG, DANIEL STERN, and ANITA MILLER SOSTEK. (Editors). *High-Risk Infants and Children: Adult and Peer Interactions*

GILBERTE PIERAUT-LE BONNIEC. *The Development of Modal Reasoning: Genesis of Necessity and Possibility Notions*

JONAS LANGER. *The Origins of Logic: Six to Twelve Months*

LYNN S. LIBEN. *Deaf Children: Developmental Perspectives*